PENGUIN

THE

Henry James was born in 1843 at Washington Place, New York, of Scottish and Irish descent. His father was a prominent theologian and philosopher, and his elder brother, William, was also famous as a philosopher. He attended schools in New York and later in London, Paris and Geneva, entering the Law School at Harvard in 1862. In 1865 he began to contribute reviews and short stories to American journals. In 1875, after two prior visits to Europe, he settled for a year in Paris, where he met Flaubert, Turgenev, and other literary figures. However, the next year he moved to London, where he became such an inveterate diner-out that in the winter of 1878–9 he confessed to accepting 107 invitations. In 1898 he left London and went to live at Lamb House, Rye, Sussex. Henry James became naturalized in 1915, was awarded the O.M., and died early in 1916.

In addition to many short stories, plays, books of criticism, autobiography, and travel he wrote some twenty novels, the first published being *Roderick Hudson* (1875). They include *The Europeans*, *The Portrait of a Lady*, *The Princess Casamassima*, *The Spoils of Poynton*, *Washington Square*, *The Bostonians*, *What Maisie Knew*, *The Golden Bowl*, *The Wings of the Dove* and *The Turn of the Screw and other Stories* (all published in Penguins).

PENGUIN MODERN CLASSICS

THE TRAGIC MUSE

HENRY JAMES
The Tragic Muse

PENGUIN BOOKS

Penguin Books Ltd, Harmondsworth, Middlesex, England
Penguin Books, 625 Madison Avenue, New York, New York 10022, U.S.A.
Penguin Books Australia Ltd, Ringwood, Victoria, Australia
Penguin Books Canada Ltd, 2801 John Street, Markham, Ontario, Canada L3R 1B4
Penguin Books (N.Z.) Ltd, 182–190 Wairau Road, Auckland 10, New Zealand

—

First published by Macmillan 1890
This edition first published by Rupert Hart-Davis 1948
Published in Penguin Books 1978

—

Made and printed in Great Britain by
Hazell Watson & Viney Ltd, Aylesbury, Bucks
Set in Linotype Times Roman

Except in the United States of America,
this book is sold subject to the condition
that it shall not, by way of trade or otherwise,
be lent, re-sold, hired out, or otherwise circulated
without the publisher's prior consent in any form of
binding or cover other than that in which it is
published and without a similar condition
including this condition being imposed
on the subsequent purchaser

NOTE ON THE TEXT

HENRY JAMES minutely revised *The Tragic Muse*, written when he was forty-six, for the New York 'Definitive' Edition of the *Novels and Tales*, published in 1908, making verbal changes in almost every paragraph. These changes are usually, by the introduction of a periphrasis, to make the sentences more allusive and less simple and direct.

To give a few examples: 'Mr Carteret' is altered to 'The rich old man ...'; 'Miriam' to 'his young friend', in a context where the phrase sounds oddly for a girl with whom the young man is passionately in love. A typical example is: 'Miriam's colour rose, through her paint, at this vivid picture ...' changed to 'Miriam's colour rose, through all her artificial surfaces, at this all but convincing appeal...'

Some critics prefer the manner of Henry James's old age to that of his maturity, and the changes in the Definitive Edition do tend to approximate the style of the earlier novels to that of the later. But they blur their sharp outlines. To read Henry James in the Definitive Edition is to obscure his development and to miss the full flavour of his middle years. Either text was equally available for this edition; after comparing them, the publishers have deliberately chosen that of the first edition (Macmillan 1890) on grounds of taste.

1

THE people of France have made it no secret that those of England, as a general thing, are, to their perception, an inexpressive and speechless race, perpendicular and unsociable, unaddicted to enriching any bareness of contact with verbal or other embroidery. This view might have derived encouragement, a few years ago, in Paris, from the manner in which four persons sat together in silence, one fine day about noon, in the garden, as it is called, of the Palais de l'Industrie – the central court of the great glazed bazaar where, among plants and parterres, gravelled walks and thin fountains, are ranged the figures and groups, the monuments and busts, which form, in the annual exhibition of the Salon, the department of statuary. The spirit of observation is naturally high at the Salon, quickened by a thousand artful or artless appeals, but no particular tension of the visual sense would have been required to embrace the character of the four persons in question. As a solicitation of the eye on definite grounds, they too constituted a successful plastic fact; and even the most superficial observer would have perceived them to be striking products of an insular neighbourhood, representatives of that tweed-and-waterproof class with which, on the recurrent occasions when the English turn out for a holiday – Christmas and Easter, Whitsuntide and the autumn – Paris besprinkles itself at a night's notice. They had about them the indefinable professional look of the British traveller abroad; that air of preparation for exposure, material and moral, which is so oddly combined with the serene revelation of security and of persistence, and which excites, according to individual susceptibility, the ire or the admiration of foreign communities. They were the more unmistakable as they illustrated very favourably the energetic race to which they had the honour to belong. The fresh, diffused light of the Salon made them clear and important; they were finished productions, in their way, and ranged there motionless, on their green bench, they were almost as much on exhibition as if they had been hung on the line.

Three ladies and a young man, they were obviously a family

– a mother, two daughters and a son – a circumstance which had the effect at once of making each member of the group doubly typical and of helping to account for their fine taciturnity. They were not, with each other, on terms of ceremony, and moreover they were probably fatigued with their course among the pictures, the rooms on the upper floor. Their attitude, on the part of visitors who had superior features, even if they might appear to some passers-by to have neglected a fine opportunity for completing these features with an expression, was after all a kind of tribute to the state of exhaustion, of bewilderment, to which the genius of France is still capable of reducing the proud.

'*En v'la des abrutis!*' more than one of their fellow-gazers might have been heard to exclaim; and certain it is that there was something depressed and discouraged in this interesting group, who sat looking vaguely before them, not noticing the life of the place, somewhat as if each had a private anxiety. A very close observer would have guessed that though on many questions they were closely united, this present anxiety was not the same for each. If they looked grave, moreover, this was doubtless partly the result of their all being dressed in mourning, as if for a recent bereavement. The eldest of the three ladies had indeed a face of a fine austere mould, which would have been moved to gaiety only by some force more insidious than any she was likely to recognize in Paris. Cold, still and considerably worn, it was neither stupid nor hard, but it was firm, narrow and sharp. This competent matron, acquainted evidently with grief, but not weakened by it, had a high forehead, to which the quality of the skin gave a singular polish – it glittered even when seen at a distance; a nose which achieved a high, free curve, and a tendency to throw back her head and carry it well above her, as if to disengage it from the possible entanglements of the rest of her person. If you had seen her walk you would have perceived that she trod the earth in a manner suggesting that in a world where she had long since discovered that one couldn't have one's own way, one could never tell what annoying aggression might take place, so that it was well, from hour to hour, to save what one could. Lady Agnes saved her head, her white triangular forehead, over which her closely crinkled flaxen hair, reproduced in different shades in her children, made a sort of looped silken canopy, like the marquee at a garden-party. Her daughters were tall, like herself – that was visible even as they sat there – and one of them, the younger evidently, was very pretty: a straight, slender, gray-eyed English girl, with a 'good' figure and a fresh complexion. The sister, who was not pretty, was also straight and slender and gray-eyed. But the gray, in this case, was not so pure, nor were the slenderness and the straightness so maidenly. The brother of these young ladies had taken off his hat, as if he felt the air of the summer day heavy in the great pavilion. He was a lean, strong, clear-faced youth, with a straight nose and light-brown hair, which lay continuously and profusely back from his forehead, so that to smooth it from the brow to the neck but a single movement of the hand was required. I cannot describe him better than by saying that he was the sort of young Englishman who looks particularly well abroad, and whose general aspect – his inches, his limbs, his friendly eyes, the modulation of his voice, the cleanness of his flesh-tints and the fashion of his garments – excites on the part of those who encounter him in far countries on the ground of a common speech a delightful sympathy of race. This sympathy is sometimes qualified by an apprehension of undue literalness, but it almost revels as soon as such a danger is dispelled. We shall see quickly enough how accurate a measure it might have taken of Nicholas Dormer. There was food for suspicion, perhaps, in the wandering blankness that sat at moments in his eyes, as if he had no attention at all, not the least in the world, at his command; but it is no more than just to add, without delay, that this discouraging symptom was known, among those who liked him, by the indulgent name of dreaminess. For his mother and sisters, for instance, his dreaminess was notorious. He is the more welcome to the benefit of such an interpretation as there is always held to be something engaging in the combination of the muscular and the musing, the mildness of strength.

After some time – a period during which these good people might have appeared to have come, individually, to the Palais de l'Industrie much less to see the works of art than to think over their domestic affairs – the young man, rousing himself from his reverie, addressed one of the girls.

'I say, Biddy, why should we sit moping here all day? Come and take a turn about with me.'

His younger sister, while he got up, leaned forward a little, looking round her, but she gave, for the moment, no further sign of complying with his invitation.

'Where shall we find you, then, if Peter comes?' inquired the other Miss Dormer, making no movement at all.

'I dare say Peter won't come. He'll leave us here to cool our heels.'

'Oh, Nick, dear!' Biddy exclaimed in a sweet little voice of protest. It was plainly her theory that Peter would come, and even, a little, her apprehension that she might miss him should she quit that spot.

'We shall come back in a quarter of an hour. Really, I must look at these things,' Nick declared, turning his face to a marble group which stood near them, on the right – a man, with the skin of a beast round his loins, tussling with a naked woman in some primitive effort of courtship or capture.

Lady Agnes followed the direction of her son's eyes, and then observed:

'Everything seems very dreadful. I should think Biddy had better sit still. Hasn't she seen enough horrors up above?'

'I dare say that if Peter comes Julia will be with him,' the elder girl remarked irrelevantly.

'Well, then, he can take Julia about. That will be more proper,' said Lady Agnes.

'Mother, dear, she doesn't care a rap about art. It's a fearful bore looking at fine things with Julia,' Nick rejoined.

'Won't you go with him, Grace?' said Biddy, appealing to her sister.

'I think she has awfully good taste!' Grace exclaimed, not answering this inquiry.

'*Don't* say nasty things about her!' Lady Agnes broke out, solemnly, to her son, after resting her eyes on him a moment with an air of reluctant reprobation.

'I say nothing but what she'd say herself,' the young man replied. 'About some things she has very good taste, but about this kind of thing she has no taste at all.'

'That's better, I think,' said Lady Agnes, turning her eyes

again to the 'kind of thing' that her son appeared to designate.

'She's awfully clever – awfully!' Grace went on, with decision.

'Awfully, awfully,' her brother repeated, standing in front of her and smiling down at her.

'You *are* nasty, Nick. You know you are,' said the young lady, but more in sorrow than in anger.

Biddy got up at this, as if the accusatory tone prompted her to place herself generously at his side. 'Mightn't you go and order lunch, in that place, you know?' she asked of her mother. 'Then we would come back when it was ready.'

'My dear child, I can't order lunch,' Lady Agnes replied, with a cold impatience which seemed to intimate that she had problems far more important than those of victualling to contend with.

'I mean Peter, if he comes. I am sure he's up in everything of that sort.'

'Oh, hang Peter!' Nick exclaimed. 'Leave him out of account, and *do* order lunch, mother; but not cold beef and pickles.'

'I must say – about him – you're not nice,' Biddy ventured to remark to her brother, hesitating, and even blushing, a little.

'You make up for it, my dear,' the young man answered, giving her chin – a very charming, rotund little chin – a friendly whisk with his forefinger.

'I can't imagine what you've got against him,' her ladyship murmured, gravely.

'Dear mother, it's a disappointed fondness,' Nick argued. 'They won't answer one's notes; they won't let one know where they are nor what to expect. "Hell has no fury like a woman scorned"; nor like a man either.'

'Peter has such a tremendous lot to do – it's a very busy time at the Embassy; there are sure to be reasons,' Biddy explained, with her pretty eyes.

'Reasons enough, no doubt!' said Lady Agnes, who accompanied these words with an ambiguous sigh, however, as if in Paris even the best reasons would naturally be bad ones.

'Doesn't Julia write to you, doesn't she answer you the very day?' Grace inquired, looking at Nick as if she were the courageous one.

He hesitated a moment, returning her glance with a certain

severity. 'What do you know about my correspondence? No doubt I ask too much,' he went on; 'I am so attached to them. Dear old Peter, dear old Julia!'

'She's younger than you, my dear!' cried the elder girl, still resolute.

'Yes, nineteen days.'

'I'm glad you know her birthday.'

'She knows yours; she always gives you something,' Lady Agnes resumed, to her son.

'Her taste is good *then*, isn't it, Nick?' Grace Dormer continued.

'She makes charming presents; but, dear mother, it isn't *her* taste. It's her husband's.'

'Her husband's?'

'The beautiful objects of which she disposes so freely are the things he collected, for years, laboriously, devotedly, poor man!'

'She disposes of them to you, but not to others,' said Lady Agnes. 'But that's all right,' she added, as if this might have been taken for a complaint of the limitations of Julia's bounty. 'She has to select, among so many, and that's a proof of taste,' her ladyship went on.

'You can't say she doesn't choose lovely ones,' Grace remarked to her brother, in a tone of some triumph.

'My dear, they are all lovely. George Dallow's judgement was so sure, he was incapable of making a mistake,' Nicholas Dormer returned.

'I don't see how you can talk of him; he was dreadful,' said Lady Agnes.

'My dear, if he was good enough for Julia to marry, he is good enough for one to talk of.'

'She did him a great honour.'

'I dare say; but he was not unworthy of it. No such intelligent collection of beautiful objects has been made in England in our time.'

'You think too much of beautiful objects,' returned her ladyship.

'I thought you were just now implying that I thought too little.'

'It's very nice – his having left Julia so well off,' Biddy interposed, soothingly, as if she foresaw a tangle.

'He treated her *en grand seigneur*, absolutely,' Nick went on.

'He used to look greasy, all the same,' Grace Dormer pursued, with a kind of dull irreconcilability. 'His name ought to have been Tallow.'

'You are not saying what Julia would like, if that's what you are trying to say,' her brother remarked.

'Don't be vulgar, Grace,' said Lady Agnes.

'I know Peter Sherringham's birthday!' Biddy broke out innocently, as a pacific diversion. She had passed her hand into her brother's arm, to signify her readiness to go with him, while she scanned the remoter portions of the garden as if it had occurred to her that to direct their steps in some such sense might after all be the shorter way to get at Peter.

'He's too much older than you, my dear,' Grace rejoined, discouragingly.

'That's why I've noticed it – he's thirty-four. Do you call that too old? I don't care for slobbering infants!' Biddy cried.

'Don't be vulgar,' Lady Agnes enjoined again.

'Come, Bid, we'll go and be vulgar together; for that's what we are, I'm afraid,' her brother said to her. 'We'll go and look at all these low works of art.'

'Do you really think it's necessary to the child's development?' Lady Agnes demanded, as the pair turned away. Nicholas Dormer was struck as by a kind of challenge, and he paused, lingering a moment, with his little sister on his arm. 'What we've been through this morning in this place, and what you've paraded before our eyes – the murders, the tortures, all kinds of disease and indecency!'

Nick looked at his mother as if this sudden protest surprised him, but as if also there were lurking explanations of it which he quickly guessed. Her resentment had the effect not so much of animating her cold face as of making it colder, less expressive, though visibly prouder. 'Ah, dear mother, don't do the British matron!' he exclaimed, good-humouredly.

'British matron is soon said! I don't know what they are coming to.'

'How odd that you should have been struck only with the disagreeable things, when, for myself, I have felt it to be the most interesting, the most suggestive morning I have passed for ever so many months!'

'Oh, Nick, Nick!' Lady Agnes murmured, with a strange depth of feeling.

'I like them better in London – they are much less unpleasant,' said Grace Dormer.

'They are things you can look at,' her ladyship went on. 'We certainly make the better show.'

'The subject doesn't matter; it's the treatment, the treatment!' Biddy announced, in a voice like the tinkle of a silver bell.

'Poor little Bid!' her brother cried, breaking into a laugh.

'How can I learn to model, mamma dear, if I don't look at things and if I don't study them?' the girl continued.

This inquiry passed unheeded, and Nicholas Dormer said to his mother, more seriously, but with a certain kind explicitness, as if he could make a particular allowance: 'This place is an immense stimulus to me; it refreshes me, excites me, it's such an exhibition of artistic life. It's full of ideas, full of refinements; it gives one such an impression of artistic experience. They try everything, they feel everything. While you were looking at the murders, apparently, I observed an immense deal of curious and interesting work. There are too many of them, poor devils; so many who must make their way, who must attract attention. Some of them can only *taper fort*, stand on their heads, turn summersaults or commit deeds of violence, to make people notice them. After that, no doubt, a good many will be quieter. But I don't know; today I'm in an appreciative mood – I feel indulgent even to them: they give me an impression of intelligence, of eager observation. All art is one – remember that, Biddy dear,' the young man continued, looking down at his sister with a smile. 'It's the same great, many-headed effort, and any ground that's gained by an individual, any spark that's struck in any province, is of use and of suggestion to all the others. We are all in the same boat.'

' "We," do you say, my dear? Are you really setting up for an artist?' Lady Agnes asked.

Nick hesitated a moment. 'I was speaking for Biddy!'

'But you *are* one, Nick – you are!' the girl cried.

Lady Agnes looked for an instant as if she were going to say once more 'Don't be vulgar!' But she suppressed these words, if she had intended them, and uttered others, few in number and not completely articulate, to the effect that she hated talking about art. While her son spoke she had watched him as if she failed to follow him; yet something in the tone of her exclamation seemed to denote that she had understood him only too well.

'We are all in the same boat,' Biddy repeated, smiling at her.

'Not me, if you please!' Lady Agnes replied. 'It's horrid, messy work, your modelling.'

'Ah, but look at the results!' said the girl, eagerly, glancing about at the monuments in the garden as if in regard even to them she were, through that unity of art that her brother had just proclaimed, in some degree an effective cause.

'There's a great deal being done here – a real vitality,' Nicholas Dormer went on, to his mother, in the same reasonable, informing way. 'Some of these fellows go very far.'

'They do, indeed!' said Lady Agnes.

'I'm fond of young schools, like this movement in sculpture,' Nick remarked, with his slightly provoking serenity.

'They're old enough to know better!'

'Mayn't I look, mamma? It *is* necessary to my development,' Biddy declared.

'You may do as you like,' said Lady Agnes, with dignity.

'She ought to see good work, you know,' the young man went on.

'I leave it to your sense of responsibility.' This statement was somewhat majestic, and for a moment, evidently, it tempted Nick, almost provoked him, or at any rate suggested to him an occasion to say something that he had on his mind. Apparently, however, he judged the occasion on the whole not good enough, and his sister Grace interposed with the inquiry –

'Please, mamma, are we *never* going to lunch?'

'Ah, mother, mother!' the young man murmured, in a troubled way, looking down at Lady Agnes with a deep fold in his forehead.

For her, also, as she returned his look, it seemed an occasion; but with this difference, that she had no hesitation in taking advantage of it. She was encouraged by his slight embarrassment; for ordinarily Nick was not embarrassed. 'You used to have so much,' she went on; 'but sometimes I don't know what has become of it – it seems all, *all* gone!'

'Ah, mother, mother!' he exclaimed again, as if there were so many things to say that it was impossible to choose. But this time he stepped closer, bent over her, and, in spite of the publicity of their situation, gave her a quick, expressive kiss. The foreign observer whom I took for granted in beginning to sketch this scene would have had to admit that the rigid English family had, after all, a capacity for emotion. Grace Dormer, indeed, looked round her to see if at this moment they were noticed. She discovered with satisfaction that they had escaped.

2

NICK DORMER walked away with Biddy, but he had not gone far before he stopped in front of a clever bust, where his mother, in the distance, saw him playing in the air with his hand, carrying out by this gesture, which presumably was applausive, some critical remark he had made to his sister. Lady Agnes raised her glass to her eyes by the long handle to which rather a clanking chain was attached, perceiving that the bust represented an ugly old man with a bald head; at which her ladyship indefinitely sighed, though it was not apparent in what way such an object could be detrimental to her daughter. Nick passed on, and quickly paused again; this time, his mother discerned, it was before the marble image of a grimacing woman. Presently she lost sight of him; he wandered behind things, looking at them all round.

'I ought to get plenty of ideas for my modelling, oughtn't I, Nick?' his sister inquired of him, after a moment.

'Ah, my poor child, what shall I say?'

'Don't you think I have any capacity for ideas?' the girl continued, ruefully.

'Lots of them, no doubt. But the capacity for applying them, for putting them into practice – how much of that have you?'

'How can I tell till I try?'

'What do you mean by trying, Biddy dear?'

'Why, you know – you've seen me.'

'Do you call that trying?' her brother asked, smiling at her.

'Ah, Nick!' murmured the girl, sensitively. Then, with more spirit, she went on: 'And please, what do you?'

'Well, this, for instance'; and her companion pointed to another bust – a head of a young man, in terra-cotta, at which they had just arrived; a modern young man, to whom, with his thick neck, his little cap and his wide ring of dense curls, the artist had given the air of a Florentine of the time of Lorenzo.

Biddy looked at the image a moment. 'Ah, that's not trying; that's succeeding.'

'Not altogether; it's only trying seriously.'

'Well, why shouldn't I be serious?'

'Mother wouldn't like it. She has inherited the queer old superstition that art is pardonable only so long as it's bad – so long as it's done at odd hours, for a little distraction, like a game of tennis or of whist. The only thing that can justify it, the effort to carry it as far as one can (which you can't do without time and singleness of purpose), she regards as just the dangerous, the criminal element. It's the oddest hind-part-before view, the drollest immorality.'

'She doesn't want one to be professional,' Biddy remarked, as if she could do justice to every system.

'Better leave it alone, then: there are duffers enough.'

'I don't want to be a duffer,' Biddy said. 'But I thought you encouraged me.'

'So I did, my poor child. It was only to encourage myself.'

'With your own work – your painting?'

'With my futile, my ill-starred endeavours. Union is strength; so that we might present a wider front, a larger surface of resistance.'

Biddy was silent a moment, while they continued their tour of observation. She noticed how her brother passed over some things quickly, his first glance sufficing to show him whether they were worth another, and recognized in a moment the figures that had something in them. His tone puzzled her, but his certainty of eye impressed her, and she felt what a difference there was yet between them – how much longer, in every case, she would have taken to discriminate. She was aware that she could rarely tell whether a picture was good or bad until she had looked at it for ten minutes; and modest little Biddy was compelled privately to add, 'And often not even then.' She was mystified, as I say (Nick was often mystifying – it was his only fault), but one thing was definite: her brother was exceedingly clever. It was the consciousness of this that made her remark at last: 'I don't so much care whether or no I please mamma, if I please you.'

'Oh, don't lean on me. I'm a wretched broken reed. I'm no use *really*!' Nick Dormer exclaimed.

'Do you mean you're a duffer?' Biddy asked, alarmed.

'Frightful, frightful!'

'So that you mean to give up your work – to let it alone, as you advise *me*?'

'It has never been my work, Biddy. If it had, it would be different. I should stick to it.'

'And you *won't* stick to it?' the girl exclaimed, standing before him, open-eyed.

Her brother looked into her eyes a moment, and she had a compunction; she feared she was indiscreet and was worrying him. 'Your questions are much simpler than the elements out of which my answer should come.'

'A great talent – what is simpler than that?'

'One thing, dear Biddy: no talent at all!'

'Well, yours is so real, you can't help it.'

'We shall see, we shall see,' said Nicholas Dormer. 'Let us go look at that big group.'

'We shall see if it's real?' Biddy went on, as she accompanied him.

'No; we shall see if I can't help it. What nonsense Paris makes one talk!' the young man added, as they stopped in front of the composition. This was true, perhaps, but not in a sense which he found himself tempted to deplore. The present was far from being his first visit to the French capital: he had often quitted England, and usually made a point of 'putting in', as he called it, a few days there on the outward journey to the Continent or on the return; but on this occasion the emotions, for the most part agreeable, attendant upon a change of air and of scene, had been more punctual and more acute than for a long time before, and stronger the sense of novelty, refreshment, amusement, of manifold suggestions looking to that quarter of thought to which, on the whole, his attention was apt most frequently, though not most confessedly, to stray. He was fonder of Paris than most of his countrymen, though not so fond, perhaps, as some other captivated aliens: the place had always had the power of quickening sensibly the life of reflection and of observation within him. It was a good while since the reflections engendered by his situation there had been so favourable to the city by the Seine; a good while, at all events, since they had ministered so to excitement, to exhilaration, to ambition, even to a restlessness which was not prevented from being agreeable by the nervous quality in it. Dormer could have given the reason of this unwonted glow; but his preference was very much to keep it to himself. Certainly, to persons not deeply knowing, or at any rate not deeply curious in relation to the young man's history, the explanation might have seemed to beg the question, consisting as it did of the simple formula that he had at last come to a crisis. Why a crisis – what was it, and why had he not come to it before? The reader shall learn these things in time, if he care enough for them.

For several years Nicholas Dormer had not omitted to see the Salon, which the general voice, this season, pronounced not particularly good. None the less, it was the exhibition of this season that, for some cause connected with his 'crisis', made him think fast, produced that effect which he had spoken of to his mother as a sense of artistic life. The precinct of the marbles and bronzes appealed to him especially today; the glazed garden, not florally rich, with its new productions alternating with perfunctory plants and its queer damp smell, partly the odour of plastic clay, of the studios of sculptors, spoke to him with the voice of old associations, of other visits, of companionships that were closed – an insinuating eloquence which was at the same time, somehow, identical with the general sharp contagion of Paris. There was youth in the air, and a multitudinous newness, for ever reviving, and the diffusion of a hundred talents, ingenuities, experiments. The summer clouds made shadows on the roof of the great building; the white images, hard in their crudity, spotted the place with provocations; the rattle of plates at the restaurant sounded sociable in the distance, and our young man congratulated himself more than ever that he had not missed the exhibition. He felt that it would help him to settle something. At the moment he made this reflection his eye fell upon a person who appeared – just in the first glimpse – to carry out the idea of help. He uttered a lively ejaculation, which, however, in its want of finish, Biddy failed to understand; so pertinent, so relevant and congruous, was the other party to this encounter.

The girl's attention followed her brother's, resting with his on a young man who faced them without seeing them, engaged as he was in imparting to two persons who were with him his ideas about one of the works exposed to view. What Biddy discerned was that this young man was fair and fat and of the middle stature; he had a round face and a short beard, and on his crown a mere reminiscence of hair, as the fact that he carried his hat in his hand permitted it to be observed. Bridget Dormer, who was quick, estimated him immediately as a gentleman, but a gentleman unlike any other gentleman she had ever seen. She would have taken him for a foreigner, but that the words proceeding from his mouth reached her ear and imposed themselves as a rare variety of English. It was not that a foreigner might not have spoken excellent English, nor yet that the English of this young man was not excellent. It had, on the contrary, a conspicuous and aggressive perfection, and Biddy was sure that no mere learner would have ventured to play such tricks with the tongue. He seemed to draw rich effects and wandering airs from it – to modulate and manipulate it as he would have done a musical instrument. Her view of the gentleman's companions was less operative, save that she made the rapid reflection that they were people whom in any country, from China to Peru, one would immediately have taken for natives. One of them was an old lady with a shawl; that was the most salient way in which she presented herself. The shawl was an ancient, voluminous fabric of embroidered cashmere, such as many ladies wore forty years ago in their walks abroad, and such as no lady wears today. It had fallen half off the back of the wearer, but at the moment Biddy permitted herself to consider her she gave it a violent jerk and brought it up to her shoulders again, where she continued to arrange and settle it, with a good deal of jauntiness and elegance, while she listened to the talk of the gentleman. Biddy guessed that this little transaction took place very frequently, and she was not unaware that it gave the old lady a droll, factitious, faded appearance, as if she were singularly out of step with the age. The other person was very much younger – she might have been a daughter – and had a pale face, a low forehead and thick, dark hair. What she chiefly had, however, Biddy rapidly discovered, was a pair of largely-gazing eyes. Our young friend was helped to the discovery by the accident of their resting at this moment, for a little while – it struck Biddy as very long – on her own. Both of these ladies were clad in light, thin, scanty gowns, giving an impression of flowered figures and odd transparencies, and in low shoes, which showed a great deal of stocking and were ornamented with large rosettes. Biddy's slightly agitated perception travelled directly to their shoes: they suggested to her vaguely that the wearers were dancers – connected possibly with the old-fashioned exhibition of the shawl-dance. By the time she had taken in so much as this the mellifluous young man had perceived and addressed himself to her brother. He came forward with an extended hand. Nick greeted him and said it was a happy chance – he was uncommonly glad to see him.

'I never come across you – I don't know why,' Nick remarked, while the two, smiling, looked each other up and down, like men reunited after a long interval.

'Oh, it seems to me there's reason enough: our paths in life are so different.' Nick's friend had a great deal of manner, as was evinced by his fashion of saluting her without knowing her.

'Different, yes, but not so different as that. Don't we both live in London, after all, and in the nineteenth century?'

'Ah, my dear Dormer, excuse me: I don't live in the nineteenth century. *Jamais de la vie!*'

'Nor in London either?'

'Yes – when I'm not in Samarcand! But surely we've diverged since the old days. I adore what you burn; you burn what I adore.' While the stranger spoke he looked cheerfully, hospitably, at Biddy; not because it was she, she easily guessed, but because it was in his nature to desire a second auditor – a kind of sympathetic gallery. Her life, somehow, was filled with shy people, and she immediately knew that she had never encountered any one who seemed so to know his part and recognize his cues.

'How do you know what I adore?' Nicholas Dormer inquired.

'I know well enough what you used to.'

'That's more than I do myself; there were so many things.'

'Yes, there are many things – many, many: that's what makes life so amusing.'

'Do you find it amusing?'

'My dear fellow, *c'est à se tordre!* Don't you think so? Ah, it was high time I should meet you – I see. I have an idea you need me.'

'Upon my word, I think I do!' Nick said, in a tone which struck his sister and made her wonder still more why, if the gentleman was so important as that, he didn't introduce him.

'There are many gods, and this is one of their temples,' the mysterious personage went on. 'It's a house of strange idols – isn't it? – and of some curious and unnatural sacrifices.'

To Biddy, as much as to her brother, this remark appeared to be offered; but the girl's eyes turned back to the ladies, who, for the moment, had lost their companion. She felt irresponsive and feared she should pass with this familiar cosmopolite for a stiff, scared English girl, which was not the type she aimed at; but there seemed an interdiction even of ocular commerce so long as she had not a sign from Nick. The elder of the strange women had turned her back and was looking at some bronze figure, losing her shawl again as she did so; but the younger stood where their escort had quitted her, giving all her attention to his sudden sociability with others. Her arms hung at her sides, her head was bent, her face lowered, so that she had an odd appearance of raising her eyes from under her brows; and in this attitude she was striking, though her air was unconciliatory, almost dangerous. Did it express resentment at having been abandoned for another girl? Biddy, who began to be frightened – there was a moment when the forsaken one resembled a tigress about to spring – was tempted to cry out that she had no wish whatever to appropriate the gentleman. Then she made the discovery that the young lady had a manner, almost as much as her cicerone, and the rapid induction that it perhaps meant no more than his. She only looked at Biddy from beneath her eyebrows, which were wonderfully arched, but there was a manner in the way she did it. Biddy had a momentary sense of being a figure in a ballet, a dramatic ballet – a subordinate, motionless figure, to be dashed at, to music, or capered up to. It would be a very dramatic ballad indeed if this young person were the heroine. She had magnificent hair, the girl reflected; and at the same moment she heard Nick say to his interlocutor: 'You're not in London – one can't meet you there?'

'I drift, I float,' was the answer; 'my feelings direct me – if such a life as mine may be said to have a direction. Where there's anything to feel I try to be there!' the young man continued with his confiding laugh.

'I should like to get hold of you,' Nick remarked.

'Well, in that case there would be something to feel. Those are the currents – any sort of personal relation – that govern my career.'

'I don't want to lose you this time,' Nick continued, in a manner that excited Biddy's surprise. A moment before, when his friend had said that he tried to be where there was anything to feel, she had wondered how he could endure him.

'Don't lose me, don't lose me!' exclaimed the stranger, with a countenance and a tone which affected the girl as the highest expression of irresponsibility that she had ever seen. 'After all, why should you? Let us remain together, unless I interfere' – and he looked, smiling and interrogative, at Biddy, who still remained blank, only observing again that Nick forbore to make them acquainted. This was an anomaly, since he prized the gentleman so; but there could be no anomaly of Nick's that would not impose itself upon his younger sister.

'Certainly, I keep you,' said Nick, 'unless, on my side, I deprive those ladies –'

'Charming women, but it's not an indissoluble union. We meet, we communicate, we part! They are going – I'm seeing them to the door. I shall come back.' With this Nick's friend rejoined his companions, who moved away with him, the strange, fine eyes of the girl lingering on Nick, as well as on Biddy, as they receded.

'Who is he – who are they?' Biddy instantly asked.

'He's a gentleman,' Nick replied, unsatisfactorily, and even, as she thought, with a shade of hesitation. He spoke as if she might have supposed he was not one; and if he was really one why didn't he introduce them? But Biddy would not for the world have put this question to her brother, who now moved to the nearest bench and dropped upon it, as if to wait for the other's return. No sooner, however, had his sister seated herself than he said: 'See here, my dear, do you think you had better stay?'

'Do you want me to go back to mother?' the girl asked, with a lengthening visage.

'Well, what do you think?' and Nick smiled down at her.

'Is your conversation to be about – about private affairs?'

'No, I can't say that. But I doubt whether mother would think it the sort of thing that's "necessary to your development".'

This assertion appeared to inspire Biddy with the eagerness with which again she broke out: 'But who are they – who are they?'

'I know nothing of the ladies. I never saw them before. The man's a fellow I knew very well at Oxford. He was thought immense fun there. We have diverged, as he says, and I had almost lost sight of him, but not so much as he thinks, because I've read him, and read him with interest. He has written a very clever book.'

'What kind of a book?'

'A sort of a novel.'

'What sort of a novel?'

'Well, I don't know – with a lot of good writing.' Biddy listened to this with so much interest that she thought it illogical her brother should add: 'I dare say Peter will have come, if you return to mother.'

'I don't care if he has. Peter's nothing to me. But I'll go if you wish it.'

Nick looked down at her again, and then said: 'It doesn't signify. We'll all go.'

'All?' Biddy echoed.

'He won't hurt us. On the contrary, he'll do us good.'

This was possible, the girl reflected in silence, but none the less the idea struck her as courageous – the idea of their taking the odd young man back to breakfast with them and with the others, especially if Peter should be there. If Peter was nothing to her, it was singular she should have attached such importance to this contingency. The odd young man reappeared, and now that she saw him without his queer female appendages he seemed personally less unusual. He struck her, moreover, as generally a good deal accounted for by the literary character, especially if it were responsible for a lot of good writing. As he took his place on the bench Nick said to him, indicating her, 'My sister Bridget,' and then mentioned his name, 'Mr Gabriel Nash.'

'You enjoy Paris – you are happy here?' Mr Nash inquired, leaning over his friend to speak to the girl.

Though his words belonged to the situation, it struck her that his tone didn't, and this made her answer him more dryly than she usually spoke. 'Oh, yes, it's very nice.'

'And French art interests you? You find things here that please?'

'Oh, yes, I like some of them.'

Mr Nash looked at her with kind eyes. 'I hoped you would say you like the Academy better.'

'She would if she didn't think you expected it,' said Nicholas Dormer.

'Oh, Nick!' Biddy protested.

'Miss Dormer is herself an English picture,' Gabriel Nash remarked, smiling like a man whose urbanity was a solvent.

'That's a compliment, if you don't like them!' Biddy exclaimed.

'Ah, some of them, some of them; there's a certain sort of thing!' Mr Nash continued. 'We must feel everything, everything that we can. We are here for that.'

'You do like English art, then?' Nick demanded, with a slight accent of surprise.

Mr Nash turned his smile upon him. 'My dear Dormer, do you remember the old complaint I used to make of you? You had formulas that were like walking in one's hat. One may see something in a case, and one may not.'

'Upon my word,' said Nick, 'I don't know any one who was fonder of a generalization than you. You turned them off as the man at the street-corner distributes hand-bills.'

'They were my wild oats. I've sown them all.'

'We shall see that!'

'Oh, they're nothing now – a tame, scanty, homely growth. My only generalizations are my actions.'

'We shall see *them*, then.'

'Ah, excuse me. You can't see them with the naked eye. Moreover, mine are principally negative. People's actions, I know, are, for the most part, the things they do, but mine are all the things I don't do. There are so many of those, so many, but they don't produce any effect. And then all the rest are shades – extremely fine shades.'

'Shades of behaviour?' Nick inquired, with an interest which surprised his sister; Mr Nash's discourse striking her mainly as the twaddle of the under-world.

'Shades of impression, of appreciation,' said the young man, with his explanatory smile. 'My only behaviour is my feelings.'

'Well, don't you show your feelings? You used to!'

'Wasn't it mainly those of disgust?' Nash asked. 'Those operate no longer. I have closed that window.'

'Do you mean you like everything?'

'Dear me, no! But I look only at what I do like.'

'Do you mean that you have lost the faculty of displeasure?'

'I haven't the least idea. I never try it. My dear fellow,' said Gabriel Nash, 'we have only one life that we know anything about: fancy taking it up with disagreeable impressions! When, then, shall we go in for the agreeable?'

'What do you mean by the agreeable?' Nick Dormer asked.

'Oh, the happy moments of our consciousness – the multiplication of those moments. We must save as many as possible from the dark gulf.'

Nick had excited a certain astonishment on the part of his

sister, but it was now Biddy's turn to make him open his eyes a little. She raised her sweet voice and inquired of Mr Nash –

'Don't you think there are any wrongs in the world – any abuses and sufferings?'

'Oh, so many, so many! That's why one must choose.'

'Choose to stop them, to reform them – isn't that the choice?' Biddy asked. 'That's Nick's,' she added, blushing and looking at this personage.

'Ah, our divergence – yes!' sighed Gabriel Nash. 'There are all kinds of machinery for that – very complicated and ingenious. Your formulas, my dear Dormer, your formulas!'

'Hang 'em, I haven't got any!' Nick exclaimed.

'To me, personally, the simplest ways are those that appeal most,' Mr Nash went on. 'We pay too much attention to the ugly; we notice it, we magnify it. The great thing is to leave it alone and encourage the beautiful.'

'You must be very sure you get hold of the beautiful,' said Nick.

'Ah, precisely, and that's just the importance of the faculty of appreciation. We must train our special sense. It is capable of extraordinary extension. Life's none too long for that.'

'But what's the good of the extraordinary extension if there is no affirmation of it, if it all goes to the negative, as you say? Where are the fine consequences?' Dormer asked.

'In one's own spirit. One is one's self a fine consequence. That's the most important one we have to do with. I am a fine consequence,' said Gabriel Nash.

Biddy rose from the bench at this, and stepped away a little, as if to look at a piece of statuary. But she had not gone far before, pausing and turning, she bent her eyes upon Mr Nash with a heightened colour, an air of hesitation and the question, after a moment: 'Are you an aesthete?'

'Ah, there's one of the formulas! That's walking in one's hat! I've *no* profession, my dear young lady. I've no *état civil*. These things are a part of the complicated, ingenious machinery. As I say, I keep to the simplest way. I find that gives one enough to do. Merely to be is such a *métier*; to live is such an art; to feel is such a career!'

Bridget Dormer turned her back and examined her statue, and her brother said to his old friend: 'And to write?'

'To write? Oh, I'll never do it again!'

'You have done it almost well enough to be inconsistent. That book of yours is anything but negative; it's complicated and ingenious.'

'My dear fellow, I'm extremely ashamed of it,' said Gabriel Nash.

'Ah, call yourself a bloated Buddhist and have done with it!' his companion exclaimed.

'Have done with it! I haven't the least desire for that. And why should one call one's self anything? One only deprives other people of their dearest occupation. Let me add that you don't *begin* to have an insight into the art of life till it ceases to be of the smallest consequence to you what you may be called. That's rudimentary.'

'But if you go in for shades, you must also go in for names. You must distinguish,' Dormer objected. 'The observer is nothing without his categories, his types and varieties.'

'Ah, trust him to distinguish!' said Gabriel Nash, sweetly. 'That's for his own convenience; he has, privately, a terminology to meet it. That's one style. But from the moment it's the convenience of others, the signs have to be grosser, the shades begin to go. That's a deplorable hour! Literature, you see, is for the convenience of others. It requires the most abject concessions. It plays such mischief with one's style that really I have had to give it up.'

'And politics?' Nick Dormer asked.

'Well, what about them?' was Mr Nash's reply, in a peculiar intonation, as he watched his friend's sister, who was still examining her statue. Biddy was divided between irritation and curiosity. She had interposed space, but she had not gone beyond earshot. Nick's question made her curiosity throb, especially in its second form, as a rejoinder to their companion's.

'That, no doubt you'll say, is still far more for the convenience of others – is still worse for one's style.'

Biddy turned round in time to hear Mr Nash exclaim: 'It has simply nothing in life to do with shades! I can't say worse for it than that.'

Biddy stepped nearer at this, and, drawing still further on her courage: 'Won't mamma be waiting? Oughtn't we to go to luncheon?' she asked.

Both the young men looked up at her, and Mr Nash remarked—

'You ought to protest! You ought to save him!'

'To save him?' said Biddy.

'He *had* a style; upon my word he had! But I've seen it go. I've read his speeches.'

'You were capable of that?' Dormer demanded.

'For you, yes. But it was like listening to a nightingale in a brass band.'

'I think they were beautiful,' Biddy declared.

Her brother got up at this tribute, and Mr Nash, rising too, said, with his bright colloquial air –

'But, Miss Dormer, he had eyes. He was made to see – to see all over, to see everything. There are so few like that.'

'I think he still sees,' Biddy rejoined, wondering a little why Nick didn't defend himself.

'He sees his side, dear young lady. Poor man, fancy your having a "side" – you, you – and spending your days and your nights looking at it! I'd as soon pass my life looking at an advertisement on a hoarding.'

'You don't see me some day a great statesman?' said Nick.

'My dear fellow, it's exactly what I've a terror of.'

'Mercy! Don't you admire them?' Biddy cried.

'It's a trade like another, and a method of making one's way which society certainly condones. But when one can be something better!'

'Dear me, what *is* better?' Biddy asked.

The young man hesitated, and Nick, replying for him, said –

'Gabriel Nash is better! You must come and lunch with us. I must keep you – I must!' he added.

'We shall save him yet,' Mr Nash observed genially to Biddy as they went; while the girl wondered still more what her mother would make of him.

3

AFTER her companions left her Lady Agnes rested for five minutes in silence with her elder daughter, at the end of which time she observed, 'I suppose one must have food, at any rate,' and, getting up, quitted the place where they had been sitting. 'And where are we to go? I hate eating out-of-doors,' she went on.

'Dear me, when one comes to Paris!' Grace rejoined, in a tone which appeared to imply that in so rash an adventure one must be prepared for compromises and concessions. The two ladies wandered to where they saw a large sign of 'Buffet' suspended in the air, entering a precinct reserved for little white-clothed tables, straw-covered chairs and long-aproned waiters. One of these functionaries approached them with eagerness and with a *'Mesdames sont seules?'* receiving in return, from her ladyship, the slightly snappish announcement, *'Non; nous sommes beaucoup!'* He introduced them to a table larger than most of the others, and under his protection they took their places at it and began, rather languidly and vaguely, to consider the question of the repast. The waiter had placed a *carte* in Lady Agnes's hands, and she studied it, through her eye-glass, with a failure of interest, while he enumerated, with professional fluency, the resources of the establishment and Grace looked at the people at the other tables. She was hungry and had already broken a morsel from a long glazed roll.

'Not cold beef and pickles, you know,' she observed to her mother. Lady Agnes gave no heed to this profane remark, but she dropped her eye-glass and laid down the greasy document. 'What does it signify? I dare say it's all nasty,' Grace continued; and she added, inconsequently: 'If Peter comes he's sure to be particular.'

'Let him be particular to come, first!' her ladyship exclaimed, turning a cold eye upon the waiter.

'Poulet chasseur, filets mignons, sauce béarnaise,' the man suggested.

'You will give us what I tell you,' said Lady Agnes, and she mentioned, with distinctness and authority, the dishes of which she desired that the meal should be composed. He interposed three or four more suggestions, but as they produced absolutely no impression on her he became silent and submissive, doing justice, apparently, to her ideas. For Lady Agnes had ideas; and though it had suited her humour, ten minutes before, to profess herself helpless in such a case, the manner in which she imposed them upon the waiter as original, practical and economical showed the high, executive woman, the mother of children, the daughter of earls, the consort of an official, the dispenser of hospitality, looking back upon a life-time of luncheons. She carried many cares, and the feeding of multitudes (she was honourably conscious of having fed them decently, as she had always done everything) had ever been one of them. 'Everything is absurdly dear,' she hinted to her daughter, as the waiter went away. To this remark Grace made no answer. She had been used, for a long time back, to hearing that everything was very dear; it was what one always expected. So she found the case herself, but she was silent and inventive about it.

Nothing further passed, in the way of conversation with her mother, while they waited for the latter's orders to be executed, till Lady Agnes reflected, audibly: 'He makes me unhappy, the way he talks about Julia!'

'Sometimes I think he does it to torment one. One can't mention her!' Grace responded.

'It's better not to mention her, but to leave it alone.'

'Yet he never mentions her of himself.'

'In some cases that is supposed to show that people like people – though of course something more than that is required,' Lady Agnes continued to meditate. 'Sometimes I think he's thinking of her; then at others, I can't fancy *what* he's thinking of.'

'It would be awfully suitable,' said Grace, biting her roll.

Her mother was silent a moment, as if she were looking for some higher ground to put it upon. Then she appeared to find this loftier level in the observation: 'Of course he must like her; he has known her always.'

'Nothing can be plainer than that she likes him,' Grace declared.

'Poor Julia!' Lady Agnes exclaimed; and her tone suggested that she knew more about that than she was ready to state.

'It isn't as if she wasn't clever and well read,' her daughter went on. 'If there were nothing else there would be a reason in her being so interested in politics, in everything that he is.'

'Ah, what he is – that's what I sometimes wonder!'

Grace Dormer looked at her mother a moment. 'Why, mother, isn't he going to be like papa?' She waited for an answer that didn't come; after which she pursued: 'I thought you thought him so like him already.'

'Well, I don't,' said Lady Agnes, quietly.

'Who is, then? Certainly Percy isn't.'

Lady Agnes was silent a moment. 'There is no one like your father.'

'Dear papa!' Grace exclaimed. Then, with a rapid transition: 'It would be so jolly for all of us; she would be so nice to us.'

'She is that already, in her way,' said Lady Agnes, conscientiously, having followed the return, quick as it was. 'Much good does it do her!' And she reproduced the note of her ejaculation of a moment before.

'It does her some, if one looks out for her. I do, and I think she knows it,' Grace declared. 'One can, at any rate, keep other women off.'

'Don't meddle! you're very clumsy,' was her mother's not particularly sympathetic rejoinder. 'There are other women who are beautiful, and there are others who are clever and rich.'

'Yes, but not all in one; that's what's so nice in Julia. Her fortune would be thrown in; he wouldn't appear to have married her for it.'

'If he does, he won't,' said Lady Agnes, a trifle obscurely.

'Yes, that's what's so charming. And he could do anything then, couldn't he?'

'Well, your father had no fortune, to speak of.'

'Yes, but didn't Uncle Percy help him?'

'His wife helped him,' said Lady Agnes.

'Dear mamma!' the girl exclaimed. 'There's one thing,' she added: 'that Mr Carteret will always help Nick.'

'What do you mean by "always"?'

'Why, whether he marries Julia or not.'

'Things are not so easy,' responded Lady Agnes. 'It will all depend on Nick's behaviour. He can stop it tomorrow.'

Grace Dormer stared; she evidently thought Mr Carteret's beneficence a part of the scheme of nature. 'How could he stop it?'

'By not being serious. It isn't so hard to prevent people giving you money.'

'Serious?' Grace repeated. 'Does he want him to be a prig, like Lord Egbert?'

'Yes, he does. And what he'll do for him he'll do for him only if he marries Julia.'

'Has he told you?' Grace inquired. And then, before her mother could answer, she exclaimed: 'I'm delighted at that!'

'He hasn't told me, but that's the way things happen.' Lady Agnes was less optimistic than her daughter, and such optimism as she cultivated was a thin tissue, with a sense of things as they are showing through it. 'If Nick becomes rich, Charles Carteret will make him more so. If he doesn't, he won't give him a shilling.'

'Oh, mamma!' Grace protested.

'It's all very well to say that in public life money isn't necessary, as it used to be,' her ladyship went on, broodingly. 'Those who say so don't know anything about it. It's always necessary.'

Her daughter was visibly affected by the gloom of her manner, and felt impelled to evoke, as a corrective, a more cheerful idea. 'I dare say; but there's the fact – isn't there? – that poor papa had so little.'

'Yes, and there's the fact that it killed him!'

These words came out with a strange, quick little flare of passion. They startled Grace Dormer, who jumped in her place and cried, 'Oh, mother!' The next instant, however, she added, in a different voice, 'Oh, Peter!' for, with an air of eagerness, a gentleman was walking up to them.

'How d'ye do, Cousin Agnes? How d'ye do, little Grace?' Peter Sherringham said, laughing and shaking hands with them; and three minutes later he was settled in his chair at their table, on which the first elements of the repast had been placed. Explanations, on one side and the other, were demanded and produced; from which it appeared that the two parties had been in some degree at cross-purposes. The day before Lady Agnes and her companions travelled to Paris, Sherringham had gone to London for forty-eight hours, on private business of the ambassador's, arriving, on his return by the night-train, only early that morning. There had accordingly been a delay in his receiving Nick Dormer's two notes. If Nick had come to the Embassy in person (he might have done him the honour to call), he would have learned that the second secretary was absent. Lady Agnes was not altogether successful in assigning a motive to her son's neglect of this courteous form; she said: 'I expected him, I wanted him, to go; and indeed, not hearing from you, he would have gone immediately – an hour or two hence, on leaving this place. But we are here so quietly, not to go out, not to seem to appeal to the ambassador. He said, "Oh, mother, we'll keep out of it; a friendly note will do." I don't know, definitely, what he wanted to keep out of, except it's anything like gaiety. The Embassy isn't gay, I know. But I'm sure his note was friendly, wasn't it? I dare say you'll see for yourself; he's different directly he gets abroad; he doesn't seem to care.' Lady Agnes paused a moment, not carrying out this particular elucidation; then she resumed: 'He said you would have seen Julia and that you would understand everything from her. And when I asked how she would know, he said, "Oh, she knows everything!"'

'He never said a word to me about Julia,' Peter Sherringham rejoined. Lady Agnes and her daughter exchanged a glance at this; the latter had already asked three times where Julia was, and her ladyship dropped that they had been hoping she would be able to come with Peter. The young man set forth that she was at that moment at an hotel in the Rue de la Paix, but had only been there since that morning: he had seen her before coming to the Champs Elysées. She had come up to Paris by an early train – she had been staying at Versailles, of all places in the world. She had been a week in Paris, on her return from Cannes (her stay *there* had been of nearly a month – fancy!) and then had gone out to Versailles to see Mrs Billinghurst. Perhaps they would remember her, poor Dallow's sister. She was staying there to teach her daughters French (she had a dozen or two!) and Julia had spent three days with her. She was to return to England about the 25th. It would make seven weeks that she would have been away from town – a rare thing for her; she usually stuck to it so in summer.

'Three days with Mrs Billinghurst – how very good-natured of her!' Lady Agnes commented.

'Oh, they're very nice to her,' Sherringham said.

'Well, I hope so!' Grace Dormer qualified. 'Why didn't you make her come here?'

'I proposed it, but she wouldn't.' Another eye-beam, at this, passed between the two ladies, and Peter went on: 'She said you must come and see her, at the Hôtel de Hollande.'

'Of course we'll do that,' Lady Agnes declared. 'Nick went to ask about her at the Westminster.'

'She gave that up; they wouldn't give her the rooms she wanted, her usual set.'

'She's delightfully particular!' Grace murmured. Then she added: 'She *does* like pictures, doesn't she?'

Peter Sherringham stared. 'Oh, I dare say. But that's not what she has in her head this morning. She has some news from London; she's immensely excited.'

'What has she in her head?' Lady Agnes asked.

'What's her news from London?' Grace demanded.

'She wants Nick to stand.'

'Nick to stand?' both the ladies cried.

'She undertakes to bring him in for Harsh. Mr Pinks is dead – the fellow, you know, that got the seat at the general election. He dropped down in London – disease of the heart, or something of that sort. Julia has her telegram, but I see it was in last night's papers.'

'Imagine, Nick never mentioned it!' said Lady Agnes.

'Don't you know, mother? – abroad he only reads foreign papers.'

'Oh, I know. I've no patience with him,' her ladyship continued. 'Dear Julia!'

'It's a nasty little place, and Pinks had a tight squeeze – 107, or something of that sort; but if it returned a Liberal a year ago, very likely it will do so again. Julia, at any rate, *se fait forte*, as they say here, to put him in.'

35

'I'm sure if she can she will,' Grace reflected.

'Dear, dear Julia! And Nick can do something for himself,' said the mother of this candidate.

'I have no doubt he can do anything,' Peter Sherringham returned, good-naturedly. Then, 'Do you mean in expenses?' he inquired.

'Ah, I'm afraid he can't do much in expenses, poor dear boy! And it's dreadful how little we can look to Percy.'

'Well, I dare say you may look to Julia. I think that's her idea.'

'Delightful Julia!' Lady Agnes ejaculated. 'If poor Sir Nicholas could have known! Of course he must go straight home,' she added.

'He won't like that,' said Grace.

'Then he'll have to go without liking it.'

'It will rather spoil *your* little excursion, if you've only just come,' Peter suggested; 'and the great Biddy's, if she's enjoying Paris.'

'We may stay, perhaps – with Julia to protect us,' said Lady Agnes.

'Ah, she won't stay; she'll go over for her man.'

'Her man?'

'The fellow that stands, whoever he is; especially if he's Nick.' These last words caused the eyes of Peter Sherringham's companions to meet again, and he went on: 'She'll go straight down to Harsh.'

'Wonderful Julia!' Lady Agnes panted. 'Of course Nick must go straight there, too.'

'Well, I suppose he must see first if they'll have him.'

'If they'll have him? Why, how can he tell till he tries?'

'I mean the people at headquarters, the fellows who arrange it.'

Lady Agnes coloured a little. 'My dear Peter, do you suppose there will be the least doubt of their "having" the son of his father?'

'Of course it's a great name, Cousin Agnes – a very great name.'

'One of the greatest, simply,' said Lady Agnes, smiling.

'It's the best name in the world!' Grace Dormer subjoined.

'All the same it didn't prevent his losing his seat.'

'By half a dozen votes: it was too odious!' her ladyship cried.

'I remember – I remember. And in such a case as that why didn't they immediately put him in somewhere else?'

'How one sees that you live abroad, Peter! There happens to have been the most extraordinary lack of openings – I never saw anything like it – for a year. They've had their hand on him, keeping him all ready. I dare say they've telegraphed to him.'

'And he hasn't told you?'

Lady Agnes hesitated. 'He's so odd when he's abroad!'

'At home, too, he lets things go,' Grace interposed. 'He does so little – takes no trouble.' Her mother suffered this statement to pass unchallenged, and she pursued, philosophically: 'I suppose it's because he knows he's so clever.'

'So he is, dear old boy. But what does he do, what has he been doing, in a positive way?'

'He has been painting.'

'Ah, not seriously!' Lady Agnes protested.

'That's the worst way,' said Peter Sherringham. 'Good things?'

Neither of the ladies made a direct response to this, but Lady Agnes said: 'He has spoken repeatedly. They are always calling on him.'

'He speaks magnificently,' Grace attested.

'That's another of the things I love, living in far countries. And he's doing the Salon now, with the great Biddy?'

'Just the things in this part. I can't think what keeps them so long,' Lady Agnes rejoined. 'Did you ever see such a dreadful place?'

Sherringham stared. 'Aren't the things good? I had an idea –'

'Good?' cried Lady Agnes. 'They're too odious, too wicked.'

'Ah,' said Peter, laughing, 'that's what people fall into, if they live abroad. The French oughtn't to live abroad.'

'Here they come,' Grace announced, at this point; 'but they've got a strange man with them.'

'That's a bore, when we want to talk!' Lady Agnes sighed.

Peter got up, in the spirit of welcome, and stood a moment watching the others approach. 'There will be no difficulty in talking, to judge by the gentleman,' he suggested; and while he remains so conspicuous our eyes may rest on him briefly. He was middling high and was visibly a representative of the nervous rather than of the phlegmatic branch of his race. He had an oval face, fine, firm features and a complexion that tended to the brown. Brown were his eyes, and women thought them soft; dark brown his hair, in which the same critics sometimes regretted the absence of a little undulation. It was perhaps to conceal this plainness that he wore it very short. His teeth were white; his moustache was pointed, and so was the small beard that adorned the extremity of his chin. His face expressed intelligence and was very much alive, and had the further distinction that it often struck superficial observers with a certain foreignness of cast. The deeper sort, however, usually perceived that it was English enough. There was an idea that, having taken up the diplomatic career and gone to live in strange lands, he cultivated the mask of an alien, an Italian or a Spaniard; of an alien in time, even – one of the wonderful ubiquitous diplomatic agents of the sixteenth century. In fact, it would have been impossible to be more modern than Peter Sherringham, and more of one's class and one's country. But this did not prevent a portion of the community – Bridget Dormer, for instance – from admiring the hue of his cheek for its olive richness and his moustache and beard for their resemblance to those of Charles I. At the same time – she rather jumbled her comparisons – she thought he looked like a Titian.

4

PETER'S meeting with Nick was of the friendliest on both sides, involving a great many 'dear fellows' and 'old boys', and his salutation to the younger of the Miss Dormers consisted of the frankest 'Delighted to see you, my dear Bid!' There was no kissing, but there was cousinship in the air, of a conscious, living kind, as Gabriel Nash no doubt quickly perceived, hovering for a moment outside the group. Biddy said nothing to

Peter Sherringham, but there was no flatness in a silence which afforded such opportunities for a pretty smile. Nick introduced Gabriel Nash to his mother and to the other two as 'a delightful old friend', whom he had just come across, and Sherringham acknowledged the act by saying to Mr Nash, but as if rather less for his sake than for that of the presenter: 'I have seen you very often before.'

'Ah, repetition – recurrence: we haven't yet, in the study of how to live, abolished that clumsiness, have we?' Mr Nash genially inquired. 'It's a poverty in the supernumeraries that we don't pass once for all, but come round and cross again, like a procession at the theatre. It's a shabby economy that ought to have been managed better. The right thing would be just *one* appearance, and the procession, regardless of expense, forever and forever different.'

The company was occupied in placing itself at table, so that the only disengaged attention, for the moment, was Grace's, to whom, as her eyes rested on him, the young man addressed these last words with a smile. 'Alas, it's a very shabby idea, isn't it? The world isn't got up regardless of expense!'

Grace looked quickly away from him, and said to her brother: 'Nick, Mr Pinks is dead.'

'Mr Pinks?' asked Gabriel Nash, appearing to wonder where he should sit.

'The member for Harsh; and Julia wants you to stand,' the girl went on.

'Mr Pinks, the member for Harsh? What names to be sure!' Gabriel mused cheerfully, still unseated.

'Julia wants me? I'm much obliged to her!' observed Nicholas Dormer. 'Nash, please sit by my mother, with Peter on her other side.'

'My dear, it isn't Julia,' Lady Agnes remarked, earnestly, to her son. 'Everyone wants you. Haven't you heard from your people? Didn't you know the seat was vacant?'

Nick was looking round the table, to see what was on it. 'Upon my word I don't remember. What else have you ordered, mother?'

'There's some *bœuf braisé*, my dear, and afterwards some galantine. Here is a dish of eggs with asparagus-tips.'

'I advise you to go in for it, Nick,' said Peter Sherringham, to whom the preparation in question was presented.

'Into the eggs with asparagus-tips? *Donnez m'en, s'il vous plaît*. My dear fellow, how can I stand? how can I sit? Where's the money to come from?'

'The money? Why, from Jul—' Grace began, but immediately caught her mother's eye.

'Poor Julia, how you do work her!' Nick exclaimed. 'Nash, I recommend the asparagus-tips. Mother, he's my best friend; do look after him.'

'I have an impression I have breakfasted – I am not sure,' Nash observed.

'With those beautiful ladies? Try again; you'll find out.'

'The money can be managed; the expenses are very small and the seat is certain,' Lady Agnes declared, not, apparently, heeding her son's injunction in respect to Nash.

'Rather – if Julia goes down!' her elder daughter exclaimed.

'Perhaps Julia won't go down!' Nick answered, humorously.

Biddy was seated next to Mr Nash, so that she could take occasion to ask, 'Who are the beautiful ladies?' as if she failed to recognize her brother's allusion. In reality this was an innocent trick: she was more curious than she could have given a suitable reason for about the odd women from whom her neighbour had separated.

'Deluded, misguided, infatuated persons!' Gabriel Nash replied, understanding that she had asked for a description. 'Strange, eccentric, almost romantic types. Predestined victims, simple-minded sacrificial lambs!'

This was copious, yet it was vague, so that Biddy could only respond, 'Oh!' But meanwhile Peter Sherringham said to Nick –

'Julia's here, you know. You must go and see her.'

Nick looked at him for an instant rather hard, as if to say, 'You too?' But Peter's eyes appeared to answer, 'No, no, not I'; upon which his cousin rejoined: 'Of course I'll go and see her. I'll go immediately. Please to thank her for thinking of me.'

'Thinking of you? There are plenty to think of you!' Lady Agnes said. 'There are sure to be telegrams at home. We must go back – we must go back!'

'We must go back to England?' Nick Dormer asked; and as his mother made no answer he continued: 'Do you mean I must go to Harsh?'

Her ladyship evaded this question, inquiring of Mr Nash if he would have a morsel of fish; but her gain was small, for this gentleman, struck again by the unhappy name of the bereaved constituency, only broke out: 'Ah, what a place to represent! How can you – how can you?'

'It's an excellent place,' said Lady Agnes, coldly. 'I imagine you have never been there. It's a very good place indeed. It belongs very largely to my cousin, Mrs Dallow.'

Gabriel partook of the fish, listening with interest. 'But I thought we had no more pocket-boroughs.'

'It's pockets we rather lack, so many of us. There are plenty of Harshes,' Nick Dormer observed.

'I don't know what you mean,' Lady Agnes said to Gabriel, with considerable majesty.

Peter Sherringham also addressed him with an 'Oh, it's all right; they come down on you like a shot!' and the young man continued, ingenuously –

'Do you mean to say you have to pay to get into that place – that it's not *you* that are paid?'

'Into that place?' Lady Agnes repeated, blankly.

'Into the House of Commons. That you don't get a high salary?'

'My dear Nash, you're delightful: don't leave me – don't leave me!' Nick cried; while his mother looked at him with an eye that demanded: 'Who is this extraordinary person?'

'What then did you think pocket-boroughs were?' Peter Sherringham asked.

Mr Nash's facial radiance rested on him. 'Why, boroughs that filled your pocket. To do that sort of thing without a bribe – *c'est trop fort!*'

'He lives at Samarcand,' Nick Dormer explained to his mother, who coloured perceptibly. 'What do you advise me? I'll do whatever you say,' he went on to his old acquaintance.

'My dear – my dear!' Lady Agnes pleaded.

'See Julia first, with all respect to Mr Nash. She's of excellent counsel,' said Peter Sherringham.

Gabriel Nash smiled across the table at Dormer. 'The lady first – the lady first! I have not a word to suggest as against any idea of hers.'

'We must not sit here too long, there will be so much to do,' said Lady Agnes, anxiously, perceiving a certain slowness in the service of the *bœuf braisé*.

Biddy had been up to this moment mainly occupied in looking, covertly and at intervals, at Peter Sherringham; as was perfectly lawful in a young lady with a handsome cousin whom she had not seen for more than a year. But her sweet voice now took licence to throw in the words: 'We know what Mr Nash thinks of politics: he told us just now he thinks they are dreadful.'

'No, not dreadful – only inferior,' the personage impugned protested. 'Everything is relative.'

'Inferior to what?' Lady Agnes demanded.

Mr Nash appeared to consider a moment. 'To anything else that may be in question.'

'Nothing else *is* in question!' said her ladyship, in a tone that would have been triumphant if it had not been dry.

'Ah, then!' And her neighbour shook his head sadly. He turned, after this, to Biddy, saying to her: 'The ladies whom I was with just now, and in whom you were so good as to express an interest?' Biddy gave a sign of assent, and he went on: 'They are persons theatrical; the younger one is trying to go upon the stage.'

'And are you assisting her?' Biddy asked, pleased that she had guessed so nearly right.

'Not in the least – I'm rather heading her off. I consider it the lowest of the arts.'

'Lower than politics?' asked Peter Sherringham, who was listening to this.

'Dear, no, I won't say that. I think the Théâtre Français a greater institution than the House of Commons.'

'I agree with you there!' laughed Sherringham; 'all the more that I don't consider the dramatic art a low one. On the contrary, it seems to me to include all the others.'

'Yes – that's a view. I think it's the view of my friends.'

'Of your friends?'

'Two ladies – old acquaintances – whom I met in Paris a week ago and whom I have just been spending an hour with in this place.'

'You should have seen them; they struck me very much,' Biddy said to her cousin.

'I should like to see them if they have really anything to say to the theatre.'

'It can easily be managed. Do you believe in the theatre?' asked Gabriel Nash.

'Passionately,' Sherringham confessed. 'Don't you?'

Before Mr Nash had had time to answer Biddy had interposed with a sigh: 'How I wish I could go – but in Paris I can't!'

'I'll take you, Biddy – I vow I'll take you.'

'But the plays, Peter,' the girl objected. 'Mamma says they're worse than the pictures.'

'Oh, we'll arrange that: they shall do one at the Français on purpose for a delightful little English girl.'

'Can you make them?'

'I can make them do anything I choose.'

'Ah, then, it's the theatre that believes in you,' said Gabriel Nash.

'It would be ungrateful if it didn't!' Peter Sherringham laughed.

Lady Agnes had withdrawn herself from between him and Mr Nash, and, to signify that she, at least, had finished eating, had gone to sit by her son, whom she held, with some importunity, in conversation. But hearing the theatre talked of, she threw across an impersonal challenge to the paradoxical young man. 'Pray, should you think it better for a gentleman to be an actor?'

'Better than being a politician? Ah, comedian for comedian, isn't the actor more honest?'

Lady Agnes turned to her son and exclaimed with spirit: 'Think of your great father, Nicholas!'

'He was an honest man; that perhaps is why he couldn't stand it.'

Peter Sherringham judged the colloquy to have taken an uncomfortable twist, though not wholly, as it seemed to him, by

43

the act of Nick's queer comrade. To draw it back to safer ground he said to this personage: 'May I ask if the ladies you just spoke of are English – Mrs and Miss Rooth: isn't that the rather odd name?'

'The very same. Only the daughter, according to her kind, desires to be known by some *nom de guerre* before she has even been able to enlist.'

'And what does she call herself?' Bridget Dormer asked.

'Maud Vavasour, or Edith Temple, or Gladys Vane – some rubbish of that sort.'

'What, then, is her own name?'

'Miriam – Miriam Rooth. I would do very well and would give her the benefit of the prepossessing fact that (to the best of my belief, at least) she is more than half a Jewess.'

'It is as good as Rachel Félix,' Sherringham said.

'The name's as good, but not the talent. The girl is magnificently stupid.'

'And more than half a Jewess? Don't you believe it!' Sherringham exclaimed.

'Don't believe she's a Jewess?' Biddy asked, still more interested in Miriam Rooth.

'No, no – that she's stupid, really. If she is, she'll be the first.'

'Ah, you may judge for yourself,' Nash rejoined, 'if you'll come tomorrow afternoon to Madame Carré, Rue de Constantinople, *à l'entresol*.'

'Madame Carré? Why, I've already a note from her – I found it this morning on my return to Paris – asking me to look in at five o'clock and listen to a *jeune Anglaise*.'

'That's my arrangement – I obtained the favour. The ladies want an opinion, and dear old Carré has consented to see them and to give one. Gladys will recite something and the venerable artist will pass judgement.'

Sherringham remembered that he had his note in his pocket, and he took it out and looked it over. 'She wishes to make her a little audience – she says she'll do better with that – and she asks me because I'm English. I shall make a point of going.'

'And bring Dormer if you can: the audience will be better. Will you come, Dormer?' Mr Nash continued, appealing to his

44

friend, – 'will you come with me to see an old French actress and to hear an English amateur recite?'

Nick looked round from his talk with his mother and Grace. 'I'l go anywhere with you, so that, as I've told you, I may not lose sight of you, may keep hold of you.'

'Poor Mr Nash, why is he so useful?' Lady Agnes demanded with a laugh.

'He steadies me, mother.'

'Oh, I wish you'd take me, Peter,' Biddy broke out, wistfully, to her cousin.

'To spend an hour with an old French actress? Do you want to go upon the stage?' the young man inquired.

'No, but I want to see something, to know something.'

'Madame Carré is wonderful in her way, but she is hardly company for a little English girl.'

'I'm not little, I'm only too big; and *she* goes, the person you speak of.'

'For a professional purpose, and with her good mother,' smiled Gabriel Nash. 'I think Lady Agnes would hardly venture –'

'Oh, I've seen her good mother!' said Biddy, as if she had an impression of what the worth of that protection might be.

'Yes, but you haven't heard her. It's then that you measure her.'

Biddy was wistful still. 'Is it the famous Honorine Carré, the great celebrity?'

'Honorine in person: the incomparable, the perfect!' said Peter Sherringham. 'The first artist of our time, taking her altogether. She and I are old pals; she has been so good as to come and "say" things, as she does sometimes still *dans le monde* as no one else does, in my rooms.'

'Make her come, then; we can go *there*!'

'One of these days!'

'And the young lady – Miriam, Edith, Gladys – make her come too.'

Sherringham looked at Nash, and the latter exclaimed: 'Oh, you'll have no difficulty; she'll jump at it!'

'Very good; I'll give a little artistic tea, with Julia, too, of course. And you must come, Mr Nash.' This gentleman promised, with an inclination, and Peter continued: 'But if, as you say, you're not for helping the young lady, how came you to arrange this interview with the great model?'

'Precisely to stop her. The great model will find her very bad. Her judgements, as you probably know, are Rhadamanthine.'

'Poor girl!' said Biddy. 'I think you're cruel.'

'Never mind; I'll look after them,' said Sherringham.

'And how can Madame Carré judge, if the girl recites English?'

'She's so intelligent that she could judge if she recited Chinese,' Peter declared.

'That's true, but the *jeune Anglaise* recites also in French,' said Gabriel Nash.

'Then she isn't stupid.'

'And in Italian, and in several more tongues, for aught I know.'

Sherringham was visibly interested. 'Very good; we'll put her through them all.'

'She must be *most* clever,' Biddy went on, yearningly.

'She has spent her life on the Continent; she has wandered about with her mother; she has picked up things.'

'And is she a lady?' Biddy asked.

'Oh, tremendous! The great ones of the earth on the mother's side. On the father's, on the other hand, I imagine, only a Jew stockbroker in the city.'

'Then they're rich – or ought to be,' Sherringham suggested.

'Ought to be – ah, there's the bitterness! The stockbroker had too short a go – he was carried off in his flower. However, he left his wife a certain property, which she appears to have muddled away, not having the safeguard of being herself a Hebrew. This is what she lived upon till today – this and another resource. Her husband, as she has often told me, had the artistic temperament; that's common, as you know, among *ces messieurs*. He made the most of his little opportunities and collected various pictures, tapestries, enamels, porcelains and similar gewgaws. He parted with them also, I gather, at a profit; in short, he carried on a neat little business as a *brocanteur*. It was nipped in the bud, but Mrs Rooth was left with a certain number of these articles in her hands; indeed they must have constituted the most palpable part of her heritage. She was not a woman of business; she turned them, no doubt, to indifferent account; but she sold them piece by piece, and they kept her going while her daughter grew up. It was to this precarious traffic, conducted with extraordinary mystery and delicacy, that, five years ago, in Florence, I was indebted for my acquaintance with her. In those days I used to collect – Heaven help me! – I used to pick up rubbish which I could ill afford. It was a little phase – we have our little phases, haven't we?' asked Gabriel Nash, with childlike trust – 'and I have come out on the other side. Mrs Rooth had an old green pot, and I heard of her old green pot. To hear of it was to long for it, so that I went to see it, under cover of night. I bought it, and a couple of years ago I overturned it and smashed it. It was the last of the little phase. It was not, however, as you have seen, the last of Mrs Rooth. I saw her afterwards in London, and I met her a year or two ago in Venice. She appears to be a great wanderer. She had other old pots, of other colours – red, yellow, black, or blue – she could produce them of any complexion you liked. I don't know whether she carried them about with her or whether she had little secret stores in the principal cities of Europe. Today, at any rate, they seem all gone. On the other hand she has her daughter, who has grown up and who is a precious vase of another kind – less fragile, I hope, than the rest. May she not be overturned and smashed!'

Peter Sherringham and Biddy Dormer listened with attention to this history, and the girl testified to the interest with which she had followed it by saying, when Mr Nash had ceased speaking: 'A Jewish stockbroker, a dealer in curiosities: what an odd person to marry – for a person who was well born! I dare say he was a German.'

'His name must have been simply Roth, and the poor lady, to smarten it up, has put in another *o*,' Sherringham ingeniously suggested.

'You are both very clever,' said Gabriel Nash, 'and Rudolf Roth, as I happen to know, was indeed the designation of Maud Vavasour's papa. But, as far as the question of derogation goes, one might as well drown as starve, for what connection is *not* a misalliance when one happens to have the cumbersome, the unaccommodating honour of being a Neville-Nugent of Castle Nugent? Such was the high lineage of Maud's mamma. I seem to have heard it mentioned that Rudolf Roth was very versatile and, like most of his species, not unacquainted with the practice of music. He had been employed to teach the harmonium to Miss Neville-Nugent and she had profited by his lessons. If his daughter is like him – and she is not like her mother – he was darkly and dangerously handsome. So I venture rapidly to reconstruct the situation.'

A silence, for the moment, had fallen upon Lady Agnes and her other two children, so that Mr Nash, with his universal urbanity, practically addressed these last remarks to them as well as to his other auditors. Lady Agnes looked as if she wondered whom he was talking about, and having caught the name of a noble residence she inquired –

'Castle Nugent – where is that?'

'It's a domain of immeasurable extent and almost inconceivable splendour, but I fear it isn't to be found in any prosaic earthly geography!' Lady Agnes rested her eyes on the tablecloth, as if she were not sure a liberty had not been taken with her, and while Mr Nash continued to abound in descriptive suppositions – 'It must be on the banks of the Manzanares or the Guadalquivir' – Peter Sherringham, whose imagination appeared to have been strongly kindled by the sketch of Miriam Rooth challenging him sociably, reminded him that he had a short time before assigned a low place to the dramatic art and had not yet answered his question as to whether he believed in the theatre. This gave Nash an opportunity to go on:

'I don't know that I understand your question; there are different ways of taking it. Do I think it's important? Is that what you mean? Important, certainly, to managers and stage-carpenters who want to make money, to ladies and gentlemen who want to produce themselves in public by lime-light, and to other ladies and gentlemen who are bored and stupid and don't know what to do with their evening. It's a commercial and social convenience which may be infinitely worked. But important artistically, intellectually? How *can* it be – so poor, so limited a form?'

'Dear me, it strikes me as so rich, so various! Do *you* think it's poor and limited, Nick?' Sherringham added, appealing to his kinsman.

'I think whatever Nash thinks. I have no opinion today but his.'

This answer of Nick Dormer's drew the eyes of his mother and sisters to him, and caused his friend to exclaim that he was not used to such responsibilities, so few people had ever tested his presence of mind by agreeing with him.

'Oh, I used to be of your way of feeling,' Nash said to Sherringham. 'I understand you perfectly. It's a phase like another. I've been through it – *j'ai été comme ça*.'

'And you went, then, very often to the Théâtre Français, and it was there I saw you. I place you now.'

'I'm afraid I noticed none of the other spectators,' Nash explained. 'I had no attention but for the great Carré – she was still on the stage. Judge of my infatuation, and how I can allow for yours, when I tell you that I sought her acquaintance, that I couldn't rest till I had told her that I hung upon her lips.'

'That's just what *I* told her,' returned Sherringham.

'She was very kind to me. She said, "*Vous me rendez des forces.*"'

'That's just what she said to me!'

'And we have remained very good friends.'

'So have *we*!' laughed Sherringham. 'And such perfect art as hers: do you mean to say you don't consider *that* important – such a rare dramatic intelligence?'

'I'm afraid you read the *feuilletons*. You catch their phrases,' Gabriel Nash blandly rejoined. 'Dramatic intelligence is never rare; nothing is more common.'

'Then why have we so many bad actors?'

'Have we? I thought they were mostly good; succeeding more easily and more completely in that business than in anything else. What could they do – those people, generally – if they didn't do that? And reflect that *that* enables them to succeed! Of course, always, there are numbers of people on the stage who are no actors at all, for it's even easier to our poor humanity to be ineffectively stupid and vulgar than to bring down the house.'

'It's not easy, by what I can see, to produce, completely, any artistic effect,' Sherringham declared; 'and those that the actor produces are among the most moving that we know. You'll not persuade me that to watch such an actress as Madame Carré was not an education of the taste, an enlargement of one's knowledge.'

'She did what she could, poor woman, but in what belittling, coarsening conditions! She had to interpret a character in a play, and a character in a play (not to say the whole piece – I speak more particularly of modern pieces) is such a wretchedly small peg to hang anything on! The dramatist shows us so little, is so hampered by his audience, is restricted to so poor an analysis.'

'I know the complaint. It's all the fashion now. The *raffinés* despise the theatre,' said Peter Sherringham, in the manner of a man abreast with the culture of his age and not to be captured by a surprise. '*Connu, connu!*'

'It will be known better yet, won't it? when the essentially brutal nature of the modern audience is still more perceived, when it has been properly analysed: the *omnium gatherum* of the population of a big commercial city, at the hour of the day when their taste is at its lowest, flocking out of hideous hotels and restaurants, gorged with food, stultified with buying and selling and with all the other sordid speculations of the day, squeezed together in a sweltering mass, disappointed in their seats, timing the author, timing the actor, wishing to get their money back on the spot, before eleven o'clock. Fancy putting the exquisite before such a tribunal as that! There's not even a question of it. The dramatist wouldn't if he could, and in nine cases out of ten he couldn't if he would. He has to make the basest concessions. One of his principal canons is that he must enable his spectators to catch the surburban trains, which stop at 11.30. What would you think of any other artist – the painter or the novelist – whose governing forces should be the dinner and the suburban trains? The old dramatists didn't defer to them (not so much, at least), and that's why they are less and less actable. If they are touched – the large fellows – it's only to be mutilated and trivialized. Besides, they had a simpler civilization to represent – societies in which the life of man was in action, in passion, in immediate and violent expression. Those things could be put upon the playhouse boards with comparatively little sacrifice of their completeness and their truth. Today we are so infinitely more reflective and complicated and diffuse that it makes all the difference. What can you do with a character, with an idea, with a feeling, between dinner and the suburban trains? You can give a gross, rough sketch of them, but how little you touch them, how bald you leave them! What crudity compared with what the novelist does!'

'Do you write novels, Mr Nash?' Peter demanded.

'No, but I read them when they are extraordinarily good, and I don't go to plays. I read Balzac, for instance – I encounter the magnificent portrait of Valérie Marneffe, in "La Cousine Bette".'

'And you contrast it with the poverty of Emile Augier's Séraphine in "Les Lionnes Pauvres"? I was awaiting you there. That's the *cheval de bataille* of you fellows.'

'What an extraordinary discussion! What dreadful authors!' Lady Agnes murmured to her son. But he was listening so attentively to the other young men that he made no response, and Peter Sherringham went on:

'I have seen Madame Carré in parts, in the modern repertory, which she has made as vivid to me, caused to abide as ineffaceably in my memory, as Valérie Marneffe. She is the Balzac, as one may say, of actresses.'

'The miniaturist, as it were, of whitewashers!' Nash rejoined, laughing.

It might have been guessed that Sherringham was irritated, but the other disputant was so good-humoured that he abundantly recognized his own obligation to appear so.

'You would be magnanimous if you thought the young lady you have introduced to our old friend would be important.'

'She might be much more so than she ever will be.'

Lady Agnes got up, to terminate the scene, and even to signify that enough had been said about people and questions she had never heard of. Everyone else rose, the waiter brought Nick the receipt of the bill, and Sherringham went on, to his interlocutor –

'Perhaps she will be more so than you think.'

'Perhaps – if *you* take an interest in her!'

'A mystic voice seems to exhort me to do so, to whisper that, though I have never seen her, I shall find something in her. What do you say, Biddy, shall I take an interest in her?'

Biddy hesitated a moment, coloured a little, felt a certain embarrassment in being publicly treated as an oracle.

'If she's not nice I don't advise it.'

'And if she *is* nice?'

'You advise it still less!' her brother exclaimed, laughing and putting his arm round her.

Lady Agnes looked sombre – she might have been saying to herself: 'Dear me, what chance has a girl of mine with a man who's so agog about actresses?' She was disconcerted and distressed; a multitude of incongruous things, all the morning, had been forced upon her attention – displeasing pictures and still more displeasing theories about them, vague portents of perversity on the part of Nicholas, and a strange eagerness on Peter's, learned apparently in Paris, to discuss, with a person who had a tone she never had been exposed to, topics irrelevant and uninteresting, the practical effect of which was to make light of her presence. 'Let us leave this – let us leave this!' she almost moaned. The party moved together towards the door of departure, and her ruffled spirit was not soothed by hearing her son remark to his terrible friend: 'You know you don't leave us – I stick to you!'

At this Lady Agnes broke out and interposed: 'Excuse me for reminding you that you are going to call on Julia.'

'Well, can't Nash also come to call on Julia? That's just what I want – that she should see him.'

Peter Sherringham came humanely to her ladyship's assistance. 'A better way, perhaps, will be for them to meet under my auspices, at my "dramatic tea". This will enable me to return one favour for another. If Mr Nash is so good as to introduce me to this aspirant for honours we estimate so differently, I will introduce him to my sister, a much more positive quantity.'

'It is easy to see who'll have the best of it!' Grace Dormer exclaimed; and Gabriel Nash stood there serenely, impartially, in a graceful, detached way which seemed characteristic of him, assenting to any decision that relieved him of the grossness of choice, and generally confident that things would turn out well for him. He was cheerfully helpless and sociably indifferent; ready to preside with a smile even at a discussion of his own admissibility.

'Nick will bring you. I have a little corner at the Embassy,' Sherringham continued.

'You are very kind. You must bring him, then, tomorrow – Rue de Constantinople.'

'At five o'clock – don't be afraid.'

'Oh dear!' said Biddy, as they went on again; and Lady Agnes, seizing his arm, marched off more quickly with her son. When they came out into the Champs Elysées Nick Dormer, looking round, saw that his friend had disappeared. Biddy had attached herself to Peter, and Grace apparently had not encouraged Mr Nash.

5

LADY AGNES'S idea had been that her son should go straight from the Palais de l'Industrie to the Hôtel de Hollande, with or without his mother and his sisters, as his humour should seem to recommend. Much as she desired to see their brilliant kinswoman and as she knew that her daughters desired it, she was quite ready to postpone their visit, if this sacrifice should contribute to a speedy confrontation for Nick. She was eager that he should talk with Mrs Dallow, and eager that he should be eager himself; but it presently appeared that he was really not anything that could impartially be called so. His view was that she and the girls should go to the Hôtel de Hollande without delay and should spend the rest of the day with Julia, if they liked. He would go later; he would go in the evening. There were lots of things he wanted to do meanwhile.

This question was discussed with some intensity, though not at length, while the little party stood on the edge of the Place de la Concorde, to which they had proceeded on foot; and Lady Agnes noticed that the 'lots of things' to which he proposed to give precedence over an urgent duty, a conference with a person who held out full hands to him, were implied somehow in the friendly glance with which he covered the great square, the opposite bank of the Seine, the steep blue roofs of the quay, the bright immensity of Paris. What in the world could be more important than making sure of his seat? – so quickly did the good lady's imagination travel. And now that idea appealed to him less than a ramble in search of old books and prints, for she was sure this was what he had in his head. Julia would be flattered if she knew it, but of course she must not know it. Lady Agnes was already thinking of the most honourable explanations she could give of the young man's want of precipitation. She would have liked to represent him as tremendously occupied, in his room at their own hotel, in getting off political letters to every one it should concern, and particularly in drawing up his address to the electors of Harsh. Fortunately she was a woman of innumerable discretions, and a part of the worn look that sat in her face came from her having schooled herself for years, in her relations with her husband and her sons, not to insist unduly. She would have liked to insist, nature had formed her to insist, and the self-control had told in more ways than one. Even now it was powerless to prevent her suggesting that before doing anything else Nick should at least repair to the inn and see if there were not some telegrams.

He freely consented to do so much as this, and having called a cab, that she might go her way with the girls, he kissed her again, as he had done at the exhibition. This was an attention that could never displease her, but somehow when he kissed her often her anxiety was apt to increase: she had come to recognize it as a sign that he was slipping away from her. She drove off with a vague sense that at any rate she and the girls might do something towards keeping the place warm for him. She had been a little vexed that Peter had not administered more of a push towards the Hôtel de Hollande, clear as it had become to her now that there was a foreignness in Peter which was not to be counted on and which made him speak of English affairs and even of English domestic politics as local. Of course they were local, and was not that the warm human comfort of them? As she left the two young men standing together in the middle of the Place de la Concorde, the grand composition of which Nick, as she looked back, appeared to have paused to admire (as if he had not seen it a thousand times!), she wished she might have thought of Peter's influence with her son as exerted a little more in favour of localism. She had a sense that he would not abbreviate the boy's ill-timed *flânerie*. However, he had been very nice: he had invited them all to dine with him that evening at a convenient restaurant, promising to bring Julia and one of his colleagues. So much as this he had been willing to do to make sure that Nick and his sister should meet. His want of localism, moreover, was not so great as that if it should turn out that there *was* anything beneath his manner towards Biddy —! The conclusion of this reflection is, perhaps, best indicated by the circumstance of her ladyship's remarking, after a minute, to her younger daughter, who sat opposite to her in the *voiture de place*, that it would do no harm if she should get a new hat, and that the article might be purchased that afternoon.

'A French hat, mamma?' said Grace. 'Oh, do wait till she gets home!'

'I think they are prettier here, you know,' Biddy rejoined; and Lady Agnes said, simply, 'I dare say they're cheaper.' What was in her mind, in fact, was, 'I dare say Peter thinks them becoming.' It will be seen that she had plenty of spiritual occupation, the sum of which was not diminished by her learning, when she reached the top of the Rue de la Paix, that Mrs Dallow had gone out half an hour before and had left no message. She was more disconcerted by this incident than she could have explained or than she thought was right, for she had taken for granted that Julia would be in a manner waiting for them. How did she know that Nick was not coming? When people were in Paris for a few days they didn't mope in the house; but Julia might have waited a little longer or might have left an explanation. Was she then not so much in earnest about Nick's standing? Didn't she recognize the importance of being there to see him about it? Lady Agnes wondered whether Julia's behaviour were a sign that she was already tired of the way this young gentleman treated her. Perhaps she had gone out because an instinct told her that its being important he should see her early would make no difference with him – told her that he wouldn't come. Her heart sank as she glanced at this possibility that Julia was already tired, for she, on her side, had an instinct there were still more tiresome things in store. She had disliked having to tell Mrs Dallow that Nick wouldn't see her till the evening, but now she disliked still more her not being there to hear it. She even resented a little her kinswoman's not having reasoned that she and the girls would come in any event, and not thought them worth staying in for. It occurred to her that she would perhaps have gone to their hotel, which was a good way up the Rue de Rivoli, near the Palais Royal, and she directed the cabman to drive to that establishment.

As he jogged along she took in some degree the measure of what that might mean, Julia's seeking a little to avoid them. Was she growing to dislike them? Did she think they kept too sharp an eye on her, so that the idea of their standing in a still closer relation to her would not be enticing? Her conduct up to this time had not worn such an appearance, unless perhaps a little, just a very little, in the matter of poor Grace. Lady Agnes knew that she was not particularly fond of poor Grace, and was even able to guess the reason – the manner in which Grace betrayed the most that they wanted to make sure of her. She remembered how long the girl had stayed the last time she had gone to Harsh. She had gone for an acceptable week, and she had been in the house a month. She took a private, heroic vow that Grace should not go near the place again for a year; that is, not unless Nick and Julia were married before this. If that were to happen she shouldn't care. She recognized that it was not absolutely everything that Julia should be in love with Nick; it was also better she should dislike his mother and sisters after than before. Lady Agnes did justice to the natural rule in virtue of which it usually comes to pass that a woman doesn't get on with her husband's female belongings, and was even willing to be sacrificed to it in her disciplined degree. But she desired not to be sacrificed for nothing: if she was to be objected to as a mother-in-law she wished to be the mother-in-law first.

At the hotel in the Rue de Rivoli she had the disappointment of finding that Mrs Dallow had not called, and also that no telegrams had come. She went in with the girls for half an hour, and then she straggled out with them again. She was undetermined and dissatisfied, and the afternoon was rather a problem; of the kind moreover that she disliked most and was least accustomed to: not a choice between different things to do (her life had been full of that), but a want of anything to do at all. Nick had said to her before they separated: 'You can knock about with the girls, you know; everything is amusing here.' That was easily said, while he sauntered and gossiped with Peter Sherringham and perhaps went to see more pictures like those in the Salon. He was usually, on such occasions, very good-natured about spending his time with them; but this episode had taken altogether a perverse, profane form. She had no desire whatever to knock about, and she was far from finding everything in Paris amusing. She had no aptitude for aimlessness, and moreover she thought it vulgar. If she had found Julia's card at the hotel (the sign of a hope of catching them just as they came back from the Salon), she would have made a second attempt to see her before the evening; but now certainly they would leave her alone. Lady Agnes wandered joylessly with the girls in the Palais Royal and the Rue de Richelieu, and emerged upon the Boulevard, where they continued their frugal prowl, as Biddy rather irritatingly called it. They went into five shops to buy a hat for Biddy, and her ladyship's presuppositions of cheapness were woefully belied.

'Who in the world is your funny friend?' Peter Sherringham meanwhile asked of his kinsman. He lost no time as they walked together.

'Ah, there's something else you lost by going to Cambridge – you lost Gabriel Nash!'

'He sounds like an Elizabethan dramatist,' Sherringham said. 'But I haven't lost him, since it appears now that I shall not be able to have you without him.'

'Oh, as for that, wait a little. I'm going to try him again, but I don't know how he wears. What I mean is that you have probably lost his freshness. I have an idea he has become conventional, or at any rate serious.'

'Bless me, do you call that serious?'

'He used to be so gay. He had a real genius for suggestive paradox. He was a wonderful talker.'

'It seems to me he does very well now,' said Peter Sherringham.

'Oh, this is nothing. He had great flights of old, very great flights; one saw him rise and rise and turn summersaults in the blue, and wondered how far he could go. He's very intelligent, and I should think it might be interesting to find out what it is that prevents the whole man from being as good as his parts. I mean in case he isn't so good.'

'I see you more than suspect that. May it not simply be that he's an ass?'

'That would be the whole — I shall see in time — but it certainly isn't one of the parts. It may be the effect, but it isn't the cause, and it's for the cause that I claim an interest. I imagine you think he's an ass on account of what he said about the theatre, his pronouncing it a coarse art.'

'To differ about him that reason will do,' said Sherringham. 'The only bad one would be one that shouldn't preserve our difference. You needn't tell me you agree with him, for frankly I don't care.'

'Then your passion still burns?' Nick Dormer asked.

'My passion?'

'I don't mean for any individual exponent of the contestable art: mark the guilty conscience, mark the rising blush, mark the confusion of mind! I mean the old sign one knew you best by: your permanent stall at the Français, your inveterate attendance at *premières*, the way you "follow" the young talents and the old.'

'Yes, it's still my little hobby: my little folly, if you like. I don't see that I get tired of it. What will you have? Strong predilections are rather a blessing; they are simplifying. I'm fond of representation — the representation of life: I like it better, I think, than the real thing. You like it, too, so you have no right to cast the stone. You like it best done one way and I another; and our preference, on either side, has a deep root in us. There is a fascination to me in the way the actor does it, when his talent (ah, he must have that!) has been highly trained (ah, it must *be* that!). The things he can do, in this effort at representation (with the dramatist to give him his lift) seem to me innumerable – he can carry it to a delicacy! – and I take great pleasure in observing them, in recognizing them and comparing them. It's an amusement like another: I don't pretend to call it by any exalted name; but in this vale of friction it will serve. One can lose one's self in it, and it has this recommendation (in common, I suppose, with the study of the other arts), that the further you go in it the more you find. So I go rather far, if you will. But is it the principal sign one knows me by?' Sherringham abruptly asked.

'Don't be ashamed of it, or it will be ashamed of you. I ought to discriminate. You are distinguished among my friends and relations by being a rising young diplomatist; but you know I always want the further distinction, the last analysis. Therefore I surmise that you are conspicuous among rising young diplomatists for the infatuation that you describe in such pretty terms.'

'You evidently believe that it will prevent me from rising very high. But pastime for pastime, is it any idler than yours?'

'Than mine?'

'Why, you have half a dozen, while I only allow myself the luxury of one. For the theatre is my sole vice, really. Is this more wanton, say, than to devote weeks to ascertaining in what particular way your friend Mr Nash may be a twaddler? That's not my ideal of choice recreation, but I would undertake to do it sooner. You're a young statesman (who happens to be *en disponibilité* for the moment), but you spend not a little of your time in besmearing canvas with bright-coloured pigments. The idea of representation fascinates you, but in your case it's representation in oils – or do you practise water-colours too? You even go much further than I, for I study my art of predilection only in the works of others. I don't aspire to leave works of my own. You're a painter, possibly a great one; but I'm not an actor.' Nick Dormer declared that he would certainly become one – he was on the way to it; and Sherringham, without heeding this charge, went on: 'Let me add that, considering you *are* a painter, your portrait of the complicated Nash is lamentably dim.'

'He's not at all complicated; he's only too simple to give

an account of. Most people have a lot of attributes and appendages that dress them up and superscribe them, and what I like him for is that he hasn't any at all. It makes him so cool.'

'By Jove, you match him there! It's an attribute to be tolerated. How does he manage it?'

'I haven't the least idea – I don't know that he *is* tolerated. I don't think any one has ever detected the process. His means, his profession, his belongings have never anything to do with the question. He doesn't shade off into other people; he's as neat as an outline cut out of paper with scissors. I like him, therefore, because in intercourse with him you know what you've got hold of. With most men you don't: to pick the flower you must break off the whole dusty, thorny, worldly branch; you find you are taking up in your grasp all sorts of other people and things, dangling accidents and conditions. Poor Nash has none of those ramifications; he's the solitary blossom.'

'My dear fellow, you would be better for a little of the same pruning!' Sherringham exclaimed; and the young men continued their walk and their gossip, jerking each other this way and that with a sociable roughness consequent on their having been boys together. Intimacy had reigned, of old, between the little Sherringhams and the little Dormers, united by country contiguity and by the circumstance that there was first-cousinship, not neglected, among the parents, Lady Agnes standing in this convertible relation to Lady Windrush, the mother of Peter and Julia as well as of other daughters and of a maturer youth who was to inherit, and who since then had inherited, the ancient barony. Since then many things had altered, but not the deep foundation of sociability. One of our young men had gone to Eton and the other to Harrow (the scattered school on the hill was the tradition of the Dormers), and the divergence had taken its course later, in university years. Bricket, however, had remained accessible to Windrush, and Windrush to Bricket, to which Percival Dormer had now succeeded, terminating the interchange a trifle rudely by letting out that pleasant white house in the midlands (its expropriated inhabitants, Lady Agnes and her daughters, adored it) to an American reputed rich, who, in the first flush of international comparison, considered that for twelve hundred a year he got it at a bargain. Bricket had come to the late Sir Nicholas from his elder brother, who died wifeless and childless. The new baronet, so different from his father (though he recalled at some points the uncle after whom he had been named) that Nick had to make it up by aspirations of resemblance, roamed about the world taking shots which excited the enthusiasm of society, when society heard of them, at the few legitimate creatures of the chase which the British rifle had spared. Lady Agnes, meanwhile, settled with her girls in a gabled, latticed house in a creditable quarter, though it was still a little raw, of the temperate zone of London. It was not into her lap, poor woman, that the revenues of Bricket were poured. There was no dower-house attached to that moderate property, and the allowance with which the estate was charged on her ladyship's behalf was not an incitement to grandeur.

Nick had a room under his mother's roof, which he mainly used to dress for dinner when he dined in Calcutta Gardens, and he had 'kept on' his chambers in the Temple; for to a young man in public life an independent address was indispensable. Moreover, he was suspected of having a studio in an out-of-the-way quarter of the town, the indistinguishable parts of South Kensington, incongruous as such a retreat might seem in the case of a member of Parliament. It was an absurd place to see his constituents, unless he wanted to paint their portraits, a kind of representation with which they scarcely would have been satisfied; and in fact the only question of portraiture had been when the wives and daughters of several of them expressed a wish for the picture of their handsome young member. Nick had not offered to paint it himself, and the studio was taken for granted rather than much looked into by the ladies in Calcutta Gardens. To express a disposition to regard whims of this sort as a pure extravagance was known by them to be open to correction; for they were not oblivious that Mr Carteret had humours which weighed against them in the shape of convenient cheques nestling between the inside pages of legible letters of advice. Mr Carteret was Nick's providence, as Nick was looked to, in a general way, to be that of his mother and sisters, especially since it had become so plain that Percy, who was ungracefully selfish, would operate, mainly with a 'six-bore', quite out of that sphere. It was not for studios, certainly, that Mr Carteret sent cheques; but they were an expression of general confidence in Nick, and a little expansion was natural to a young man enjoying such a luxury as that. It was sufficiently felt, in Calcutta Gardens, that Nick could be looked to not to betray such a confidence; for Mr Carteret's behaviour could have no name at all unless one were prepared to call it encouraging. He had never promised anything, but he was one of the delightful persons with whom the redemption precedes or dispenses with the vow. He had been an early and lifelong friend of the late right honourable gentleman, a political follower, a devoted admirer, a stanch supporter in difficult hours. He had never married, espousing nothing more reproductive than Sir Nicholas's views (he used to write letters to *The Times* in favour of them), and had, so far as was known, neither chick nor child; nothing but an amiable little family of eccentricities, the flower of which was his odd taste for living in a small, steep, clean country town, all green gardens and red walls, with a girdle of hedge-rows, clustering about an immense brown old abbey. When Lady Agnes's imagination rested upon the future of her second son she liked to remember that Mr Carteret had nothing to 'keep-up': the inference seemed so direct that he would keep up Nick.

The most important event in the life of this young man had been incomparably his victory, under his father's eyes, more than two years before, in the sharp contest for Crockhurst – a victory which his consecrated name, his extreme youth, his ardour in the fray, the general personal sympathy of the party and the attention excited by the fresh cleverness of his speeches, tinted with young idealism and yet sticking sufficiently to the question (the burning question, it has since burnt out), had rendered almost brilliant. There had been leaders in the newspapers about it, half in compliment to her husband, who was known to be failing so prematurely (he was almost as young to die, and to die famous – Lady Agnes regarded it as famous – as his son had been to stand), which the boy's mother religiously preserved, cut out and tied together with a ribbon, in the innermost drawer of a favourite cabinet. But it had been a barren, or almost a barren triumph, for in the order of importance in Nick's history another incident had run it, as the phrase is, very close: nothing less than the quick dissolution of the Parliament in which he was so manifestly destined to give symptoms of a future. He had not recovered his seat at the general election, for the second contest was even sharper than the first, and the Tories had put forward a loud, vulgar, rattling, almost bullying man. It was to a certain extent a comfort that poor Sir Nicholas, who had been witness of the bright hour, passed away before the darkness. He died, with all his hopes on his second son's head, unconscious of near disaster, handing on the torch and the tradition, after a long, supreme interview with Nick, at which Lady Agnes had not been present but which she knew to have been a sort of paternal dedication, a solemn communication of ideas on the highest national questions (she had reason to believe he had touched on those of external as well as of domestic and of colonial policy), leaving on the boy's nature and manner from that moment the most unmistakable traces. If his tendency to reverie increased, it was because he had so much to think over in what his pale father had said to him in the hushed, dim chamber, laying upon him the great mission of carrying out the unachieved and reviving a silent voice. It was work cut out for a lifetime, and that 'co-ordinating power in relation to detail', which was one of the great characteristics of Sir Nicholas's high distinction (the most analytic of the weekly papers was always talking about it), had enabled him to rescue the prospect from any shade of vagueness or of ambiguity.

Five years before Nick Dormer went up to be questioned by the electors of Crockhurst, Peter Sherringham appeared before a board of examiners who let him off much less easily, though there were also some flattering prejudices in his favour; such influences being a part of the copious, light, unembarrassing baggage with which each of the young men began life. Peter passed, however, passed high, and had his reward in prompt assignment to small, subordinate diplomatic duties in Germany. Since then he had had his professional adventures, which need not arrest us, inasmuch as they had all paled in the light of his appointment, nearly three years previous to the

moment of our making his acquaintance, to a secretaryship of
embassy in Paris. He had done well and had gone fast, and
for the present he was willing enough to rest. It pleased him
better to remain in Paris as a subordinate than to go to Honduras as a principal, and Nick Dormer had not put a false
colour on the matter in speaking of his stall at the Théâtre
Français as a sedative to his ambition. Nick's inferiority in age
to his cousin sat on him more lightly than when they had been
in their teens; and indeed no one can very well be much older
than a young man who has figured for a year, however imperceptibly, in the House of Commons. Separation and diversity
had made them strange enough to each other to give a taste
to what they shared; they were friends without being particular
friends; that further degree could always hang before them
as a suitable but not oppressive contingency, and they were both
conscious that it was in their interest to keep certain differences
to 'chaff' each other about — so possible was it that they might
have quarrelled if they had only agreed. Peter, as being wide-
minded, was a little irritated to find his cousin always so intensely British, while Nick Dormer made him the object of the
same compassionate criticism, recognized that he had a rare
knack with foreign tongues, but reflected, and even with extravagance declared, that it was a pity to have gone so far from
home only to remain so homely. Moreover, Nick had his ideas
about the diplomatic mind; it was the moral type of which, on
the whole, he thought least favourably. Dry, narrow, barren,
poor, he pronounced it in familiar conversation with the clever
secretary; wanting in imagination, in generosity, in the finest
perceptions and the highest courage. This served as well as
anything else to keep the peace between them; it was a necessity
of their friendly intercourse that they should scuffle a little,
and it scarcely mattered what they scuffled about. Nick Dormer's express enjoyment of Paris, the shop-windows on the
quays, the old books on the parapet, the gaiety of the river, the
grandeur of the Louvre, all the amusing tints and tones, struck
his companion as a sign of insularity; the appreciation of such
things having become with Sherringham an unconscious habit,
a contented assimilation. If poor Nick, for the hour, was demonstrative and lyrical, it was because he had no other way of

sounding the note of farewell to the independent life of which the term seemed now definitely in sight; the sense pressed upon him that these were the last moments of his freedom. He would waste time till half-past seven, because half-past seven meant dinner, and dinner meant his mother, solemnly attended by the strenuous shade of his father and reinforced by Julia.

6

WHEN Nick arrived with the three members of his family, Peter Sherringham was seated, in the restaurant at which the tryst had been taken, at a small but immaculate table; but Mrs Dallow was not yet on the scene, and they had time for a sociable settlement – time to take their places and unfold their napkins, crunch their rolls, breathe the savoury air and watch the door, before the usual raising of heads and suspension of forks, the sort of stir that accompanied most of this lady's movements, announced her entrance. The *dame de comptoir* ducked and re-ducked, the people looked round, Peter and Nick got up, there was a shuffling of chairs – Julia was there. Peter had related how he had stopped at her hotel to bring her with him and had found her, according to her custom, by no means ready; on which, fearing that his guests would come first to the rendezvous and find no proper welcome, he had come off without her, leaving her to follow. He had not brought a friend, as he intended, having divined that Julia would prefer a pure family party, if she wanted to talk about her candidate. Now she stood there, looking down at the table and her expectant kinsfolk, drawing off her gloves, letting her brother draw off her jacket, lifting her hands for some rearrangement of her bonnet. She looked at Nick last, smiling, but only for a moment. She said to Peter, 'Are we going to dine here? Oh dear, why didn't you have a private room?'

Nick had not seen her at all for several weeks, and had seen her but little for a year, but her off-hand, cursory manner had not altered in the interval. She spoke remarkably fast, as if speech were not in itself a pleasure – to have it over as soon as possible; and her *brusquerie* was of the kind that friendly critics account for by pleading shyness. Shyness had never appeared to him an ultimate quality or a real explanation of anything; it only explained an effect by another effect, giving a bad fault another name. What he suspected in Julia was that her mind was less graceful than her person; an ugly, a really damnatory idea, which as yet he had only half accepted. It was a case in which she was entitled to the benefit of every doubt and ought not to be judged without a complete trial. Dormer, meanwhile, was afraid of the trial (this was partly why, of late, he had been to see her so little), because he was afraid of the sentence, afraid of anything happening which should lessen the pleasure it was actually in the power of her beauty to give. There were people who thought her rude, and he hated rude women. If he should fasten on that view, or rather if that view should fasten on him, what could still please and what he admired in her would lose too much of its sweetness. If it be thought odd that he had not yet been able to read the character of a woman he had known since childhood, the answer is that that character had grown faster than Nick Dormer's observation. The growth was constant, whereas the observation was but occasional, though it had begun early. If he had attempted to phrase the matter to himself, as he probably had not, he might have said that the effect she produced upon him was too much a compulsion; not the coercion of design, of importunity, nor the vulgar pressure of family expectation, a suspected desire that he should like her enough to marry her, but something that was a mixture of diverse things, of the sense that she was imperious and generous – but probably more the former than the latter – and of a certain prevision of doom, the influence of the idea that he should come to it, that he was predestined.

This had made him shrink from knowing the worst about her; the desire, not to get used to it in time, but what was more characteristic of him, to interpose a temporary illusion. Illusions and realities and hopes and fears, however, fell into confusion whenever he met her after a separation. The separation, so far as seeing her alone or as continuous talk was concerned, had now been tolerably long; had lasted really ever since his failure to regain his seat. An impression had come to him that she judged that failure rather harshly, had thought he ought to have done better.

This was a part of her imperious strain, and a part to which it was not easy to accommodate one's self on a present basis. If he were to marry her he should come to an understanding with her: he should give her his own measure as well as take hers. But the understanding, in the actual case, might suggest too much that he *was* to marry her. You could quarrel with your wife, because there were compensations – for her; but you might not be prepared to offer these compensations as prepayment for the luxury of quarrelling.

It was not that such a luxury would not be considerable, Nick Dormer thought, as Julia Dallow's fine head poised itself before him again; a high spirit was a better thing than a poor one to be mismated with, any day in the year. She had much the same colouring as her brother, but as nothing else in her face was the same the resemblance was not striking. Her hair was of so dark a brown that it was commonly regarded as black, and so abundant that a plain arrangement was required to keep it in discreet relation to the rest of her person. Her eyes were of a gray tint that was sometimes pronounced too light; and they were not sunken in her face, but placed well on the surface. Her nose was perfect, but her mouth was too small; and Nick Dormer, and doubtless other persons as well, had sometimes wondered how, with such a mouth, her face could have expressed decision. Her figure helped it, for she looked tall (being extremely slender), though she was not; and her head took turns and positions which, though they were a matter of but half an inch out of the common, this way or that, somehow contributed to the air of resolution and temper. If it had not been for her extreme delicacy of line and surface she might have been called bold; but as it was she looked refined and quiet – refined by tradition and quiet for a purpose. And altogether she was beautiful, with the pure style of her capable head, her hair like darkness, her eyes like early twilight, her mouth like a rare pink flower.

Peter said that he had not taken a private room because he knew Biddy's tastes; she liked to see the world (she had told him so), the curious people, the coming and going of Paris. 'Oh, anything for Biddy!' Julia replied, smiling at the girl and taking her place. Lady Agnes and her elder daughter exchanged one of their looks, and Nick exclaimed jocosely that he didn't see why the whole party should be sacrificed to a presumptuous child. The presumptuous child blushingly protested she had never expressed any such wish to Peter, upon which Nick, with broader humour, revealed that Peter had served them so out of stinginess: he had pitchforked them together in the public room because he wouldn't go to the expense of a *cabinet*. He had brought no guest, no foreigner of distinction nor diplomatic swell, to honour them, and now they would see what a paltry dinner he would give them. Peter stabbed him indignantly with a long roll, and Lady Agnes, who seemed to be waiting for some manifestation on Mrs Dallow's part which didn't come, concluded, with a certain coldness, that they quite sufficed to themselves for privacy as well as for society. Nick called attention to this fine phrase of his mother's, said it was awfully neat, while Grace and Biddy looked harmoniously at Julia's clothes. Nick felt nervous, and joked a good deal to carry it off – a levity that didn't prevent Julia's saying to him, after a moment: 'You might have come to see me today, you know. Didn't you get my message from Peter?'

'Scold him, Julia – scold him well. I begged him to go,' said Lady Agnes; and to this Grace added her voice with an 'Oh, Julia, do give it to him!' These words, however, had not the effect they suggested, for Mrs Dallow only murmured, with an ejaculation, in her quick, curt way, that that would be making far too much of him. It was one of the things in her which Nick Dormer mentally pronounced ungraceful that a perversity of pride or shyness always made her disappoint you a little if she saw you expected a thing. She was certain to snub effusiveness. This vice, however, was the last thing of which Lady Agnes would have consented to being accused; and Nick, while he replied to Julia that he was certain he shouldn't have found her, was not unable to perceive the operation on his mother of that shade of manner. 'He ought to have gone; he owed you that,' she went on; 'but it's very true he would have had the same luck as we. I went with the girls directly after luncheon. I suppose you got our card.'

'He might have come after I came in,' said Mrs Dallow.

'Dear Julia, I'm going to see you tonight. I've been waiting for that,' Nick rejoined.

'Of course *we* had no idea when you would come in,' said Lady Agnes.

'I'm so sorry. You must come tomorrow. I hate calls at night,' Julia remarked.

'Well, then, will you roam with me? Will you wander through Paris on my arm?' Nick asked, smiling. 'Will you take a drive with me?'

'Oh, that would be perfection!' cried Grace.

'I thought we were all going somewhere – to the Hippodrome, Peter,' said Biddy.

'Oh, not all; just you and me!' laughed Peter.

'I am going home to my bed. I've earned my rest,' Lady Agnes sighed.

'Can't Peter take *us*?' asked Grace. 'Nick can take you home, mamma, if Julia won't receive him, and I can look perfectly after Peter and Biddy.'

'Take them to something amusing; please take them,' Mrs Dallow said to her brother. Her voice was kind, but had the expectation of assent in it, and Nick observed both the indulgence and the pressure. 'You're tired, poor dear,' she continued to Lady Agnes. 'Fancy your being dragged about so! What did you come over for?'

'My mother came because I brought her,' Nick said. 'It's I who have dragged her about. I brought her for a little change. I thought it would do her good. I wanted to see the Salon.'

'It isn't a bad time. I have a carriage, and you must use it; you must use nothing else. It will take you everywhere. I will drive you about tomorrow.' Julia dropped these words in the same perfunctory, casual way as any others; but Nick had already noted, and he noted now afresh, with pleasure, that her abruptness was perfectly capable of conveying a benevolence. It was quite sufficiently manifest to him that for the rest of the time she might be near his mother she would do her numberless good turns. She would give things to the girls – he had a private adumbration of that; expensive Parisian, perhaps not perfectly useful things.

Lady Agnes was a woman who measured reciprocities and distances; but she was both too subtle and too just not to recognize the smallest manifestation that might count, either technically or essentially, as a service. 'Dear Julia!' she exclaimed, responsively; and her tone made this brevity of acknowledgement sufficient. What Julia had said was all she wanted. 'It's so interesting about Harsh,' she added. 'We're immensely excited.'

'Yes, Nick looks it. *Merci, pas de vin*. It's just the thing for you, you know.'

'To be sure he knows it. He's immensely grateful. It's really very kind of you.'

'You do me a very great honour, Julia,' said Nick.

'Don't be tiresome!' exclaimed Mrs Dallow.

'We'll talk about it later. Of course there are lots of points,' Nick pursued. 'At present let us be purely convivial. Somehow Harsh is such a false note here. *A tout à l'heure!*'

'My dear fellow, you've caught exactly the tone of Mr Gabriel Nash,' Peter Sherringham observed.

'Who is Mr Gabriel Nash?' Mrs Dallow asked.

'Nick, *is* he a gentleman? Biddy says so,' Grace Dormer interposed before this inquiry was answered.

'It is to be supposed that any one Nick brings to lunch with us –' Lady Agnes murmured.

'Ah, Grace, with your tremendous standard!' her brother said; while Peter Sherringham replied to Julia that Mr Nash was Nick's new Mentor or oracle; whom moreover she should see if she would come and have tea with him.

'I haven't the least desire to see him,' Julia declared, 'any more than I have to talk about Harsh and bore poor Peter.'

'Oh, certainly, dear, you would bore me,' said Sherringham.

'One thing at a time, then. Let us by all means be convivial. Only you must show me how,' Mrs Dallow went on to Nick. 'What does he mean, Cousin Agnes? Does he want us to drain the wine-cup, to flash with repartee?'

'You'll do very well,' said Nick. 'You are charming, this evening.'

'Do go to Peter's, Julia, if you want something exciting. You'll see a marvellous girl,' Biddy broke in, with her smile on Peter.

'Marvellous for what?'

'For thinking she can act, when she can't,' said the roguish Biddy.

'Dear me, what people you all know! I hate Peter's theatrical people.'

'And aren't you going home, Julia?' Lady Agnes inquired.

'Home to the hotel?'

'Dear, no, to Harsh, to see about everything.'

'I'm in the midst of telegrams. I don't know yet.'

'I suppose there's no doubt they'll have him,' Lady Agnes decided to pursue.

'Who will have whom?'

'Why, the local people – and the party; those who invite a gentleman to stand. I'm speaking of my son.'

'They'll have the person I want them to have, I dare say. There are so many people in it, in one way or another, it's dreadful. I like the way you sit there,' Mrs Dallow added to Nick Dormer.

'So do I,' he smiled back at her; and he thought she *was* charming now, because she was gay and easy and willing really, though she might plead incompetence, to understand how jocose a dinner in a pot-house in a foreign town might be. She was in good-humour, or she was going to be, and not grand, nor stiff, nor indifferent, nor haughty, nor any of the things that people who disliked her usually found her and sometimes even a little made him believe her. The spirit of mirth, in some cold natures, manifests itself not altogether happily; their effort of recreation resembles too much the bath of the hippopotamus. But when Mrs Dallow put her elbows on the table one felt she could be trusted to get them safely off again.

For a family in mourning the dinner was lively; the more so that before it was half over Julia had arranged that her brother, eschewing the inferior spectacle, should take the girls to the Théâtre Français. It was her idea, and Nick had a chance to observe how an idea was apt not to be successfully controverted when it was Julia's. Even the programme appeared to have been pre-arranged to suit it, just the thing for the cheek of the young person – *Il ne Faut Jurer de Rien* and *Mademoiselle de la Seiglière*. Peter was all willingness, but it was Julia who settled it, even to sending for the newspaper (her brother, by a rare accident, was unconscious of the evening's bill), and to reassuring Biddy, who was happy but anxious, on the article of their not getting places, their being too late. Peter could always get places: a word from him and the best box was at his disposal. She made him write the word on a card, and saw that a messenger was dispatched with it to the Rue de Richelieu; and all this was done without loudness or insistence, parenthetically and authoritatively. The box was bespoken; the carriage, as soon as they had had their coffee was found to be there; Peter drove off in it with the girls, with the understanding that he was to send it back; Nick sat waiting for it, over the finished repast, with the two ladies, and then his mother was relegated to it and conveyed to her apartments: and all the while it was Julia who governed the succession of events. 'Do be nice to her,' Lady Agnes murmured to him, as he placed her in the vehicle at the door of the restaurant; and he guessed that it gave her a comfort to have left him sitting there with Mrs Dallow.

Nick had every disposition to be nice to her; if things went as she liked them it was an acknowledgement of a certain force that was in her – the force of assuming that they would. Julia had her differences – some of them were much for the better; and when she was in a mood like this evening's, liberally dominant, he was ready to encourage her assumptions. While they waited for the return of the carriage, which had rolled away with his mother, she sat opposite to him, with her elbows on the table, playing first with one and then with another of the objects that encumbered it: after five minutes of which she exclaimed, 'Oh, I say, we'll go!' and got up abruptly, asking for her jacket. He said something about the carriage's having had orders to come back for them, and she replied: 'Well, it can go away again!' She added: 'I don't want a carriage; I want to walk'; and in a moment she was out of the place, with the people at the tables turning round again and the *cassière* swaying in her high seat. On the pavement of the boulevard she looked up and down: there were people at little tables, at the door; there were people all over the broad expanse of the asphalt; there was a profusion of light and a pervasion of sound; and everywhere, though the establishment at which they had been dining was not in the thick of the fray, the tokens of a great traffic of pleasure, that night-aspect of Paris which represents it as a huge market for sensations. Beyond the Boulevard des Capucines it flared through the warm evening like a vast bazaar; and opposite the Café Durand the Madeleine rose theatrical, a high, clever *décor*, before the footlights of the Rue Royale. 'Where shall we go, what shall we do?' Mrs Dallow asked, looking at her companion and somewhat to his surprise, as he had supposed she only wanted to go home.

'Anywhere you like. It's so warm we might drive, instead of going indoors. We might go to the Bois. That would be agreeable.'

'Yes, but it wouldn't be walking. However, that doesn't matter. It's mild enough for anything – for sitting out, like all these people. And I've never walked in Paris at night: it would amuse me.'

Nick hesitated. 'So it might, but it isn't particularly recommended to ladies.'

'I don't care, if it happens to suit me.'

'Very well, then, we'll walk to the Bastille, if you like.'

Julia hesitated, on her side, still looking round her.

'It's too far; I'm tired; we'll sit here.' And she dropped beside an empty table, on the 'terrace' of M. Durand. 'This will do; it's amusing enough, and we can look at the Madeleine; that's respectable. If we must have something we'll have a *madère*: is that respectable? Not particularly? So much the better. What are those people having? *Bocks*? Couldn't we have bocks? Are they very low? Then I shall have one. I've been so wonderfully good – I've been staying at Versailles: *je me dois bien cela*.'

She insisted, but pronounced the thin liquid in the tall glass very disgusting when it was brought. Nick was amazed, reflecting that it was not for such a discussion as this that his mother had left him with such complacency; and indeed he too had, as she would have had, his share of perplexity, observing that nearly half an hour passed without his cousin's saying anything about Harsh.

Mrs Dallow leaned back against the lighted glass of the café, comfortable and beguiled, watching the passers, the opposite shops, the movement of the square in front of them. She talked about London, about the news written to her in her absence, about Cannes and the people she had seen there, about her poor sister-in-law and her numerous progeny and two or three droll things that had happened at Versailles. She discoursed considerably about herself, mentioning certain things she meant to do on her return to town, her plans for the rest of the season. Her carriage came and stood there, and Nick asked if he should send it away; to which she said: 'No, let it stand a bit.' She let it stand a long time, and then she told him to dismiss it: they would walk home. She took his arm and they went along the boulevard, on the right hand side, to the Rue de la Paix, saying little to each other during the transit; and then they passed into the hotel and up to her rooms. All she had said on the way was that she was very tired of Paris. There was a shaded lamp in her *salon*, but the windows were open and the light of the street, with its undisturbing murmur, as if everything ran on india-rubber, came up through the interstices of the balcony and made a vague glow and a flitting of shadows on the ceiling. Her maid appeared, busying herself a moment; and when she had gone out Julia said suddenly to her companion: 'Should you mind telling me what's the matter with you?'

'The matter with me?'

'Don't you want to stand?'

'I'll do anything to oblige you.'

'Why should you oblige me?'

'Why, isn't that the way people treat you?' asked Nick.

'They treat me best when they are a little serious.'

'My dear Julia, it seems to me I'm serious enough. Surely it isn't an occasion to be so very solemn, the idea of going down into a stodgy little country town and talking a lot of rot.'

'Why do you call it "rot"?'

'Because I can think of no other name that, on the whole, describes it so well. You know the sort of thing. Come! you've listened to enough of it, first and last. One blushes for it when one sees it in print, in the local papers. The local papers – ah, the thought of them makes me want to stay in Paris.'

'If you don't speak well it's your own fault: you know how to, perfectly. And you usually do.'

'I always do, and that's what I'm ashamed of. I've got the cursed humbugging trick of it. I speak beautifully. I can turn it on, a fine flood of it, at the shortest notice. The better it is the worse it is, the kind is so inferior. It has nothing to do with the truth or the search for it; nothing to do with intelligence, or candour, or honour. It's an appeal to everything that for one's self one despises,' the young man went on – 'to stupidity, to ignorance, to density, to the love of names and phrases, the love of hollow, idiotic words, of shutting the eyes tight and making a noise. Do men who respect each other or themselves talk to each other that way? They know they would deserve kicking if they were to attempt it. A man would blush to say to himself in the darkness of the night the things he stands up on a platform in the garish light of day to stuff into the ears of a multitude whose intelligence he pretends that he esteems.' Nick Dormer stood at one of the windows, with his hands in his pockets. He had been looking out, but as his words followed each other faster he turned towards Mrs Dallow, who had dropped upon a sofa with her face to the window. She had given her jacket and gloves to her maid, but had kept on her bonnet; and she leaned forward a little as she sat, with her hands clasped together in her lap and her eyes upon her companion. The lamp, in a corner, was so thickly veiled that the room was in tempered obscurity, lighted almost equally from the street, from the brilliant shop-fronts opposite. 'Therefore, why be sapient and solemn about it, like an editorial in a newspaper?' Nick added, with a smile.

She continued to look at him for a moment after he had spoken; then she said: 'If you don't want to stand, you have only to say so. You needn't give your reasons.'

'It's too kind of you to let me off that! And then I'm a tremendous fellow for reasons; that's my strong point, don't you know? I've a lot more besides those I've mentioned, done up and ready for delivery. The odd thing is that they don't always govern my behaviour. I rather think I do want to stand.'

'Then what you said just now was a speech,' Mrs Dallow rejoined.

'A speech?'

'The "rot", the humbug of the hustings.'

'No, those great truths remain, and a good many others. But an inner voice tells me I'm in for it. And it will be much more graceful to embrace this opportunity, accepting your co-operation, than to wait for some other and forfeit that advantage.'

'I shall be very glad to help you anywhere,' said Mrs Dallow.

'Thanks, awfully,' murmured the young man, still standing there with his hands in his pockets. 'You would do it best in your own place, and I have no right to deny myself such a help.'

Julia smiled at him for an instant. 'I don't do it badly.'

'Ah, you're so political!'

'Of course I am; it's the only decent thing to be. But I can only help you if you'll help yourself. I can do a good deal but I can't do everything. If you'll work I'll work with you; but if you are going into it with your hands in your pockets I'll have nothing to do with you.' Nick instantly changed the position of these members, and sank into a seat with his elbows on his knees. 'You're very clever, but you must really take a little trouble. Things don't drop into people's mouths.'

'I'll try – I'll try. I have a great incentive,' Nick said.

'Of course you have.'

'My mother, my poor mother.' Mrs Dallow made a slight exclamation, and he went on: 'And of course, always, my father, dear man. My mother's even more political than you.'

'I dare say she is, and quite right!' said Mrs Dallow.

'And she can't tell me a bit more than you can what she thinks, what she believes, what she desires.'

'Excuse me, I can tell you perfectly. There's one thing I always desire – to keep out a Tory.'

'I see; that's a great philosophy.'

'It will do very well. And I desire the good of the country. I'm not ashamed of that.'

'And can you give me an idea of what it is – the good of the country?'

'I know perfectly what it isn't. It isn't what the Tories want to do.'

'What do they want to do?'

'Oh, it would take me long to tell you. All sorts of trash.'

'It would take you long, and it would take them longer! All they want to do is to prevent *us* from doing. On our side, we want to prevent them from preventing us. That's about as clearly as we all see it. So, on one side and the other, it's a beautiful, lucid, inspiring programme.'

'I don't believe in you,' Mrs Dallow replied to this, leaning back on her sofa.

'I hope not, Julia, indeed!' He paused a moment, still with his face toward her and his elbows on his knees; then he pursued: 'You are a very accomplished woman and a very zealous one; but you haven't an idea, you know – to call an idea. What you mainly want is to be at the head of a political salon; to start one, to keep it up, to make it a success.'

'Much you know me!' Julia exclaimed; but he could see through the dimness that she had coloured a little.

'You'll have it, in time, but I won't come to it,' Nick went on.

'You can't come less than you do.'

'When I say you'll have it, I mean you've already got it. That's why I don't come.'

'I don't think you know what you mean,' said Mrs Dallow. 'I have an idea that's as good as any of yours, any of those you have treated me to this evening, it seems to me – the simple idea that one ought to do something or other for one's country.'

' "Something or other" certainly covers all the ground. There is one thing one can always do for one's country, which is not to be afraid.'

'Afraid of what?'

Nick Dormer hesitated a moment, laughing; then he said: 'I'll tell you another time. It's very well to talk so glibly of standing,' he added; 'but it isn't absolutely foreign to the question that I haven't got the cash.'

'What did you do before?' asked Mrs Dallow.

'The first time, my father paid.'

'And the other time?'

'Oh, Mr Carteret.'

'Your expenses won't be at all large; on the contrary,' said Julia.

'They shan't be; I shall look out sharp for that. I shall have the great Hutchby.'

'Of course; but, you know, I want you to do it well.' She paused an instant, and then: 'Of course you can send the bill to me.'

'Thanks, awfully; you're tremendously kind. I shouldn't think of that.' Nick Dormer got up as he said these words, and walked to the window again, his companion's eyes resting upon him as he stood for a moment with his back to her. 'I shall manage it somehow,' he went on.

'Mr Carteret will be delighted,' said Julia.

'I dare say, but I hate taking people's money.'

'That's nonsense, when it's for the country. Isn't it for *them*?'

'When they get it back!' Nick replied, turning round and looking for his hat. 'It's startlingly late; you must be tired.' Mrs Dallow made no response to this, and he pursued his quest, successful only when he reached a duskier corner of the room, to which the hat had been relegated by his cousin's maid. 'Mr Carteret will expect so much, if he pays. And so would you.'

'Yes, I'm bound to say I should!' And Mrs Dallow emphasized this assertion by the way she rose erect. 'If you're only going in to lose, you had better stay out.'

'How can I lose, with you?' the young man asked, smiling. She uttered a word, impatiently but indistinguishably, and he continued: 'And even if I do, it will have been immense fun.'

'It *is* immense fun,' said Julia. 'But the best fun is to win. If you don't –'

'If I don't?' he repeated, as she hesitated.

'I'll never speak to you again.'

'How much you expect, even when you don't pay!'

Mrs Dallow's rejoinder was a justification of this remark, embodying as it did the fact that if they should receive on the morrow certain information on which she believed herself entitled to count, information tending to show that the Tories meant to fight the seat hard, not to lose it again, she should look to him to be in the field as early as herself. Sunday was a lost day; she should leave Paris on Monday.

'Oh, they'll fight it hard; they'll put up Kingsbury,' said Nick, smoothing his hat. 'They'll all come down – all that can get away. And Kingsbury has a very handsome wife.'

'She is not so handsome as your cousin,' Mrs Dallow hazarded.

'Oh dear, no – a cousin sooner than a wife, any day!' Nick

laughed as soon as he had said this, as if the speech had an awkward side; but the reparation, perhaps, scarcely mended it, the exaggerated mock-meekness with which he added: 'I'll do any blessed thing you tell me.'

'Come here tomorrow, then, as early as ten.' She turned round, moving to the door with him; but before they reached it she demanded, abruptly: 'Pray, isn't a gentleman to do anything, to be anything?'

'To be anything?'

'If he doesn't aspire to serve the State.'

'To make his political fortune, do you mean? Oh, bless me, yes, there are other things.'

'What other things, that can compare with that?'

'Well, I, for instance, I'm very fond of the arts.'

'Of the arts?'

'Did you never hear of them? I'm awfully fond of painting.'

At this Mrs Dallow stopped short, and her fine gray eyes had for a moment the air of being set further forward in her head. 'Don't be odious! Good-night,' she said, turning away and leaving him to go.

7

PETER SHERRINGHAM, the next day, reminded Nick that he had promised to be present with him at Madame Carré's interview with the ladies introduced to her by Gabriel Nash; and in the afternoon, in accordance with this arrangement, the two men took their way to the Rue de Constantinople. They found Mr Nash and his friends in the small beflounced drawing-room of the old actress, who, as they learned, had sent in a request for ten minutes' grace, having been detained at a lesson – a rehearsal of a *comédie de salon*, to be given, for a charity, by a fine lady, at which she had consented to be present as an adviser. Mrs Rooth sat on a black satin sofa, with her daughter beside her, and Gabriel Nash wandered about the room, looking at the votive offerings which converted the little panelled box, decorated in sallow white and gold, into a theatrical museum: the presents, the portraits, the wreaths, the diadems, the letters, framed and glazed, the trophies and tributes and relics collected by Madame Carré during half a century of renown. The profusion of this testimony was hardly more striking than the confession of something missed, something hushed, which seemed to rise from it all and make it melancholy, like a reference to clappings which, in the nature of things, could now only be present as a silence: so that if the place was full of history, it was the form without the fact, or at the most a redundancy of the one to a pinch of the other – the history of a mask, of a squeak, a record of movements in the air.

Some of the objects exhibited by the distinguished artist, her early portraits, in lithograph or miniature, represented the costume and embodied the manner of a period so remote that Nick Dormer, as he glanced at them, felt a quickened curiosity to look at the woman who reconciled being alive today with having been alive so long ago. Peter Sherringham already knew how she managed this miracle, but every visit he paid to her added to his amused, charmed sense that it *was* a miracle, that his extraordinary old friend had seen things that he should never, never see. Those were just the things he wanted to see most, and her duration, her survival, cheated him agreeably and helped him a little to guess them. His appreciation of the actor's art was so systematic that it had an antiquarian side, and at the risk of representing him as attached to a futility, it must be said that he had as yet hardly known a keener regret for anything than for the loss of that antecedent world, and in particular for his having come too late for the great *comédienne*, the light of the French stage in the early years of the century, of whose example and instruction Madame Carré had had the inestimable benefit. She had often described to him her rare predecessor, straight from whose hands she had received her most celebrated parts, and of whom her own manner was often a religious imitation; but her descriptions troubled him more than they consoled, only confirming his theory, to which so much of his observation had already ministered, that the actor's art, in general, is going down and down, descending a slope with abysses of vulgarity at its foot, after having reached its perfection, more than fifty years ago, in the talent of the lady in question. He would have liked to dwell for an hour beneath the meridian.

Gabriel Nash introduced the new-comers to his companions; but the younger of the two ladies gave no sign of lending herself to this transaction. The girl was very white; she huddled there, silent and rigid, frightened to death, staring, expressionless. If Bridget Dormer had seen her at this moment she might have felt avenged for the discomfiture she had suffered the day before, at the Salon, under the challenging eyes of Maud Vavasour. It was plain at the present hour, that Miss Vavasour would have run away had she not felt that the persons present would prevent her escape. Her aspect made Nick Dormer feel as if the little temple of art in which they were collected had been the waiting-room of a dentist. Sherringham had seen a great many nervous girls trembling before the same ordeal, and he liked to be kind to them, to say things that would help them to do themselves justice. The probability, in a given case, was almost overwhelmingly in favour of their having any other talent one could think of in a higher degree than the dramatic; but he could rarely forbear to interpose, even as against his conscience, to keep the occasion from being too cruel. There were occasions indeed that could scarcely be too cruel to punish properly certain examples of presumptuous ineptitude. He remembered what Mr Nash had said about this blighted maiden, and perceived that though she might be inept she was now anything but presumptuous. Gabriel fell to talking with Nick Dormer, and Peter addressed himself to Mrs Rooth. There was no use as yet in saying anything to the girl; she was too scared even to hear. Mrs Rooth, with her shawl fluttering about her, nestled against her daughter, putting out her hand to take one of Miriam's, soothingly. She had pretty, silly, near-sighted eyes, a long, thin nose and an upper lip which projected over the under as an ornamental cornice rests on its support. 'So much depends – really everything!' she said in answer to some sociable observation of Sherringham's. 'It's either this,' and she rolled her eyes expressively about the room, 'or it's – I don't know too much what!'

'Perhaps we're too many,' Peter hazarded, to her daughter. 'But really, you'll find, after you fairly begin, that you'll do better with four or five.'

Before she answered she turned her head and lifted her fine

eyes. The next instant he saw they were full of tears. The word she spoke, however, though uttered in a deep, serious tone, had not the note of sensibility: 'Oh, I don't care for *you*!' He laughed at this, declared it was very well said and that if she could give Madame Carré such a specimen as that —! The actress came in before he had finished his phrase, and he observed the way the girl slowly got up to meet her, hanging her head a little and looking at her from under her brows. There was no sentiment in her face — only a kind of vacancy of terror which had not even the merit of being fine of its kind, for it seemed stupid and superstitious. Yet the head was good, he perceived at the same moment; it was strong and salient and made to tell at a distance. Madame Carré scarcely noticed her at first, greeting her only in her order, with the others, and pointing to seats, composing the circle with smiles and gestures, as if they were all before the prompter's box. The old actress presented herself to a casual glance as a red-faced woman in a wig, with beady eyes, a hooked nose and pretty hands; but Nick Dormer, who had a perception of physiognomy, speedily observed that these free characteristics included a great deal of delicate detail — an eyebrow, a nostril, a flitting of expressions, as if a multitude of little facial wires were pulled from within. This accomplished artist had in particular a mouth which was visibly a rare instrument, a pair of lips whose curves and fine corners spoke of a lifetime of 'points' unerringly made and verses exquisitely spoken, helping to explain the purity of the sound that issued from them. Her whole countenance had the look of long service — of a thing infinitely worn and used, drawn and stretched to excess, with its elasticity overdone and its springs relaxed, yet religiously preserved and kept in repair, like an old valuable time-piece, which might have quivered and rumbled, but could be trusted to strike the hour. At the first words she spoke Gabriel Nash exclaimed, endearingly: '*Ah, la voix de Célimène!*' Célimène, who wore a big red flower on the summit of her dense wig, had a very grand air, a toss of the head and sundry little majesties of manner; in addition to which she was strange, almost grotesque, and to some people would have been even terrifying, capable of reappearing, with her hard eyes, as a queer vision in the darkness. She excused herself for having made the company wait, and mouthed and mimicked in the drollest way, with intonations as fine as a flute, the performance and the pretensions of the *belles dames* to whom she had just been endeavouring to communicate a few of the rudiments. '*Mais celles-là, c'est une plaisanterie,*' she went on, to Mrs Rooth; 'whereas you and your daughter, *chère madame* – I am sure that you are quite another matter.'

The girl had got rid of her tears and was gazing at her, and Mrs Rooth leaned forward and said insinuatingly: 'She knows four languages.'

Madame Carré gave one of her histrionic stares, throwing back her head. 'That's three too many. The thing is to do something with one of them.'

'We're very much in earnest,' continued Mrs Rooth, who spoke excellent French.

'I'm glad to hear it – *il n'y a que ça. La tête est bien* – the head is very good,' she said, looking at the girl. 'But let us see, my dear child, what you've got in it!' The young lady was still powerless to speak; she opened her lips, but nothing came. With the failure of this effort she turned her deep, sombre eyes upon the three men. '*Un beau regard* – it carries well,' Madame Carré hinted. But even as she spoke Miss Rooth's fine gaze was suffused again, and the next moment she had begun to weep. Nick Dormer sprung up; he felt embarrassed and intrusive – there was such an indelicacy in sitting there to watch a poor girl's struggle with timidity. There was a momentary confusion; Mrs Rooth's tears were seen also to flow; Gabriel Nash began to laugh, addressing however at the same time the friendliest, most familiar encouragement to his companions, and Peter Sherringham offered to retire with Nick on the spot, if their presence was oppressive to the young lady. But the agitation was over in a minute; Madame Carré motioned Mrs Rooth out of her seat and took her place beside the girl, and Gabriel Nash explained judiciously to the other men that she would be worse if they were to go away. Her mother begged them to remain, 'so that there should be at least some English'; she spoke as if the old actress were an army of Frenchwomen. The girl was quickly better, and Madame Carré, on the sofa beside her, held her hand and emitted a perfect music of reassurance. 'The nerves, the nerves – they are half of our trade. Have as many as you like, if you've got something else too. *Voyons* – do you know anything?'

'I know some pieces.'

'Some pieces of the *répertoire*?'

Miriam Rooth stared as if she didn't understand. 'I know some poetry.'

'English, French, Italian, German,' said her mother.

Madame Carré gave Mrs Rooth a look which expressed irritation at the recurrence of this announcement. 'Does she wish to act in all those tongues? The phrase-book isn't the comedy.'

'It is only to show you how she has been educated.'

'Ah, *chère madame*, there is no education that matters! I mean save the right one. Your daughter must have a language, like me, like *ces messieurs*.'

'You see if I can speak French,' said the girl, smiling dimly at her hostess. She appeared now almost to have collected herself.

'You speak it in perfection.'

'And English just as well,' said Miss Rooth.

'You oughtn't to be an actress; you ought to be a governess.'

'Oh, don't tell us that: it's to escape from that!' pleaded Mrs Rooth.

'I'm very sure your daughter will escape from that,' Peter Sherringham was moved to remark.

'Oh, if *you* could help her!' the lady exclaimed, pathetically.

'She has certainly all the qualities that strike the eye,' said Peter.

'You are *most* kind, sir!' Mrs Rooth declared, elegantly draping herself.

'She knows Célimène; I have heard her do Célimène,' Gabriel Nash said to Madame Carré.

'And she knows Juliet, and Lady Macbeth, and Cleopatra,' added Mrs Rooth.

'*Voyons*, my dear child, do you wish to work for the French stage or for the English?' the old actress demanded.

'Ours would have sore need of you, Miss Rooth,' Sherringham gallantly interposed.

'Could you speak to any one in London – could you introduce her?' her mother eagerly asked.

'Dear madam, I must hear her first, and hear what Madame Carré says.'

'She has a voice of rare beauty, and I understand voices,' said Mrs Rooth.

'Ah, then, if she has intelligence she has every gift.'

'She has a most poetic mind,' the old lady went on.

'I should like to paint her portrait; she's made for that,' Nick Dormer ventured to observe to Mrs Rooth; partly because he was struck with the girl's capacity as a model, partly to mitigate the crudity of inexpressive spectatorship.

'So all the artists say. I have had three or four heads of her, if you would like to see them: she has been done in several styles. If you were to do her I am sure it would make her celebrated.'

'And me too,' said Nick, laughing.

'It would indeed, a member of Parliament!' Nash declared.

'Ah, I have the honour –?' murmured Mrs Rooth, looking gratified and mystified.

Nick explained that she had no honour at all, and meanwhile Madame Carré had been questioning the girl. '*Chère madame*, I can do nothing with your daughter: she knows too much!' she broke out. 'It's a pity, because I like to catch them wild.'

'Oh, she's wild enough, if that's all! And that's the very point, the question of where to try,' Mrs Rooth went on. 'Into what do I launch her – upon what dangerous, stormy sea? I've thought of it so anxiously.'

'Try here – try the French public: they're so much the most serious,' said Gabriel Nash.

'Ah, no, try the English: there's such a rare opening!' Sherringham exclaimed, in quick opposition.

'Ah, it isn't the public, dear gentlemen. It's the private side, the other people – it's the life – it's the moral atmosphere.'

'*Je ne connais qu'une scène – la nôtre*,' Madame Carré asserted. 'I have been informed there is no other.'

'And very correctly,' said Gabriel Nash. 'The theatre in our countries is puerile and barbarous.'

'There is something to be done for it, and perhaps *mademoiselle* is the person to do it,' Sherringham suggested, contentiously.

'Ah, but, *en attendant*, what can it do for her?' Madame Carré asked.

'Well, anything that I can help it to do,' said Peter Sherringham, who was more and more struck with the girl's rich type. Miriam Rooth sat in silence, while this discussion went on, looking from one speaker to the other with a suspended, literal air.

'Ah, if your part is marked out, I congratulate you, *mademoiselle*!' said the old actress, underlining the words as she had often underlined such words on the stage. She smiled with large permissiveness on the young aspirant, who appeared not to understand her. Her tone penetrated, however, to certain depths in the mother's nature, adding another stir to agitated waters.

'I feel the responsibility of what she shall find in the life, the standards, of the theatre,' Mrs Rooth explained. 'Where is the purest tone – where are the highest standards? that's what I ask,' the good lady continued, with a persistent candour which elicited a peal of unceremonious but sociable laughter from Gabriel Nash.

'The purest tone – *qu'est-ce-que-c'est que ça?*' Madame Carré demanded, in the finest manner of modern comedy.

'We are very, *very* respectable,' Mrs Rooth went on, smiling and achieving lightness, too. 'What I want to do is to place my daughter where the conduct – and the picture of conduct, in which she should take part – wouldn't be absolutely dreadful. Now, *chère madame*, how about all that? how about the conduct in the French theatre – the things she should see, the things she should hear?'

'I don't think I know what you are talking about. They are the things she may see and hear everywhere; only they are better done, they are better said. The only conduct that concerns an actress, it seems to me, is her own, and the only way for her to behave herself is not to be a stick. I know no other conduct.'

'But there are characters, there are situations, which I don't think I should like to see *her* undertake.'

'There are many, no doubt, which she would do well to leave alone!' laughed the Frenchwoman.

'I shouldn't like to see her represent a very bad woman – a *really* bad one,' Mrs Rooth serenely pursued.

'Ah, in England, then, and in your theatre, every one is good? Your plays must be even more ingenious than I supposed.'

'We haven't any plays,' said Gabriel Nash.

'People will write them for Miss Rooth – it will be a new era,' Peter Sherringham rejoined, with wanton, or at any rate combative, optimism.

'Will *you*, sir – will you do something? A sketch of some truly noble female type?' the old lady asked, engagingly.

'Oh, I know what you do with our pieces – to show your superior virtue!' Madame Carré broke in, before he had time to reply that he wrote nothing but diplomatic memoranda. 'Bad women? *Je n'ai joué que ça, madame*. "Really" bad? I tried to make them real!'

'I can say "L'Aventurière",' Miriam interrupted, in a cold voice which seemed to hint at a want of participation in the maternal solicitudes.

'Confer on us the pleasure of hearing you, then, Madame Carré will give you the *réplique*,' said Peter Sherringham.

'Certainly, my child; I can say it without the book,' Madame Carré responded. 'Put yourself there – move that chair a little away.' She patted her young visitor, encouraging her to rise, settling with her the scene they should take, while the three men sprang up to arrange a place for the performance. Miriam left her seat and looked vaguely round her; then, having taken off her hat and given it to her mother, she stood on the designated spot with her eyes on the ground. Abruptly, however, instead of beginning the scene, Madame Carré turned to the elder lady with an air which showed that a rejoinder to this visitor's remarks of a moment before had been gathering force in her breast.

'You mix things up, *chère madame*, and I have it on my heart to tell you so. I believe it's rather the case with you other English, and I have never been able to learn that either your morality or your talent is the gainer by it. To be too respectable to go where things are done best is, in my opinion, to be very vicious indeed; and to do them badly in order to preserve your virtue is to fall into a grossness more shocking than any other. To do them well is virtue enough, and not to make a mess of it the only respectability. That's hard enough to merit Paradise.

Everything else is base humbug! *Voilà, chère madame*, the answer I have for your scruples!'

'It's admirable – admirable; and I am glad my friend Dormer here has had the great advantage of hearing you utter it!' Gabriel Nash exclaimed, looking at Nick.

Nick thought it, in effect, a speech denoting an intelligence of the question, but he rather resented the idea that Nash should assume that it would strike him as a revelation; and to show his familiarity with the line of thought it indicated, as well as to play his part appreciatively in the little circle, he observed to Mrs Rooth, as if they might take many things for granted: 'In other words, your daughter must find her safeguard in the artistic conscience.' But he had no sooner spoken than he was struck with the oddity of their discussing so publicly, and under the poor girl's nose, the conditions which Miss Rooth might find the best for the preservation of her personal integrity. However, the anomaly was light and unoppressive – the echoes of a public discussion of delicate questions seemed to linger so familiarly in the egotistical little room. Moreover the heroine of the occasion evidently was losing her embarrassment; she was the priestess on the tripod, awaiting the afflatus and thinking only of that. Her bared head, of which she had changed the position, holding it erect, while her arms hung at her sides, was admirable; and her eyes gazed straight out of the window, at the houses on the opposite side in the Rue de Constantinople.

Mrs Rooth had listened to Madame Carré with startled, respectful attention, but Nick, considering her, was very sure that she had not understood her hostess's little lesson. Yet this did not prevent her from exclaiming in answer to him: 'Oh, a fine artistic life – what indeed is more beautiful?'

Peter Sherringham had said nothing; he was watching Miriam and her attitude. She wore a black dress, which fell in straight folds; her face, under her mobile brows, was pale and regular, with a strange, strong, tragic beauty. 'I don't know what's in her,' he said to himself; 'nothing, it would seem, from her persistent vacancy. But such a face as that, such a head, is a fortune!' Madame Carré made her commence, giving her the first line of the speech of Clorinde: '*Vous ne me fuyez pas, mon enfant, aujourd'hui.*' But still the girl hesitated, and for an instant she appeared to make a vain, convulsive effort. In this effort she frowned portentously; her low forehead overhung her eyes; the eyes themselves, in shadow, stared, splendid and cold, and her hands clinched themselves at her sides. She looked austere and terrible, and during this moment she was an incarnation the vividness of which drew from Sherringham a stifled cry. '*Elle est bien belle – ah, ça!*' murmured the old actress; and in the pause which still preceded the issue of sound from the girl's lips Peter turned to his kinsman and said in a low tone:

'You must paint her just like that.'

'Like that?'

'As the Tragic Muse.'

She began to speak; a long, strong, colourless voice came quavering from her young throat. She delivered the lines of Clorinde, in the fine interview with Célie, in the third act of the play, with a rude monotony, and then, gaining confidence, with an effort at modulation which was not altogether successful and which evidently she felt not to be so. Madame Carré sent back the ball without raising her hand, repeating the speeches of Célie, which her memory possessed from their having so often been addressed to her, and uttering the verses with soft, communicative art. So they went on through the scene, and when it was over it had not precisely been a triumph for Miriam Rooth. Sherringham forbore to look at Gabriel Nash, and Madame Carré said: 'I think you have a voice, *ma fille*, somewhere or other. We must try and put our hand on it.' Then she asked her what instruction she had had, and the girl, lifting her eyebrows, looked at her mother, while her mother prompted her.

'Mrs Delamere, in London; she was once an ornament of the English stage. She gives lessons just to a very few; it's a great favour. Such a very nice person! But above all, Signor Ruggieri – I think he taught us most.' Mrs Rooth explained that this gentleman was an Italian tragedian, in Rome, who instructed Miriam in the proper manner of pronouncing his language, and also in the art of declaiming and gesticulating.

'Gesticulating, I'll warrant,' said their hostess. 'They mimic as if for the deaf, they emphasize as if for the blind. Mrs Delamere is doubtless an epitome of all the virtues, but I never heard of her. You travel too much,' Madame Carré went on; 'that's very amusing, but the way to study is to stay at home, to shut yourself up and hammer at your scales.' Mrs Rooth complained that they had no home to stay at; in rejoinder to which the old actress exclaimed: 'Oh, you English, you are *d'une légèreté à faire rougir*. If you haven't a home you must make one. In our profession it's the first requisite.'

'But where? That's what I ask!' said Mrs Rooth.

'Why not here?' Sherringham inquired.

'Oh, here!' And the good lady shook her head, with a world of suggestions.

'Come and live in London, and then I shall be able to paint your daughter,' Nick Dormer interposed.

'Is that all that it will take, my dear fellow?' asked Gabriel Nash.

'Ah, London is full of memories,' Mrs Rooth went on. 'My father had a great house there – we always came up. But all that's over.'

'Study here, and go to London to appear,' said Peter Sherringham, feeling frivolous even as he spoke.

'To appear in French?'

'No, in the language of Shakespeare.'

'But we can't study that here.'

'Monsieur Sherringham means that he will give you lessons,' Madame Carré explained. 'Let me not fail to say it – he's an excellent critic.'

'How do you know that – you who are perfect?' asked Sherringham: an inquiry to which the answer was forestalled by the girl's rousing herself to make it public that she could recite the 'Nights' of Alfred de Musset.

'*Diable!*' said the actress, 'that's more than I can! But by all means give us a specimen.'

The girl again placed herself in position and rolled out a fragment of one of the splendid conversations of Musset's poet with his muse – rolled it loudly and proudly, tossed it and tumbled it about the room. Madame Carré watched her at first, but after a few moments she shut her eyes, though the best part of the business was to look. Sherringham had supposed Miriam was abashed by the flatness of her first performance, but now he perceived that she could not have been conscious of this; she was rather exhilarated and emboldened. She made a muddle of the divine verses, which, in spite of certain sonorities and cadences, an evident effort to imitate a celebrated actress, a comrade of Madame Carré, whom she had heard declaim them, she produced as if she had but a dim idea of their meaning. When she had finished Madame Carré passed no judgement; she only said, 'Perhaps you had better say something English.' She suggested some little piece of verse – some fable, if there were fables in English. She appeared but scantily surprised to hear that there were not – it was a language of which one expected so little. Mrs Rooth said, 'She knows her Tennyson by heart. I think he's more profound than La Fontaine'; and after some deliberation and delay Miriam broke into 'The Lotos-Eaters', from which she passed directly, almost breathlessly, to 'Edward Gray'. Sherringham had by this time heard her make four different attempts, and the only generalization which could be very present to him was that she uttered these dissimilar compositions in exactly the same tone – a solemn, droning, dragging measure, adopted with an intention of pathos, a crude idea of 'style'. It was funereal, and at the same time it was rough and childish. Sherringham thought her English performance less futile than her French, but he could see that Madame Carré listened to it with even less pleasure. In the way the girl wailed forth some of her Tennysonian lines he detected a possibility of a thrill. But the further she went, the more violently she acted on the nerves of Mr Gabriel Nash: that also he could discover, from the way this gentleman ended by slipping discreetly to the window and leaning there, with his head out and his back to the exhibition. He had the art of mute expression; his attitude said, as clearly as possible: 'No, no, you can't call me either ill-mannered or ill-natured. I'm the showman of the occasion, moreover, and I avert myself, leaving you to judge. If there's a thing in life I hate, it's this idiotic new fashion of the drawing-room recitation, and the insufferable creatures who practise it, who prevent conversation and whom, as they are beneath it, you can't punish by criticism. Therefore what I am is only too magnanimous – bringing these benighted women here, paying with my person, stifling my just repugnance.'

At the same time that Sherringham pronounced privately that

the manner in which Miss Rooth had acquitted herself offered no element of interest, he remained conscious that something surmounted and survived her failure, something that would perhaps be worth taking hold of. It was the element of outline and attitude, the way she stood, the way she turned her eyes, her head, and moved her limbs. These things held the attention; they had a natural felicity and, in spite of their suggesting too much the school-girl in the *tableau-vivant*, a sort of grandeur. Her face, moreover, grew as he watched it; something delicate dawned in it, a dim promise of variety and a touching plea for patience, as if it were conscious of being able to show in time more expressions than the simple and striking gloom which, as yet, had mainly graced it. In short the plastic quality of her person was the only definite sign of a vocation. He almost hated to have to recognize this; he had seen that quality so often when it meant nothing at all that he had come at last to regard it as almost a guarantee of incompetence. He knew Madame Carré valued it, by itself, so little that she counted it out in measuring an histrionic nature; when it was not accompanied with other properties which helped and completed it she was near considering it as a positive hindrance to success – success of the only kind that she esteemed. Far oftener than he, she had sat in judgement on young women for whom hair and eyebrows and a disposition for the statuesque would have worked the miracle of attenuating their stupidity if the miracle were workable. But that particular miracle never was. The qualities she deemed most interesting were not the gifts, but the conquests – the effects the actor had worked hard for, had wrested by unwearying study. Sherringham remembered to have had, in the early part of their acquaintance, a friendly dispute with her on this subject; he having been moved at that time to defend the cause of the gifts. She had gone so far as to say that a serious comedian ought to be ashamed of them – ashamed of resting his case on them; and when Sherringham had cited Mademoiselle Rachel as a great artist whose natural endowment was rich and who had owed her highest triumphs to it, she had declared that Rachel was the very instance that proved her point – a talent embodying one or two primary aids, a voice and an eye, but essentially formed by work, unremitting and ferocious work. 'I don't care a straw for your handsome girls,' she said; 'but bring me the one who is ready to drudge the tenth part of the way Rachel drudged, and I'll forgive her her beauty. Of course, *notez bien*, Rachel wasn't a *bête*: that's a gift, if you like!'

Mrs Rooth, who was evidently very proud of the figure her daughter had made, appealed to Madame Carré, rashly and serenely, for a verdict; but fortunately this lady's voluble *bonne* came rattling in at the same moment with the tea-tray. The old actress busied herself in dispensing this refreshment, an hospitable attention to her English visitors, and under cover of the diversion thus obtained, while the others talked together, Sherringham said to his hostess: 'Well, is there anything in her?'

'Nothing that I can see. She's loud and coarse.'

'She's very much afraid; you must allow for that.'

'Afraid of me, immensely, but not a bit afraid of her authors – nor of you!' added Madame Carré, smiling.

'Aren't you prejudiced by what Mr Nash has told you?'

'Why prejudiced? He only told me she was very handsome.'

'And don't you think she is?'

'Admirable. But I'm not a photographer nor a dressmaker. I can't do anything with that.'

'The head is very noble,' said Peter Sherringham. 'And the voice, when she spoke English, had some sweet tones.'

'Ah, your English – possibly! All I can say is that I listened to her conscientiously, and I didn't perceive in what she did a single *nuance*, a single inflection or intention. But not one, *mon cher*. I don't think she's intelligent.'

'But don't they often seem stupid at first?'

'Say always!'

'Then don't some succeed – even when they are handsome?'

'When they are handsome they always succeed – in one way or another.'

'You don't understand us English,' said Peter Sherringham.

Madame Carré drank her tea; then she replied: 'Marry her, my son, and give her diamonds. Make her an ambassadress; she will look very well.'

'She interests you so little that you don't care to do anything for her?'

'To do anything?'

'To give her a few lessons.'

The old actress looked at him a moment: after which, rising from her place near the table on which the tea had been served, she said to Miriam Rooth: 'My dear child, I give my voice for the *scène anglaise*. You did the English things best.'

'Did I do them well?' asked the girl.

'You have a great deal to learn; but you have force. The principal things *sont encore à dégager*, but they will come. You must work.'

'I think she has ideas,' said Mrs Rooth.

'She gets them from you,' Madame Carré replied.

'I must say, if it's to be *our* theatre I'm relieved. I think it's safer,' the good lady continued.

'Ours is dangerous, no doubt.'

'You mean you are more severe,' said the girl.

'Your mother is right,' the actress smiled; 'you have ideas.'

'But what shall we do then – how shall we proceed?' Mrs Rooth inquired.

She made this appeal, plaintively and vaguely, to the three gentlemen; but they had collected a few steps off and were talking together, so that it failed to reach them.

'Work – work – work!' exclaimed the actress.

'In English I can play Shakespeare. I want to play Shakespeare,' Miriam remarked.

'That's fortunate, as in English you haven't anyone else to play.'

'But he's so great – and he's so pure!' said Mrs Rooth.

'That also seems very fortunate for you!' Madame Carré phrased.

'You think me actually pretty bad, don't you?' the girl demanded, with her serious face.

'*Mon Dieu, que vous dirai-je?* Of course you're rough; but so was I, at your age. And if you find your voice it may carry you far. Besides, what does it matter what I think? How can I judge for your English public?'

'How shall I find my voice?' asked Miriam Rooth.

'By trying. *Il n'y a que ça*. Work like a horse, night and day. Besides, Monsieur Sherringham, as he says, will help you.'

Sherringham, hearing his name, turned round, and the girl appealed to him. 'Will you help me, really?'

'To find her voice,' Madame Carré interposed.

'The voice, when it's worth anything, comes from the heart; so I suppose that's where to look for it,' Gabriel Nash suggested.

'Much you know; you haven't got any!' Miriam retorted, with the first scintillation of gaiety she had shown on this occasion.

'Any voice, my child?' Mr Nash inquired.

'Any heart – or any manners!'

Peter Sherringham made the secret reflection that he liked her better when she was lugubrious; for the note of pertness was not totally absent from her mode of emitting these few words. He was irritated, moreover, for in the brief conference he had just had with the young lady's introducer he had had to face the necessity of saying something optimistic about her, which was not particularly easy. Mr Nash had said with his bland smile, 'And what impression does my young friend make?' to which it appeared to Sherringham that uncomfortable consistency compelled him to reply that there was evidently a good deal in her. He was far from being sure of that. At the same time the young lady, both with the exaggerated 'points' of her person and the poverty of her instinct of expression, constituted a kind of challenge – presented herself to him as a subject for inquiry, a problem, a piece of work, an explorable country. She was too bad to jump at, and yet she was too individual to overlook, especially when she rested her tragic eyes on him with the appeal of her deep 'Really?' This appeal sounded as if it were in a certain way to his honour, giving him a chance to brave verisimilitude, to brave ridicule even, a little, in order to show, in a special case, what he had always maintained in general, that the direction of a young person's studies for the stage may be an interest of as high an order as any other artistic consideration.

'Mr Nash has rendered us the great service of introducing us to Madame Carré, and I'm sure we're immensely indebted to him,' Mrs Rooth said to her daughter, with an air affectionately corrective.

'But what good does that do us?' the girl asked, smiling at

the actress and gently laying her finger-tips upon her hand. 'Madame Carré listens to me with adorable patience and then sends me about my business – in the prettiest way in the world.'

'*Mademoiselle*, you are not so rough; the tone of that is very *juste. A la bonne heure*; work – work!' The actress exclaimed. 'There was an inflection there, or very nearly. Practise it till you've got it.'

'Come and practise it to me, if your mother will be so kind as to bring you,' said Peter Sherringham.

'Do you give lessons – do you understand?' Miriam asked.

'I'm an old playgoer, and I have unbounded belief in my own judgement.'

'"Old", sir, is too much to say,' Mrs Rooth remonstrated. 'My daughter knows your high position, but she is very direct. You will always find her so. Perhaps you'll say there are less honourable faults. We'll come to see you with pleasure. Oh, I've been at the Embassy, when I was her age. Therefore why shouldn't she go today? That was in Lord Davenant's time.'

'A few people are coming to tea with me tomorrow. Perhaps you will come then, at five o'clock.'

'It will remind me of the dear old times,' said Mrs Rooth.

'Thank you; I'll try and do better tomorrow,' Miriam remarked, very sweetly.

'You do better every minute!' Sherringham exclaimed, looking at Madame Carré in emphasis of this declaration.

'She is finding her voice,' the actress responded.

'She is finding a friend!' cried Mrs Rooth.

'And don't forget, when you come to London, my hope that you'll come and see *me*,' Nick Dormer said to the girl. 'To try and paint you – that would do me good!'

'She is finding even two,' said Madame Carré.

'It's to make up for one I've lost!' And Miriam looked with very good stage-scorn at Gabriel Nash. 'It's he that thinks I'm bad.'

'You say that to make me drive you home; you know it will,' Nash returned.

'We'll all take you home; why not?' Sherringham asked.

Madame Carré looked at the handsome girl, handsomer than ever at this moment, and at the three young men who had taken their hats and stood ready to accompany her. A deeper expression came for an instant into her hard, bright eyes, while she sighed: '*Ah, la jeunesse!* you'd always have that, my child, if you were the greatest goose on earth!'

8

AT Peter Sherringham's, the next day, Miriam Rooth had so evidently come with the expectation of 'saying' something that it was impossible such a patron of the drama should forbear to invite her, little as the exhibition at Madame Carré's could have contributed to render the invitation prompt. His curiosity had been more appeased than stimulated, but he felt none the less that he had 'taken up' the dark-browed girl and her reminiscential mother and must face the immediate consequences of the act. This responsibility weighed upon him during the twenty-four hours that followed the ultimate dispersal of the little party at the door of the Hôtel de la Garonne.

On quitting Madame Carré's the two ladies had gracefully declined Mr Nash's offered cab and had taken their way homeward on foot, with the gentlemen in attendance. The streets of Paris at that hour were bright and episodical, and Sherringham trod them good-humouredly enough, and not too fast, leaning a little to talk to the young lady as he went. Their pace was regulated by the mother's, who walked in advance, on the arm of Gabriel Nash (Nick Dormer was on her other side), in refined deprecation. Her sloping back was before them, exempt from retentive stiffness in spite of her rigid principles, with the little drama of her lost and recovered shawl perpetually going on.

Sherringham said nothing to the girl about her performance or her powers; their talk was only of her manner of life with her mother – their travels, their *pensions*, their economies, their want of a home, the many cities she knew well, the foreign tongues and the wide view of the world she had acquired. He guessed easily enough the dolorous type of exile of the two ladies, wanderers in search of Continental cheapness, inured to queer contacts and compromises, 'remarkably well connected' in England, but going out for their meals. The girl was but indirectly communicative, not, apparently, from any intention of concealment, but from the habit of associating with people whom she didn't honour with her confidence. She was fragmentary and abrupt, as well as not in the least shy, subdued to dread of Madame Carré as she had been for the time. She gave Sherringham a reason for this fear, and he thought her reason innocently pretentious. 'She admired a great artist more than anything in the world; and in the presence of art, of *great* art, her heart beat so fast.' Her manners were not perfect, and the friction of a varied experience had rather roughened than smoothed her. She said nothing that showed that she was clever, though he guessed that this was the intention of two or three of her remarks; but he parted from her with the suspicion that she was, according to the contemporary French phrase, a 'nature'.

The Hôtel de la Garonne was in a small, unrenovated street, in which the cobble-stones of old Paris still flourished, lying between the Avenue de l'Opéra and the Place de la Bourse. Sherringham had occasionally passed through this dim byway, but he had never noticed the tall, stale *maison meublée*, whose aspect, that of a third-rate provincial inn, was an illustration of Mrs Rooth's shrunken standard.

'We would ask you to come up, but it's quite at the top and we haven't a sitting-room,' the poor lady bravely explained. 'We had to receive Mr Nash at a café.'

Nick Dormer declared that he liked cafés, and Miriam, looking at his cousin, dropped with a flash of passion the demand: 'Do you wonder that I should want to do something, so that we can stop living like pigs?'

Sherringham recognized eventually, the next day, that though it might be rather painful to listen to her it was better to make her recite than to let her do nothing, so effectually did the presence of his sister and that of Lady Agnes, and even of Grace and Biddy, appear, by a sort of tacit opposition, to deprive hers, ornamental as it was, of a reason. He had only to see them all together to perceive that she couldn't pass for having come to 'meet' them – even her mother's insinuating gentility failed to put the occasion on that footing – and that she must therefore be assumed to have been brought to show them something. She was not subdued nor colourless enough to sit there for nothing, or even for conversation (the sort of conversation that was likely to come off), so that it was inevitable to treat her position as connected with the principal place on the carpet, with silence and attention and the pulling together of chairs. Even when so established it struck him at first as precarious, in the light or the darkness of the inexpressive faces of the other ladies, sitting in couples and rows on sofas (there were several in addition to Julia and the Dormers; mainly the wives, with their husbands, of Sherringham's fellow-secretaries), scarcely one of whom he felt that he might count upon to say something gushing when the girl should have finished.

Miss Rooth gave a representation of Juliet drinking her potion, according to the system, as her mother explained, of the famous Signor Ruggieri – a scene of high, fierce sound, of many cries and contortions: she shook her hair (which proved magnificent) half down before the performance was over. Then she declaimed several short poems by Victor Hugo, selected, among many hundred, by Mrs Rooth, as the good lady was careful to make known. After this she jumped to the American lyre, regaling the company with specimens, both familiar and fresh, of Longfellow, Lowell, Whittier, Holmes, and of two or three poetesses revealed to Sherringham on this occasion. She flowed so copiously, keeping the floor and rejoicing visibly in her opportunity, that Sherringham was mainly occupied with wondering how he could make her leave off. He was surprised at the extent of her repertory, which, in view of the circumstance that she could never have received much encouragement – it must have come mainly from her mother, and he didn't believe in Signor Ruggieri – denoted a very stiff ambition and a kind of misplaced perseverance. It was her mother who checked her at last, and he found himself suspecting that Gabriel Nash had intimated to the old woman that interference was necessary. For himself he was chiefly glad that Madame Carré was not there. It was present to him that she would have deemed the exhibition, with its badness, its assurance, the absence of criticism, almost indecent.

His only new impression of the girl was that of this same high assurance – her coolness, her complacency, her eagerness to go on. She had been deadly afraid of the old actress, but she was not a bit afraid of a cluster of *femmes du monde*, of Julia, of Lady Agnes, of the smart women of the Embassy. It was positively these personages who were rather frightened; there was certainly a moment when even Julia was scared, for the first time that he had ever seen her. The space was too small; the cries, the rushes of the dishevelled girl were too near. Lady Agnes, much of the time, wore the countenance she might have worn at the theatre during a play in which pistols were fired; and indeed the manner of the young reciter had become more spasmodic, more explosive. It appeared, however, that the company in general thought her very clever and successful; which showed, to Sherringham's sense, how little they understood the matter. Poor Biddy was immensely struck, and grew flushed and absorbed as Miriam, at her best moments, became pale and fatal. It was she who spoke to her first, after it was agreed that they had better not fatigue her any more; she advanced a few steps, happening to be near her, murmuring: 'Oh, thank you, thank you so much. I never saw anything so beautiful, so grand.'

She looked very red and very pretty as she said this. Peter Sherringham liked her enough to notice and to like her better when she looked prettier than usual. As he turned away he heard Miriam answer, with rather an ungracious irrelevance: 'I have seen you before, three days ago, at the Salon, with Mr Dormer. Yes, I know he's your brother. I have made his acquaintance since. He wants to paint my portrait. Do you think he'll do it well?' He was afraid Miriam was something of a brute, and also somewhat grossly vain. This impression would perhaps have been confirmed if a part of the rest of the short conversation of the two girls had reached his ear. Biddy ventured to remark that she herself had studied modelling a little and that she could understand how any artist would think Miss Rooth a splendid subject. If, indeed, *she* could attempt her head, that would be a chance to do something.

'Thank you,' said Miriam, with a laugh. 'I think I had rather not *passer par toute la famille*!' Then she added: 'If your brother's an artist, I don't understand how he's in Parliament.'

'Oh, he isn't in Parliament now; we only hope he will be.'

'Oh, I see.'

'And he isn't an artist, either,' Biddy felt herself conscientiously bound to subjoin.

'Then he isn't anything,' said Miss Rooth.

'Well – he's immensely clever.'

'Oh, I see,' Miss Rooth again replied. 'Mr Nash has puffed him up so.'

'I don't know Mr Nash,' said Biddy, guilty of a little dryness, and also of a little misrepresentation, and feeling rather snubbed.

'Well, you needn't wish to.'

Biddy stood with her a moment longer, still looking at her and not knowing what to say next, but not finding her any less handsome because she had such odd manners. Biddy had an ingenious little mind, which always tried as much as possible to keep different things separate. It was pervaded now by the observation, made with a certain relief, that if the girl spoke to her with such unexpected familiarity of Nick she said nothing at all about Peter. Two gentlemen came up, two of Peter's friends, and made speeches to Miss Rooth of the kind, Biddy supposed, that people learned to make in Paris. It was also doubtless in Paris, the girl privately reasoned, that they learned to listen to them as this striking performer listened. She received their advances very differently from the way she had received Biddy's. Sherringham noticed his young kinswoman turn away, still blushing, to go and sit near her mother again, leaving Miriam engaged with the two men. It appeared to have come over Biddy that for a moment she had been strangely spontaneous and bold and had paid a little of the penalty. The seat next her mother was occupied by Mrs Rooth, towards whom Lady Agnes's head had inclined itself with a preoccupied air of benevolence. He had an idea that Mrs Rooth was telling her about the Neville-Nugents of Castle Nugent, and that Lady Agnes was thinking it odd she never had heard of them. He said to himself that Biddy was generous. She had urged Julia to come, in order that they might see how bad the strange young woman would be; but now that she turned out so dazzling she forgot this calculation and rejoiced in what she innocently supposed to be her triumph. She kept away from Julia, however; she didn't even look at her to invite her also to confess that, in vulgar parlance, they had been sold. He himself spoke to his sister, who was leaning back, in rather a detached way, in the corner of a sofa, saying something which led her to remark in reply: 'Ah, I dare say it's extremely fine, but I don't care for tragedy when it treads on one's toes. She's like a cow who has kicked over the milking-pail. She ought to be tied up!'

'My poor Julia, it isn't extremely fine; it isn't fine at all,' Sherringham rejoined, with some irritation.

'Excuse me. I thought that was why you invited us.'

'I thought she was different,' Sherringham said.

'Ah, if you don't care for her, so much the better. It has always seemed to me that you make too much of those people.'

'Oh, I do care for her in a way, too. She's interesting.' His sister gave him a momentary mystified glance, and he added, 'And she's awful!' He felt stupidly annoyed, and he was ashamed of his annoyance, for he could have assigned no reason for it. It didn't make it less, for the moment, to see Gabriel Nash approach Mrs Dallow, introduced by Nick Dormer. He gave place to the two young men with a certain alacrity, for he had a sense of being put in the wrong, in respect to the heroine of the occasion, by Nash's very presence. He remembered that it had been part of their bargain, as it were, that he should present that gentleman to his sister. He was not sorry to be relieved of the office by Nick, and he even, tacitly and ironically, wished his cousin's friend joy of a colloquy with Mrs Dallow. Sherringham's life was spent with people, he was used to people, and both as a host and as a guest he carried them, in general, lightly. He could observe, especially in the former capacity, without uneasiness, take the temperature without anxiety. But at present his company oppressed him; he felt himself nervous, which was the thing in the world that he had always held to be least an honour to a gentleman dedicated to diplomacy. He was vexed with the levity in himself which had made him call them together on so poor a pretext, and yet he was vexed with the stupidity in them which made them think, as they evidently did, that the pretext was sufficient. He inwardly groaned at the precipitancy with which he had saddled himself with the Tragic Muse (a tragic muse who was noisy and pert), and yet he wished his visitors would go away and leave him alone with her.

Nick Dormer said to Mrs Dallow that he wanted her to know

an old friend of his, one of the cleverest men he knew; and he added the hope that she would be gentle and encouraging with him: he was so timid and so easily disconcerted.

Gabriel Nash dropped into a chair by the arm of Julia's sofa, Nick Dormer went away, and Mrs Dallow turned her glance upon her new acquaintance without a perceptible change of position. Then she emitted, with rapidity, the remark: 'It's very awkward when people are told one is clever.'

'It's awkward if one isn't,' said Mr Nash, smiling.

'Yes, but so few people are – enough to be talked about.'

'Isn't that just the reason why such a matter, such an exception, ought to be mentioned to them?' asked Gabriel Nash. 'They mightn't find it out for themselves. Of course, however, as you say, there ought to be a certainty; then they are surer to know it. Dormer's a dear fellow, but he's rash and superficial.'

Mrs Dallow, at this, turned her glance a second time upon her interlocutor; but during the rest of the conversation she rarely repeated the movement. If she liked Nick Dormer extremely (and it may without further delay be communicated to the reader that she did), her liking was of a kind that opposed no difficulty whatever to her not liking (in case of such a complication) a person attached or otherwise belonging to him. It was not in her nature to extend tolerances to others for the sake of an individual she loved: the tolerance was usually consumed in the loving; there was nothing left over. If the affection that isolates and simplifies its object may be distinguished from the affection that seeks communications and contacts for it, Julia Dallow's belonged wholly to the former class. She was not so much jealous as rigidly direct. She desired no experience for the familiar and yet partly mysterious kinsman in whom she took an interest that she would not have desired for herself; and, indeed, the cause of her interest in him was partly the vision of his helping her to the particular emotion that she did desire – the emotion of great affairs and of public action. To have such ambitions for him appeared to her the greatest honour she could do him; her conscience was in it as well as her inclination, and her scheme, in her conception, was noble enough to varnish over any disdain she might feel for forces drawing him another way. She had a prejudice, in general, against his connections, a suspicion of them and a supply of unwrought contempt ready for them. It was a singular circumstance that she was sceptical even when, knowing her as well as he did, he thought them worth recommending to her: the recommendation indeed inveterately confirmed the suspicion.

This was a law from which Gabriel Nash was condemned to suffer, if suffering could on any occasion be predicated of Gabriel Nash. His pretension was, in truth, that he had purged his life of such incongruities, though probably he would have admitted that if a sore spot remained the hand of a woman would be sure to touch it. In dining with her brother and with the Dormers, two evenings before, Mrs Dallow had been moved to exclaim that Peter and Nick knew the most extraordinary people. As regards Peter the attitudinizing girl and her mother now pointed that moral with sufficient vividness; so that there was little arrogance in taking a similar quality for granted in the conceited man at her elbow, who sat there as if he would be capable, from one moment to another, of leaning over the arm of her sofa. She had not the slightest wish to talk with him about himself, and was afraid, for an instant, that he was on the point of passing from the chapter of his cleverness to that of his timidity. It was a false alarm, however, for instead of this he said something about the pleasures of the monologue, as the distraction that had just been offered was called by the French. He intimated that in his opinion these pleasures were mainly for the performers. They had all, at any rate, given Miss Rooth a charming afternoon; that, of course, was what Mrs Dallow's kind brother had mainly intended in arranging the little party. (Mrs Dallow hated to hear him call her brother 'kind'; the term seemed offensively patronizing.) But he himself, he related, was now constantly employed in the same beneficence, listening, two-thirds of his time, to 'intonations' and shrieks. She had doubtless observed it herself, how the great current of the age, the adoration of the mime, was almost too strong for any individual; how it swept one along and hurled one against the rocks. As she made no response to this proposition Gabriel Nash asked her if she had not been struck with the main sign of the time, the preponderance of the mountebank, the glory and renown, the personal favour, that he enjoyed. Hadn't she noticed what an immense part of the public attention he held, in London at least? For in Paris society was not so pervaded with him, and the women of the profession, in particular, were not in every drawing-room.

'I don't know what you mean,' Mrs Dallow said. 'I know nothing of any such people.'

'Aren't they under your feet wherever you turn – their performances, their portraits, their speeches, their autobiographies, their names, their manners, their ugly mugs, as the people say, and their idiotic pretensions?'

'I dare say it depends on the places one goes to. If they're everywhere' – and Mrs Dallow paused a moment – 'I don't go everywhere.'

'I don't go anywhere, but they mount on my back, at home, like the Old Man of the Sea. Just observe a little when you return to London,' Nash continued, with friendly instructiveness. Mrs Dallow got up at this – she didn't like receiving directions; but no other corner of the room appeared to offer her any particular reason for crossing to it: she never did such a thing without a great inducement. So she remained standing there, as if she were quitting the place in a moment, which indeed she now determined to do; and her interlocutor, rising also, lingered beside her, unencouraged but unperturbed. He went on to remark that Mr Sherringham was quite right to offer Miss Rooth an afternoon's sport; she deserved it as a fine, brave, amiable girl. She was highly educated, knew a dozen languages, was of illustrious lineage and was immensely particular.

'Immensely particular?' Mrs Dallow repeated.

'Perhaps I should say that her mother is, on her behalf. Particular about the sort of people they meet – the tone, the standard. I'm bound to say they're like you: they don't go everywhere. That spirit is meritorious; it should be recognized and rewarded.'

Mrs Dallow said nothing for a moment; she looked vaguely round the room, but not at Miriam Rooth. Nevertheless she presently dropped, in allusion to her, the words: 'She's dreadfully vulgar.'

'Ah, don't say that to my friend Dormer!' Gabriel Nash exclaimed.

'Are you and he such great friends?' Mrs Dallow asked, looking at him.

'Great enough to make me hope we shall be greater.'

Again, for a moment, she said nothing; then she went on –

'Why shouldn't I say to him that she's vulgar?'

'Because he admires her so much; he wants to paint her.'

'To paint her?'

'To paint her portrait.'

'Oh, I see. I dare say she'd do for that.'

Gabriel Nash laughed gaily. 'If that's your opinion of her you are not very complimentary to the art he aspires to practise.'

'He aspires to practise?' Mrs Dallow repeated.

'Haven't you talked with him about it? Ah, you must keep him up to it!'

Julia Dallow was conscious, for a moment, of looking uncomfortable; but it relieved her to demand of her neighbour, in a certain tone, 'Are you an artist?'

'I try to be,' Nash replied, smiling; 'but I work in such a difficult material.'

He spoke this with such a clever suggestion of unexpected reference that, in spite of herself, Mrs Dallow said after him –

'Difficult material?'

'I work in life!'

At this Mrs Dallow turned away, leaving Nash the impression that she probably misunderstood his speech, thinking he meant that he drew from the living model, or some such platitude: as if there could have been any likelihood that he drew from the dead one. This, indeed, would not fully have explained the abruptness with which she dropped their conversation. Gabriel Nash, however, was used to sudden collapses, and even to sudden ruptures, on the part of his interlocutors, and no man had more the secret of remaining gracefully with his ideas on his hands. He saw Mrs Dallow approach Nick Dormer, who was talking with one of the ladies of the Embassy, and apparently signify to him that she wished to speak to him. He got up, they had a minute's conversation, and then he turned and took leave of his fellow-visitor. Mrs Dallow said a word to her brother, Dormer joined her, and then they came together to the door. In this movement they had to pass near Nash, and it gave her an opportunity to nod good-bye to him, which he was by no means sure she would have done if Nick had not been with her. The young man stopped a moment; he said to Nash: 'I should like to see you this evening, late; you must meet me somewhere.'

'We'll take a walk – I should like that,' Nash replied. 'I shall smoke a cigar at the café on the corner of the Place de l'Opéra; you'll find me there.' Gabriel prepared to compass his own departure, but before doing so he addressed himself to the duty of saying a few words of civility to Lady Agnes. This proved difficult, for on one side she was defended by the wall of the room and on the other rendered inaccessible by Miriam's mother, who clung to her with a quickly-rooted fidelity, showing no symptom of desistance. Gabriel compromised on her daughter Grace, who said to him:

'You were talking with my cousin, Mrs Dallow.'

'*To* her rather than with her,' Nash smiled.

'Ah, she's very charming,' said Grace.

'She's very beautiful,' Nash rejoined.

'And very clever,' Miss Dormer continued.

'Very, very intelligent.' His conversation with the young lady went little further than this, and he presently took leave of Peter Sherringham; remarking to him, as he shook hands, that he was very sorry for him. But he had courted his fate.

'What do you mean by my fate?' Sherringham asked.

'You've got them for life.'

'Why for life, when I now lucidly and courageously recognize that she isn't good?'

'Ah, but she'll become so,' said Gabriel Nash.

'Do you think that?' Sherringham inquired, with a candour which made his visitor laugh.

'*You* will – that's more to the purpose!' Gabriel exclaimed, as he went away.

Ten minutes later Lady Agnes substituted a general vague assent to all further particular ones and, with her daughters, withdrew from Mrs Rooth and from the rest of the company. Peter had had very little talk with Biddy, but the girl kept her disappointment out of her pretty eyes and said to him:

'You told us she didn't know how – but she does!' There was no suggestion of disappointment in this.

Sherringham held her hand a moment. 'Ah, it's you who know how, dear Biddy!' he answered; and he was conscious that if the occasion had been more private he would lawfully have kissed her.

Presently three others of his guests departed, and Mr Nash's assurance that he had them for life recurred to him as he observed that Mrs Rooth and her daughter quite failed to profit by so many examples. The Lovicks remained – a colleague and his sociable wife – and Peter gave them a hint that they were not to leave him absolutely alone with the two ladies. Miriam quitted Mrs Lovick, who had attempted, with no great subtlety, to engage her, and came up to Sherringham as if she suspected him of a design of stealing from the room and had the idea of preventing it.

'I want some more tea: will you give me some more? I feel quite faint. You don't seem to suspect how that sort of thing takes it out of you.'

Sherringham apologized, extravagantly, for not having seen that she had the proper quantity of refreshment, and took her to the round table, in a corner, on which the little collation had been served. He poured out tea for her and pressed bread-and-butter on her, and *petits fours*, of all which she profusely and methodically partook. It was late; the afternoon had faded and a lamp had been brought in, the wide shade of which shed a fair glow upon the tea-service, the little plates of comestibles. The Lovicks sat with Mrs Rooth at the other end of the room, and the girl stood at the table drinking her tea and eating her bread-and-butter. She consumed these articles so freely that he wondered if she had been in serious want of food – if they were so poor as to have to count with that sort of privation. This supposition was softening, but still not so much so as to make him ask her to sit down. She appeared indeed to prefer to stand: she looked better so, as if the freedom, the conspicuity of being on her feet and treading a stage were agreeable to her. While Sherringham lingered near her, vaguely, with his hands in his pockets, not knowing exactly what to say and instinctively avoiding, now, the theatrical question (there were moments when he was plentifully tired of it), she broke out abruptly: 'Confess that you think me intolerably bad!'

'Intolerably – no.'

'Only tolerably! I think that's worse.'

'Every now and then you do something very clever,' Sherringham said.

'How many such things did I do today?'

'Oh, three or four. I don't know that I counted very carefully.'

She raised her cup to her lips, looking at him over the rim of it – a proceeding which gave her eyes a strange expression. 'It bores you, and you think it disagreeable,' she said in a moment – 'a girl always talking about herself.' He protested that she could never bore him, and she went on: 'Oh, I don't want compliments – I want the truth. An actress has to talk about herself; what else can she talk about, poor vain thing?'

'She can talk sometimes about other actresses.'

'That comes to the same thing. You won't be serious. I'm awfully serious.' There was something that caught his attention in the way she said this – a longing, half-hopeless, half-argumentative, to be believed in. 'If one really wants to do anything one must worry it out; of course everything doesn't come the first day,' she pursued. 'I can't see everything at once; but I can see a little more – step by step – as I go: can't I?'

'That's the way – that's the way,' said Sherringham. 'If you see the things to do, the art of doing them will come, if you hammer away. The great point is to see them.'

'Yes; and you don't think me clever enough for that.'

'Why do you say so, when I've asked you to come here on purpose?'

'You've asked me to come, but I've had no success.'

'On the contrary; every one thought you wonderful.'

'Oh, they don't know!' said Miriam Rooth. 'You've not said a word to me. I don't mind your not having praised me; that would be too *banal*. But if I'm bad – and I know I'm dreadful – I wish you would talk to me about it.'

'It's delightful to talk to you,' Sherringham said.

'No, it isn't, but it's kind,' she answered, looking away from him.

Her voice had a quality, as she uttered these words, which made him exclaim, 'Every now and then you say something –!'

She turned her eyes back to him, smiling. 'I don't want it to come by accident.' Then she added: 'If there's any good to be got from trying, from showing one's self, how can it come unless one hears the simple truth, the truth that turns one inside out? It's all for that – to know what one is, if one's a stick!'

'You have great courage, you have rare qualities,' said Sherringham. She had begun to touch him, to seem different: he was glad she had not gone.

For a moment she made no response to this, putting down her empty cup and looking vaguely over the table, as if to select something more to eat. Suddenly she raised her head and broke out with vehemence: 'I will, I will, I will!'

'You'll do what you want, evidently.'

'I will succeed – I will be great. Of course I know too little, I've seen too little. But I've always liked it; I've never liked anything else. I used to learn things, and to do scenes, and to rant about the room, when I was five years old.' She went on, communicative, persuasive, familiar, egotistical (as was necessary), and slightly common, or perhaps only natural; with reminiscences, reasons and anecdotes, an unexpected profusion, and with an air of comradeship, of freedom of intercourse, which appeared to plead that she was capable at least of embracing that side of the profession she desired to adopt. He perceived that if she had seen very little, as she said, she had also seen a great deal; but both her experience and her innocence had been accidental and irregular. She had seen very little acting – the theatre was always too expensive. If she could only go often – in Paris, for instance, every night for six months – to see the best, the worst, everything, she would make things out, she would observe and learn what to do, what not to do: it would be a kind of school. But she couldn't, without selling the clothes off her back. It was vile and disgusting to be poor; and if ever she were to know the bliss of having a few francs in her pocket she would make up for it – that she could promise! She had never been acquainted with any one who could tell her anything – if it was good or bad, right or wrong – except Mrs Delamere and poor Ruggieri. She supposed they had told her a great deal, but perhaps they hadn't, and she was perfectly willing to give it up if it was bad. Evidently Madame Carré thought so;

she thought it was horrid. Wasn't it perfectly divine, the way the old woman had said those verses, those speeches of Célie? If she would only let her come and listen to her once in a while, like that, it was all she would ask. She had got lots of ideas, just from that; she had practised them over, over and over again, the moment she got home. He might ask her mother – he might ask the people next door. If Madame Carré didn't think she could work she might have heard something that would show her. But she didn't think her even good enough to criticize; for that wasn't criticism, telling her her head was good. Of course her head was good; she didn't need travel up to the *quartiers excentriques* to find that out. It was her mother – the way she talked – who gave that idea, that she wanted to be elegant, and very moral, and a *femme du monde*, and all that sort of trash. Of course that put people off, when they were only thinking of the right way. Didn't *she* know, Miriam herself, that that was the only thing to think of? But any one would be kind to her mother who knew what a dear she was. 'She doesn't know when it's right or wrong, but she's a perfect saint,' said the girl, obscuring considerably her vindication. 'She doesn't mind when I say things over by the hour, dinning them into her ears while she sits there and reads. She's a tremendous reader; she's awfully up in literature. She taught me everything herself – I mean all that sort of thing. Of course I'm not so fond of reading; I go in for the book of life.' Sherringham wondered whether her mother had not, at any rate, taught her that phrase, and thought it highly probable. 'It would give on *my* nerves, the life I lead her,' Miriam continued; 'but she's really a delicious woman.'

The oddity of this epithet made Sherringham laugh, and altogether, in a few minutes, which is perhaps a sign that he abused his right to be a man of moods, the young lady had produced a revolution of curiosity in him, re-awakened his sympathy. Her mixture, as it spread itself before one, was a quickening spectacle: she was intelligent and clumsy – she was underbred and fine. Certainly she was very various, and that was rare; not at all at this moment the heavy-eyed, frightened creature who had pulled herself together with such an effort at Madame Carré's, nor the elated 'phenomenon' who had just been declaiming, nor the rather affected and contradictious young person with whom he had walked home from the Rue de Constantinople. Was this succession of phases a sign that she really possessed the celebrated artistic temperament, the nature that made people provoking and interesting? That Sherringham himself was of that shifting complexion is perhaps proved by his odd capacity for being of two different minds at very nearly the same time. Miriam was pretty now, with likeable looks and charming usual eyes. Yes, there were things he could do for her; he had already forgotten the chill of Mr Nash's irony, of his prophecy. He was even scarcely conscious how much, in general, he detested hints, insinuations, favours asked obliquely and plaintively: that was doubtless also because the girl was so pretty and so fraternizing. Perhaps indeed it was unjust to qualify it as roundabout, the manner in which Miss Rooth conveyed to him that it was open to him not only to pay for lessons for her, but to meet the expense of her nightly attendance, with her mother, at instructive exhibitions of theatrical art. It was a large order, sending the pair to all the plays; but what Sherringham now found himself thinking about was not so much its largeness as that it would be rather interesting to go with them sometimes and point the moral (the technical one), showing her the things he liked, the things he disapproved. She repeated her declaration that she recognized the fallacy of her mother's views about 'noble' heroines and about the importance of her looking out for such tremendously proper people. 'One must let her talk, but of course it creates a prejudice,' she said, with her eyes on Mr and Mrs Lovick, who had got up, terminating their communion with Mrs Rooth. 'It's a great muddle, I know, but she can't bear anything coarse – and quite right, too. I shouldn't, either, if I didn't have to. But I don't care where I go if I can act, or who they are if they'll help me. I want to act – that's what I want to do; I don't want to meddle in people's affairs. I can look out for myself – *I'm* all right!' the girl exclaimed, roundly, frankly, with a ring of honesty which made her crude and pure. 'As for doing the bad ones, I'm not afraid of that.'

'The bad ones?'

'The bad women, in the plays – like Madame Carré. I'll do anything.'

'I think you'll do best what you are,' remarked Sherringham, laughing. 'You're a strange girl.'

'*Je crois bien!* Doesn't one have to be, to want to go and exhibit one's self to a loathsome crowd, on a platform, with trumpets and a big drum, for money – to parade one's body and one's soul?'

Sherringham looked at her a moment: her face changed constantly; now there was a little flush and a noble delicacy in it.

'Give it up; you're too good for it,' he said, abruptly.

'Never, never – never till I'm pelted!'

'Then stay on here a bit; I'll take you to the theatres.'

'Oh, you dear!' Miriam delightedly exclaimed. Mr and Mrs Lovick, accompanied by Mrs Rooth, now crossed the room to them, and the girl went on, in the same tone: 'Mamma, dear, he's the best friend we've ever had; he's a great deal nicer than I thought.'

'So are you, *mademoiselle*,' said Peter Sherringham.

'Oh, I trust Mr Sherringham – I trust him infinitely,' Mrs Rooth returned, covering him with her mild, respectable, wheedling eyes. 'The kindness of every one has been beyond everything. Mr and Mrs Lovick can't say enough. They make the most obliging offers; they want you to know their brother.'

'Oh, I say, he's no brother of mine,' Mr Lovick protested, good-naturedly.

'They think he'll be so suggestive, he'll put us up to the right things,' Mrs Rooth went on.

'It's just a little brother of mine – such a dear, clever boy,' Mrs Lovick explained.

'Do you know she has got nine? Upon my honour she has!' said her husband. 'This one is the sixth. Fancy if I had to take them over!'

'Yes, it makes it rather awkward,' Mrs Lovick amiably conceded. 'He has gone on the stage, poor dear boy; he acts rather well.'

'He tried for the diplomatic service, but he didn't precisely dazzle his examiners,' Mr Lovick remarked.

113

'Edmund's very nasty about him. There are lots of gentlemen on the stage; he's not the first.'

'It's such a comfort to hear that,' said Mrs Rooth.

'I'm much obliged to you. Has he got a theatre?' Miriam asked.

'My dear young lady, he hasn't even got an engagement,' replied the young man's unsympathizing brother-in-law.

'He hasn't been at it very long, but I'm sure he'll get on. He's immensely in earnest, and he's very good-looking. I just said that if he should come over to see us you might rather like to meet him. He might give you some tips, as my husband says.'

'I don't care for his looks, but I *should* like his tips,' said Miriam, smiling.

'And *is* he coming over to see you?' asked Sherringham, to whom, while this exchange of remarks, which he had not lost, was going on, Mrs Rooth had, in lowered accents, addressed herself.

'Not if I can help it, I think!' Mr Lovick declared, but so jocosely that it was not embarrassing.

'Oh, sir, I'm sure you're fond of him,' Mrs Rooth remonstrated, as the party passed together into the ante-chamber.

'No, really, I like some of the others – four or five of them; but I don't like Arty.'

'We'll make it up to him, then; *we'll* like him,' Miriam declared, gaily; and her voice rang in the staircase (Sherringham went a little way with them), with a charm which her host had not perceived in her sportive note the day before.

9

NICK DORMER found his friend Nash, that evening, on the spot he had designated, smoking a cigar in the warm, bright night, in front of the café at the corner of the square before the Opera. He sat down with him, but at the end of five minutes he uttered a protest against the crush and confusion, the publicity and vulgarity, of the place, the shuffling procession of the crowd, the jostle of fellow-customers, the perpetual brush of waiters. 'Come away. I want to talk to you, and I can't talk here,' he said to his companion. 'I don't care where we go. It will be pleasant to walk; we'll stroll away to the *quartiers sérieux*. Each time I come to Paris, at the end of three days, I take the boulevard, with its conventional grimace, into greater disfavour. I hate even to cross it, I go half a mile round to avoid it.'

The young men took their course together down the Rue de la Paix to the Rue de Rivoli, which they crossed, passing beside the gilded railing of the Tuileries. The beauty of the night – the only defect of which was that the immense illumination of Paris kept it from being quite night enough, made it a sort of bedizened, rejuvenated day – gave a charm to the quieter streets, drew our friends away to the right, to the river and the bridges, the older, duskier city. The pale ghost of the palace that had died by fire hung over them awhile, and, by the passage now open at all times across the garden of the Tuileries, they came out upon the Seine. They kept on and on, moving slowly, smoking, talking, pausing, stopping to look, to emphasize, to compare. They fell into discussion, into confidence, into inquiry, sympathetic or satiric, and into explanation which needed in turn to be explained. The balmy night, the time for talk, the amusement of Paris, the memory of young confabulations gave a quality to the occasion. Nick had already forgotten the little brush he had had with Mrs Dallow, when they quitted Peter's tea-party together, and that he had been almost disconcerted by the manner in which she characterized the odious man he had taken it into his head to present to her. Impertinent and fatuous she had called him; and when Nick began to explain that he was really neither of these things, though he could imagine his manner might sometimes suggest them, she had declared that she didn't wish to argue about him or even to hear of him again. Nick had not counted on her liking Gabriel Nash, but he had thought it wouldn't matter much if she should dislike him a little. He had given himself the diversion, which he had not dreamed would be cruel to any one concerned, of seeing what she would make of a type she had never encountered before. She had made even less than he expected, and her implication that he had played her a trick had been irritating enough to prevent him from reflecting that the fault might have been in some degree with Nash. But he had recovered from his resentment sufficiently to ask this personage, with every possible circumstance of implied consideration for the lady, what *he*, on his side, had made of his charming cousin.

'Upon my word, my dear fellow, I don't regard that as a fair question,' was the answer. 'Besides, if you think Mrs Dallow charming, what on earth need it matter to you what I think? The superiority of one man's opinion over another's is never so great as when the opinion is about a woman.'

'It was to help me to find out what I think of yourself,' said Nick Dormer.

'Oh, that you'll never do. I shall bother you to the end. The lady with whom you were so good as to make me acquainted is a beautiful specimen of the English garden-flower, the product of high cultivation and much tending; a tall, delicate stem, with the head set upon it in a manner which, as I recall it, is distinctly so much to the good in my day. She's the perfect type of the object *raised*, or bred, and everything about her is homogeneous, from the angle of her elbow to the way she drops that vague, conventional, dry little "Oh!" which dispenses with all further performance. That sort of completeness is always satisfying. But I didn't satisfy her, and she didn't understand me. I don't think they usually understand.'

'She's no worse than I, then.'

'Ah, she didn't try.'

'No, she doesn't try. But she probably thought you conceited, and she would think so still more if she were to hear you talk about her trying.'

'Very likely – very likely,' said Gabriel Nash. 'I have an idea a good many people think that. It appears to me so droll. I suppose it's a result of my little system.'

'Your little system?'

'Oh, it's nothing wonderful. Only the idea of being just the same to every one. People have so bemuddled themselves that the last thing they can conceive is that one should be simple.'

'Lord, do you call yourself simple?' Nick ejaculated.

'Absolutely; in the sense of having no interest of my own to push, no nostrum to advertise, no power to conciliate, no axe to grind. I'm not a savage – ah, far from it – but I really think I'm perfectly independent.'

'Oh, that's always provoking!' laughed Nick.

'So it would appear, to the great majority of one's fellow-mortals; and I well remember the pang with which I originally made that discovery. It darkened my spirit, at a time when I had no thought of evil. What we like, when we are unregenerate, is that a new-comer should give us a password, come over to our side, join our little camp or religion, get into our little boat, in short, whatever it is, and help us to row it. It's natural enough; we are mostly in different tubs and cockles, paddling for life. Our opinions, our convictions and doctrines and standards, are simply the particular thing that will make the boat go – *our* boat, naturally, for they may very often be just the thing that will sink another. If you won't get in, people generally hate you.'

'Your metaphor is very lame,' said Nick; 'it's the overcrowded boat that goes to the bottom.'

'Oh, I'll give it another leg or two! Boats can be big, in the infinite of space, and a doctrine is a raft that floats the better the more passengers it carries. A passenger jumps over from time to time, not so much from fear of sinking as from a want of interest in the course or the company. He swims, he plunges, he dives, he dips down and visits the fishes and the mermaids and the submarine caves; he goes from craft to craft and splashes about, on his own account, in the blue, cool water. The regenerate, as I call them, are the passengers who jump over in search of better fun. I turned my summersault long ago.'

'And now, of course, you're at the head of the regenerate; for, in your turn, you all form a select school of porpoises.'

'Not a bit, and I know nothing about heads, in the sense you mean. I've grown a tail, if you will; I'm the merman wandering free. It's a delightful trade!'

Before they had gone many steps further Nick Dormer stopped short and said to his companion: 'I say, my dear fellow, do you mind mentioning to me whether you are the greatest humbug and charlatan on earth, or a genuine intelligence, one that has sifted things for itself?'

'I do puzzle you – I'm so sorry,' Nash replied, benignly. 'But I'm very sincere. And I *have* tried to straighten out things a bit for myself.'

'Then why do you give people such a handle?'

'Such a handle?'

'For thinking you're an – for thinking you're not wise.'

'I dare say it's my manner; they're so unused to candour.'

'Why don't you try another?' Nick inquired.

'One has the manner that one can; and mine, moreover, is a part of my little system.'

'Ah, if you've got a little system you're no better than anyone else,' said Nick, going on.

'I don't pretend to be better, for we are all miserable sinners; I only pretend to be bad in a pleasanter, brighter way, by what I can see. It's the simplest thing in the world; I just take for granted a certain brightness in life, a certain frankness. What is essentially kinder than that, what is more harmless? But the tradition of dreariness, of stodginess, of dull, dense, literal prose, has so sealed people's eyes that they have ended by thinking the most normal thing in the world the most fantastic. Why be dreary, in our little day? No one can tell me why, and almost every one calls me names for simply asking the question. But I keep on, for I believe one can do a little good by it. I want so much to do a little good,' Gabriel Nash continued, taking his companion's arm. 'My persistence is systematic: don't you see what I mean? I won't be dreary – no, no, no; and I won't recognize the necessity, or even, if there is any way out of it, the accident of dreariness in the life that surrounds me. That's enough to make people stare: they're so stupid!'

'They think you're impertinent,' Dormer remarked.

At this his companion stopped him short, with an ejaculation of pain, and, turning his eyes, Nick saw under the lamps of the quay that he had brought a vivid blush into Nash's face. 'I don't strike *you* that way?' Gabriel asked, reproachfully.

'Oh, me! Wasn't it just admitted that I don't in the least make you out?'

'That's the last thing!' Nash murmured, as if he were thinking the idea over, with an air of genuine distress. 'But with a little patience we'll clear it up together, if you care enough about it,' he added, more cheerfully. He let his friend go on again and he continued: 'Heaven help us all! what do people mean by impertinence? There are many, I think, who don't understand its nature or its limits; and upon my word I have literally seen mere quickness of intelligence or of perception, the jump of a step or two, a little whirr of the wings of talk, mistaken for it. Yes, I have encountered men and women who thought you were impertinent if you were not so stupid as they. The only impertinence is aggression, and I indignantly protest that I am never guilty of *that* clumsiness. Ah, for what do they take one, with *their* presumptions? Even to defend myself, sometimes, I have to make believe to myself that I care. I always feel as if I didn't successfully make others think so. Perhaps they see an impertinence in that. But I dare say the offence is in the things that I take, as I say, for granted; for if one tries to be pleased one passes, perhaps inevitably, for being pleased above all with one's self. That's really not my case, for I find my capacity for pleasure deplorably below the mark I've set. That's why, as I have told you, I cultivate it, I try to bring it up. And I am actuated by positive benevolence; I have that pretension. That's what I mean by being the same to every one, by having only one manner. If one is conscious and ingenious to that end, what's the harm, when one's motives are so pure? By never, *never* making the concession, one may end by becoming a perceptible force for good.'

'What concession are you talking about?' asked Nick Dormer.

'Why, that we are only here for dreariness. It's impossible to grant it sometimes, if you wish to withhold it ever.'

'And what do you mean by dreariness? That's modern slang, and it's terribly vague. Many good things are dreary – virtue and decency and charity and perseverance and courage and honour.'

'Say at once that life is dreary, my dear fellow!' Gabriel Nash exclaimed.

'That's on the whole my usual impression.'

'*C'est là que je vous attends!* I'm precisely engaged in trying what can be done in taking it the other way. It's my little personal experiment. Life consists of the personal experiments of each of us, and the point of an experiment is that it shall succeed. What we contribute is our treatment of the material, our rendering of the text, our style. A sense of the qualities of a style is so rare that many persons should doubtless be forgiven for not being able to read, or at all events to enjoy us; but is that a reason for giving it up – for not being, in this other sphere, if one possibly can, a Macaulay, a Ruskin, a Renan? Ah, we must write our best; it's the great thing we can do in the world, on the right side. One has one's form, *que diable*, and a mighty good thing that one has. I'm not afraid of putting all life into mine, without unduly squeezing it. I'm not afraid of putting in honour and courage and charity, without spoiling them: on the contrary, I'll only do them good. People may not read you at sight, may not like you, but there's a chance they'll come round; and the only way to court the chance is to keep it up – always to keep it up. That's what I do, my dear fellow, if you don't think I've perseverance. If someone likes it here and there, if you give a little impression of solidity, that's your reward; besides, of course, the pleasure for yourself.'

'Don't you think your style is a little affected?' Nick asked, laughing, as they proceeded.

'That's always the charge against a personal manner; if you have any at all people think you have too much. Perhaps, perhaps – who can say? Of course one isn't perfect; but that's the delightful thing about art, that there is always more to learn and more to do; one can polish and polish and refine and refine. No doubt I'm rough still, but I'm in the right direction: I make it my business to take for granted an interest in the beautiful.'

'Ah, the beautiful – there it stands, over there!' said Nick Dormer. 'I am not so sure about yours – I don't know what I've got hold of. But Notre Dame *is* solid; Notre Dame *is* wise; on Notre Dame the distracted mind can rest. Come over and look at her!'

They had come abreast of the low island from which the great cathedral, disengaged today from her old contacts and adhesions, rises high and fair, with her front of beauty and her majestic mass, darkened at that hour, or at least simplified, under the stars, but only more serene and sublime for her happy union, far aloft, with the cool distance and the night. Our young men, gossiping as profitably as I leave the reader to estimate, crossed the wide, short bridge which made them face towards the monuments of old Paris – the Palais de Justice, the Conciergerie, the holy chapel of Saint Louis. They came out before the church, which looks down on a square where the past, once so thick in the very heart of Paris, has been made rather a blank, pervaded, however, by the everlasting freshness of the great cathedral-face. It greeted Nick Dormer and Gabriel Nash with a kindness which the centuries had done nothing to dim. The lamplight of the great city washed its foundations, but the towers and buttresses, the arches, the galleries, the statues, the vast rose-window, the large, full composition, seemed to grow clearer as they climbed higher, as if they had a conscious benevolent answer for the upward gaze of men.

'How it straightens things out and blows away one's vapours – anything that's *done*!' said Nick; while his companion exclaimed, blandly and affectionately:

'The dear old thing!'

'The great point is to do something, instead of standing muddling and questioning; and, by Jove, it makes me want to!'

'Want to build a cathedral?' Nash inquired.

'Yes, just that.'

'It's you who puzzle *me*, then, my dear fellow. You can't build them out of words.'

'What is it the great poets do?' asked Nick.

'*Their* words are ideas – their words are images, enchanting collocations and unforgettable signs. But the verbiage of parliamentary speeches!'

'Well,' said Nick, with a candid, reflective sigh, 'you can rear a great structure of many things – not only of stones and timbers and painted glass.' They walked round Notre Dame, pausing, criticizing, admiring and discussing; mingling the grave with the gay and paradox with contemplation. Behind and at the sides the huge dusky vessel of the church seemed to dip into the Seine, or rise out of it, floating expansively – a ship of stone, with its flying buttresses thrown forth like an array of mighty oars. Nick Dormer lingered near it with joy, with a certain soothing content; as if it had been the temple of a faith so dear to him that there was peace and security in its precinct. And there was comfort too, and consolation of the same sort, in the company, at this moment, of Nash's equal response, of his appreciation, exhibited by his own signs, of the great effect. He felt it so freely and uttered his impression with such vividness that Nick was reminded of the luminosity his boyish admiration had found in him of old, the natural intelligence of everything of that kind. 'Everything of that kind' was, in Nick's mind, the description of a wide and bright domain.

They crossed to the further side of the river, where the influence of the Gothic monument threw a distinction even over the Parisian smartnesses – the municipal rule and measure, the importunate symmetries, the 'handsomeness' of everything, the extravagance of gaslight, the perpetual click on the neat bridges. In front of a quiet little café on the right bank Gabriel Nash said, 'Let's sit down' – he was always ready to sit down. It was a friendly establishment and an unfashionable quarter, far away from the Grand Hotel; there were the usual little tables and chairs on the quay, the muslin curtains behind the glazed front, the general sense of sawdust and of drippings of watery beer. The place was subdued to stillness, but not extinguished, by the lateness of the hour; no vehicles passed, but only now and then a light Parisian foot. Beyond the parapet they could hear the flow of the Seine. Nick Dormer said it made him think of the old Paris, of the great Revolution, of Madame Roland, *quoi*! Gabriel Nash said they could have watery beer but were not obliged to drink it. They sat a long time; they talked a great deal, and the more they said the more the unsaid came up. Presently Nash found occasion to remark: 'I go about my business, like any good citizen – that's all.'

'And what *is* your business?'

'The spectacle of the world.'

Nick laughed out. 'And what do you do with that?'

'What does any one do with a spectacle? I look at it.'

'You are full of contradictions and inconsistencies. You described yourself to me half an hour ago as an apostle of beauty.'

'Where is the inconsistency? I do it in the broad light of day, whatever I do: that's virtually what I meant. If I look at the spectacle of the world I look in preference at what is charming in it. Sometimes I have to go far to find it – very likely; but that's just what I do. I go far – as far as my means permit me. Last year I heard of such a delightful little spot: a place where a wild fig-tree grows in the south wall, the outer side, of an old Spanish city. I was told it was a deliciously brown corner, with the sun making it warm in winter! As soon as I could I went there.'

'And what did you do?'

'I lay on the first green grass – I liked it.'

'If that sort of thing is all you accomplish you are not encouraging.'

'I accomplish my happiness – it seems to me that's something. I have feelings, I have sensations: let me tell you that's not so common. It's rare to have them; and if you chance to have them it's rare not to be ashamed of them. I go after them – when I judge they won't hurt any one.'

'You're lucky to have money for your travelling expenses,' said Nick.

'No doubt, no doubt; but I do it very cheap. I take my stand on my nature, on my disposition. I'm not ashamed of it. I don't think it's so horrible, my disposition. But we've befogged and befouled so the whole question of liberty, of spontaneity, of good-humour and inclination and enjoyment, that there's nothing that makes people stare so as to see one natural.'

'You are always thinking too much of "people".'

'They say I think too little,' Gabriel smiled.

'Well, I've agreed to stand for Harsh,' said Nick, with a roundabout transition.

'It's you then who are lucky to have money.'

'I haven't,' Nick replied. 'My expenses are to be paid.'

'Then you too must think of "people".'

Nick made no answer to this, but after a moment he said: 'I wish very much you had more to show for it.'

'To show for what?'

'Your little system – the aesthetic life.'

Nash hesitated, tolerantly, gaily, as he often did, with an air of being embarrassed to choose between several answers, any one of them would be so right. 'Oh, having something to show is such a poor business. It's a kind of confession of failure.'

'Yes, you're more affected than anything else,' said Nick, impatiently.

'No, my dear boy, I'm more good-natured: don't I prove it? I'm rather disappointed to find that you are not worthy of the esoteric doctrine. But there is, I confess, another plane of intelligence, honourable, and very honourable in its way, from which it *may* legitimately appear important to have something to show. If you *must* confine yourself to that plane I won't refuse you my sympathy. After all, that's what *I* have to show! But the degree of my sympathy must of course depend on the nature of the manifestation that you wish to make.'

'You know it very well – you've guessed it,' Nick rejoined, looking before him in a conscious, modest way which, if he had been a few years younger, would have been called sheepish.

'Ah, you've broken the scent with telling me you are going to return to the House of Commons,' said Nash.

'No wonder you don't make it out! My situation is certainly absurd enough. What I really want to do is to be a painter. That's the abject, crude, ridiculous fact. In this out-of-the-way corner, at the dead of night, in lowered tones, I venture to disclose it to you. Isn't that the aesthetic life?'

'Do you know how to paint?' asked Nash.

'Not in the least. No element of burlesque is therefore wanting to my position.'

'That makes no difference. I'm so glad!'

'So glad I don't know how?'

'So glad of it all. Yes, that only makes it better. You're a delightful case, and I like delightful cases. We must see it through. I rejoice that I met you.'

'Do you think I can do anything?' Nick inquired.

'Paint good pictures? How can I tell till I've seen some of your work? Doesn't it come back to me that at Oxford you used to sketch very prettily? But that's the last thing that matters.'

'What does matter, then?' Nick demanded, turning his eye on his companion.

'To be on the right side – on the side of beauty.'

'There will be precious little beauty if I produce nothing but daubs.'

'Ah, you cling to the old false measure of success. I must cure you of that. There will be the beauty of having been disinterested and independent; of having taken the world in the free, brave, personal way.'

'I shall nevertheless paint decently if I can,' Nick declared.

'I'm almost sorry! It will make your case less clear, your example less grand.'

'My example will be grand enough, with the fight I shall have to make.'

'The fight – with whom?'

'With myself, first of all. I'm awfully against it.'

'Ah, but you'll have me on the other side,' smiled Nash.

'Well, you'll have more than a handful to meet – everything, every one that belongs to me, that touches me, near or far: my family, my blood, my heredity, my traditions, my promises, my circumstances, my prejudices; my little past, such as it is; my great future, such as it has been supposed it may be.'

'I see, I see; it's admirable!' Nash exclaimed. 'And Mrs Dallow into the bargain,' he added.

'Yes, Mrs Dallow, if you like.'

'Are you in love with her?'

'Not in the least.'

'Well, she is with you – so I perceived.'

'Don't say that,' said Nick Dormer, with sudden sternness.

'Ah, you are, you are!' his companion rejoined, judging apparently from this accent.

'I don't know what I am – heaven help me!' Nick broke out, tossing his hat down on his little tin table with vehemence. 'I'm a freak of nature and a sport of the mocking gods! Why should they go out of their way to worry me? Why should they do anything so inconsequent, so improbable, so preposterous? It's the vulgarest practical joke. There has never been anything of the sort among us; we are all Philistines to the core, with about as much aesthetic sense as that hat. It's excellent soil – I don't complain of it – but not a soil to grow that flower. From where the devil, then, has the seed been dropped? I look back from generation to generation; I scour our annals without finding the least little sketching grandmother, any sign of a building, or versifying, or collecting, or even tulip-raising ancestor. They were all as blind as bats and none the less happy for that. I'm a wanton variation, an unaccountable monster. My dear father, rest his soul, went through life without a suspicion that there is anything in it that can't be boiled into blue-books; and he became, in that conviction, a very distinguished person. He brought me up in the same simplicity and in the hope of the same eminence. It would have been better if I had remained so. I think it's partly your fault that I haven't,' Nick went on. 'At Oxford you were very bad company for me, my evil genius; you opened my eyes, you communicated the poison. Since then, little by little, it has been working within me; vaguely, covertly, insensibly at first, but during the last year or two with violence, pertinacity, cruelty. I have taken every antidote in life; but it's no use – I'm stricken. It tears me to pieces, as I may say.'

'I see, I follow you,' said Nash, who had listened to this recital with radiant interest and curiosity. 'And that's why you are going to stand.'

'Precisely – it's an antidote. And, at present, you're another.'

'Another?'

'That's why I jumped at you. A bigger dose of you may disagree with me to that extent that I shall either die or get better.'

'I shall control the dilution,' said Nash. 'Poor fellow – if you're elected!' he added.

'Poor fellow either way. You don't know the atmosphere in which I live, the horror, the scandal that my apostasy would inspire, the injury and suffering that it would inflict. I believe it would kill my mother. She thinks my father is watching me from the skies.'

'Jolly to make him jump!' Nash exclaimed.

'He would jump indeed; he would come straight down on top of me. And then the grotesqueness of it – to *begin*, all of a sudden, at my age.'

'It's perfect indeed; it's a magnificent case,' Nash went on.

'Think how it sounds – a paragraph in the London papers: "Mr Nicholas Dormer, M.P. for Harsh and son of the late Right Honourable, and so forth and so forth, is about to give up his seat and withdraw from public life in order to devote himself to the practice of portrait-painting. Orders respectfully solicited." '

'The nineteenth century is better than I thought,' said Nash. 'It's the portrait that preoccupies you?'

'I wish you could see; you must come immediately to my place in London.'

'You wretch, you're capable of having talent!' cried Nash.

'No, I'm too old, too old. It's too late to go through the mill.'

'You make *me* young! Don't miss your election, at your peril. Think of the edification.'

'The edification?'

'Of your throwing it all up the next moment.'

'That would be pleasant for Mr Carteret,' Nick observed.

'Mr Carteret?'

'A dear old fellow who will wish to pay my agent's bill.'

'Serve him right, for such depraved tastes.'

'You do me good,' said Nick, getting up and turning away.

'Don't call me useless then.'

'Ah, but not in the way you mean. It's only if I don't get in that I shall perhaps console myself with the brush,' Nick continued, as they retraced their steps.

'In the name of all the muses, then, don't stand. For you *will* get in.'

'Very likely. At any rate I've promised.'

'You've promised Mrs Dallow?'

'It's her place; she'll put me in,' Nick said.

'Baleful woman! But I'll pull you out!'

10

FOR several days Peter Sherringham had business in hand which left him neither time nor freedom of mind to occupy himself actively with the ladies of the Hôtel de la Garonne. There were moments when they brushed across his memory, but their passage was rapid and not lighted up with any particular complacency of attention; for he shrank considerably from bringing it to the proof – the question of whether Miriam would be an interest or only a bore. She had left him, after their second meeting, with a quickened expectation, but in the course of a few hours that flame had burned dim. Like many other men Sherringham was a mixture of impulse and reflection; but he was peculiar in this, that thinking things over almost always made him think less well of them. He found illusions necessary, so that in order to keep an adequate number going he often earnestly forbade himself that exercise. Mrs Rooth and her daughter were there and could certainly be trusted to make themselves felt. He was conscious of their anxiety, their calculations, as of a kind of oppression; he knew that, whatever results might ensue, he should have to do something positive for them. An idea of tenacity, of worrying feminine duration, associated itself with their presence; he would have assented with a silent nod to the proposition (enunciated by Gabriel Nash) that he was saddled with them. Remedies hovered before him, but they figured also at the same time as complications; ranging vaguely from the expenditure of money to the discovery that he was in love. This latter accident would be particularly tedious; he had a full perception of the arts by which the girl's mother might succeed in making it so. It would not be a compensation for trouble, but a trouble which in itself would require compensation. Would that balm spring from the spectacle of the young lady's genius? The genius would have to be very great to justify a rising young diplomatist in making a fool of himself.

With the excuse of pressing work he put off his young pupil from day to day, and from day to day he expected to hear her knock at his door. It would be time enough when they came after him; and he was unable to see how, after all, he could serve them even then. He had proposed impetuously a course of theatres; but that would be a considerable personal effort, now that the summer was about to begin, with bad air, stale pieces, tired actors. When, however, more than a week had elapsed without a reminder of his neglected promise, it came over him that he must himself in honour give a sign. There was a delicacy in such discretion – he was touched by being let alone. The flurry of work at the Embassy was over, and he had time to ask himself what, in especial, he should do. He wished to have something definite to suggest before communicating with the Hôtel de la Garonne.

As a consequence of this speculation he went back to Madame Carré, to ask her to reconsider her unfavourable judgement and give the young English lady – to oblige him – a dozen lessons of the sort that she knew how to give. He was aware that this request scarcely stood on its feet; for in the first place Madame Carré never reconsidered, when once she had got her impression, and in the second she never wasted herself on subjects whom nature had not formed to do her honour. He knew that his asking her to strain a point to please him would give her a false idea (for that matter, she had it already) of his relations, actual or prospective, with the girl; but he reflected that he needn't care for that, as Miriam herself probably wouldn't care. What he had mainly in mind was to say to the old actress that she had been mistaken – the *jeune Anglaise* was not such a duffer. This would take some courage, but it would also add to the amusement of his visit.

He found her at home, but as soon as he had expressed the conviction I have mentioned she exclaimed: 'Oh, your *jeune Anglaise*, I know a great deal more about her than you! She has been back to see me twice; she doesn't go the longest way round. She charges me like a grenadier, and she asks me to give her – guess a little what! – private recitations, all to herself. If she doesn't succeed it won't be for want of knowing how to thump at doors. The other day, when I came in, she was waiting for me; she had been there for an hour. My private recitations – have you an idea what people pay for them?'

'Between artists, you know, there are easier conditions,' Sherringham laughed.

'How do I know if she's an artist? She won't open her mouth to me; what she wants is to make me say things to her. She does make me – I don't know how – and she sits there gaping at me with her big eyes. They look like open pockets!'

'I dare say she'll profit by it,' said Sherringham.

'I dare say *you* will! Her face is stupid while she watches me, and when she has tired me out she simply walks away. However, as she comes back –' Madame Carré paused a moment, listened, and then exclaimed: 'Didn't I tell you?'

Sherringham heard a parley of voices in the little antechamber and the next moment the door was pushed open and Miriam Rooth bounded into the room. She was flushed and breathless, without a smile, very direct.

'Will you hear me today? I know four things,' she immediately began. Then, perceiving Sherringham, she added in the same brisk, earnest tone, as if the matter were of the highest importance: 'Oh, how d'ye do? I'm very glad you are here.' She said nothing else to him than this, appealed to him in no way, made no allusion to his having neglected her, but addressed herself entirely to Madame Carré, as if he had not been there; making no excuses and using no flattery; taking rather a tone of equal authority, as if she considered that the celebrated artist had a sacred duty toward her. This was another variation, Sherringham thought; it differed from each of the attitudes in which he had previously seen her. It came over him suddenly that so far from there being any question of her having the histrionic nature, she simply had it in such perfection that she was always acting; that her existence was a series of parts assumed for the moment, each changed for the next, before the perpetual mirror of some curiosity or admiration or wonder – some spectatorship that she perceived or imagined in the people about her. Interested as he had ever been in the profession of which she was potentially an ornament, this idea startled him by its novelty and even lent, on the spot, a formidable, a really appalling character to Miriam Rooth. It struck him abruptly that a woman whose only being was to 'make believe', to make believe that she had any and every being that you liked, that would serve a purpose, produce a certain effect, and whose identity resided in the continuity of her personations, so that she had no moral privacy, as he phrased it to himself, but lived in a high wind of exhibition, of figuration – such a woman was a kind of monster, in whom of necessity there would be nothing to like, because there would be nothing to take hold of. He felt for a moment that he had been very simple not before to have achieved that analysis of the actress. The girl's very face made it vivid to him now – the discovery that she positively had no countenance of her own, but only the countenance of the occasion, a sequence, a variety (capable possibly of becoming immense), of representative movements. She was always trying them, practising them for her amusement or profit, jumping from one to the other and extending her range; and this would doubtless be her occupation more and more as she acquired ease and confidence. The expression that came nearest to belonging to her, as it were, was the one that came nearest to being a blank – an air of inanity when she forgot herself, watching something. Then her eye was heavy and her mouth rather common; though it was perhaps just at such a moment that the fine line of her head told most. She had looked slightly *bête* even when Sherringham, on their first meeting at Madame Carré's, said to Nick Dormer that she was the image of the Tragic Muse.

Now, at any rate, he had the apprehension that she might do what she liked with her face. It was an elastic substance, an element of gutta-percha, like the flexibility of the gymnast, the lady who, at a music-hall, is shot from the mouth of a cannon. He coloured a little at this quickened view of the actress; he had always looked more poetically, somehow, at that priestess of art. But what was she, the priestess, when one came to think of it, but a female gymnast, a mountebank at higher wages? She didn't literally hang by her heels from a trapeze, holding a fat man in her teeth, but she made the same use of her tongue, of her eyes, of the imitative trick, that her muscular sister made of leg and jaw. It was an odd circumstance that Miriam Rooth's face seemed to him today a finer instrument than old Madame Carré's. It was doubtless that the girl's face was fresh and strong, with a future in it, while poor Madame Carré's was worn and weary, with only a past.

The old woman said something, half in jest, half in real resentment, about the brutality of youth, as Miriam went to a mirror and quickly took off her hat, patting and arranging her hair as a preliminary to making herself heard. Sherringham saw with surprise and amusement that the clever Frenchwoman, who had in her long life exhausted every adroitness, was in a manner helpless, condemned, both protesting and consenting. Miriam had taken but a few days and a couple of visits to become a successful force; she had imposed herself, and Madame Carré, while she laughed (yet looked terrible too, with artifices of eye and gesture), was reduced to the last line of defence – that of declaring her coarse and clumsy, saying she might knock her down, but that proved nothing. She spoke jestingly enough not to offend Miriam, but her manner betrayed the irritation of an intelligent woman who, at an advanced age, found herself for the first time failing to understand. What she didn't understand was the kind of social product that had been presented to her by Gabriel Nash; and this suggested to Sherringham that the *jeune Anglaise* was perhaps indeed rare, a new type, as Madame Carré must have seen innumerable varieties. He guessed that the girl was perfectly prepared to be abused and that her indifference to what might be thought of her discretion was a proof of life, health and spirit, the insolence of conscious power.

When she had given herself a touch at the glass she turned round, with a rapid '*Ecoutez maintenant!*' and stood leaning a moment, slightly lowered and inclined backward, with her hands behind her and supporting her, on the table in front of the mirror. She waited an instant, turning her eyes from one of her companions to the other as if she were taking possession of them (an eminently conscious, intentional proceeding, which made Sherringham ask himself what had become of her former terror and whether that and her tears had all been a comedy): after which, abruptly straightening herself, she began to repeat a short French poem, a composition modern and delicate, one of the things she had induced Madame Carré to say over to her. She had learned it, practised it, rehearsed it to her mother, and now she had been childishly eager to show what she could do with it. What she mainly did was to reproduce with a crude fidelity, but with extraordinary memory, the intonations, the personal quavers and cadences of her model.

'How bad you make me seem to myself, and if I were you how much better I should say it!' was Madame Carré's first criticism.

Miriam allowed her little time to develop this idea, for she broke out, at the shortest intervals, with the five other specimens of verse to which the old actress had handed her the key. They were all delicate lyrics, of tender or pathetic intention, by contemporary poets – all things demanding perfect taste and art, a mastery of tone, of insinuation, in the interpreter. Miriam had gobbled them up, she gave them forth in the same way as the first, with close, rude, audacious mimicry. There was a moment when Sherringham was afraid Madame Carré would think she was making fun of her manner, her celebrated simpers and grimaces, so extravagant did the girl's performance cause these refinements to appear.

When she had finished, the old woman said: 'Should you like now to hear how *you* do it?' and, without waiting for an answer, phrased and trilled the last of the pieces, from beginning to end, exactly as Miriam had done, making this imitation of an imitation the drollest thing conceivable. If she had been annoyed it was a perfect revenge. Miriam had dropped on a sofa, exhausted, and she stared at first, looking flushed and wild; then she gave way to merriment, laughing with a high sense of comedy. She said afterwards, to defend herself, that the verses in question, and indeed all those she had recited, were of the most difficult sort: you had to do them; they didn't do themselves – they were things in which the *gros moyens* were of no avail. 'Ah, my poor child, your means are all *gros moyens*; you appear to have no others,' Madame Carré replied. 'You do what you can, but there are people like that; it's the way they are made. They can never come nearer to the delicate; shades don't exist for them, they don't see certain differences. It was to show you a difference that I repeated that thing as you repeat it, as you represent my doing it. If you are struck with the little the two ways have in common, so much the better. But you seem to me to coarsen everything you touch.'

Sherringham thought this judgement harsh to cruelty, and perceived that Miss Rooth had the power to set the teeth of her instructress on edge. She acted on her nerves; she was made of a thick, rough substance which the old woman was not accustomed to manipulate. This exasperation, however, was a kind of flattery; it was neither indifference nor simple contempt; it acknowledged a mystifying reality in the girl and even a degree of importance. Miriam remarked, serenely enough, that the things she wanted most to do were just those that were not for the *gros moyens*, the vulgar obvious dodges, the starts and shouts that any one could think of and that the *gros public* liked. She wanted to do what was most difficult and to plunge into it from the first; and she explained, as if it were a discovery of her own, that there were two kinds of scenes and speeches: those which acted themselves, of which the treatment was plain, the only way, so that you had just to take it; and those which were open to interpretation, with which you had to fight every step, rendering, arranging, doing it according to your idea. Some of the most effective things, and the most celebrated and admired, like the frenzy of Juliet with her potion, were of the former sort; but it was the others she liked best.

Madame Carré received this revelation good-naturedly enough, considering its want of freshness, and only laughed at the young lady for looking so nobly patronizing while she gave it. It was clear that her laughter was partly dedicated to the good faith with which Miriam described herself as preponderantly interested in the subtler problems of her art. Sherringham was charmed with the girl's pluck – if it was pluck and not mere density – the brightness with which she submitted, for a purpose, to the old woman's rough usage. He wanted to take her away, to give her a friendly caution, to advise her not to become a bore, not to expose herself. But she held up her beautiful head in a way that showed she didn't care at present how she exposed herself, and that (it was half coarseness – Madame Carré was so far right – and half fortitude) she had no intention of coming away so long as there was anything to be picked up. She sat, and still she sat, challenging her hostess with every sort of question – some reasonable, some ingenious, some strangely futile and some highly indiscreet; but all with the effect that, contrary to Sherringham's expectation, Madame Carré warmed to the work of answering and explaining, became interested, was content to keep her and to talk. Yet she took her ease; she relieved herself, with the rare cynicism of the artist, all the crudity, the irony and intensity of a discussion of esoteric things, of personal mysteries, of methods and secrets. It was the oddest hour Sherringham had ever spent, even in the course of investigation which had often led him into the *cuisine*, as the French called it, the distillery or back-shop of the admired profession. He got up several times to come away; then he remained, partly in order not to leave Miriam alone with her terrible initiatress, partly because he was both amused and edified, and partly because Madame Carré held him by the appeal of her sharp, confidential old eyes, addressing her talk to him, with Miriam as a subject, a vile illustration. She undressed this young lady, as it were, from head to foot, turned her inside out, weighed and measured and sounded her: it was all, for Sherringham, a new revelation of the point to which, in her profession and nation, a ferocious analysis had been carried, with an intelligence of the business and a special vocabulary. What struck him above all was the way she knew her reasons and everything was sharp and clear in her mind and lay under her hand. If she had rare perceptions she had traced them to their source; she could give an account of what she did; she knew perfectly why; she could explain it, defend it, amplify it, fight for it: and all this was an intellectual joy to her, allowing her a chance to abound and insist and be clever. There was a kind of cruelty, or at least of hardness in it all, to Sherringham's English sense, that sense which can never really reconcile itself to the question of execution and has extraneous sentiments to placate with compromises and superficialities, frivolities that have often a pleasant moral fragrance. In theory there was nothing that he valued more than just such a logical passion as Madame Carré's; but in fact, when he found himself in close quarters with it, it was apt to seem to him an ado about nothing.

If the old woman was hard, it was not that many of her present conclusions, as regards Miriam, were not indulgent, but that she had a vision of the great manner, of right and wrong, of the just and the false, so high and religious that the individual was nothing before it – a prompt and easy sacrifice. It made Sherringham uncomfortable, as he had been made uncomfortable by certain *feuilletons*, reviews of the theatres in the Paris newspapers, which he was committed to thinking important, but of which, when they were very good, he was rather ashamed. When they were very good, that is when they were very thorough, they were very personal, as was inevitable in dealing with the most personal of the arts: they went into details; they put the dots on the *i*'s; they discussed impartially the qualities of appearance, the physical gifts of the actor or actress, finding them in some cases reprehensibly inadequate. Sherringham could not rid himself of a prejudice against these pronouncements; in the case of the actresses especially they appeared to him brutal and indelicate – unmanly as coming from a critic sitting smoking in his chair. At the same time he was aware of the dilemma (he hated it; it made him blush still more) in which his objection lodged him. If one was right in liking the actor's art one ought to have been interested in every candid criticism of it, which, given the peculiar conditions, would be legitimate in proportion as it should be minute. If the criticism that recognized frankly these conditions seemed an inferior or an offensive thing, then what was to be said for the art itself? What an implication, if the criticism was tolerable only so long as it was worthless – so long as it remained vague and timid! This was a knot which Sherringham had never straightened out: he contented himself with saying that there was no reason a theatrical critic shouldn't be a gentleman, at the same time that he often remarked that it was an odious trade, which no gentleman could possibly follow. The best of the fraternity, so conspicuous in Paris, were those who didn't follow it – those who, while pretending to write about the stage, wrote about everything else.

It was as if Madame Carré, in pursuance of her inflamed sense that the art was everything and the individual nothing, save as he happened to serve it, had said: 'Well, if she *will* have it she shall; she shall know what she is in for, what I went through, battered and broken in as we all have been – all who are worthy, who have had the honour. She shall know the real point of view.' It was as if she were still haunted with Mrs Rooth's nonsense, her hypocrisy, her scruples – something she felt a need to belabour, to trample on. Miriam took it all as a bath, a baptism, with passive exhilaration and gleeful shivers; staring, wondering, sometimes blushing and failing to follow, but not shrinking nor wounded; laughing, when it was necessary, at her own expense, and feeling evidently that this at last was the air of the profession, an initiation which nothing could undo. Sherringham said to her that he would see her home – that he wanted to talk to her and she must walk away with him. 'And it's understood, then, she may come back,' he added to Madame Carré. 'It's my affair, of course. You'll take an interest in her for a month or two; she will sit at your feet.'

'Oh, I'll knock her about; she seems stout enough!' said the old actress.

11

WHEN she had descended into the street with Sherringham Miriam informed him that she was thirsty, dying to drink something: upon which he asked her if she would have any objection to going with him to a café.

'Objection? I have spent my life in cafés!' she exclaimed. 'They are warm in winter and are full of gaslight. Mamma and I have sat in them for hours, many a time, with a *consommation* of three sous, to save fire and candles at home. We have lived in places we couldn't sit in, if you want to know – where there was only really room if we were in bed. Mamma's money is sent out from England, and sometimes it didn't come. Once it didn't come for months – for months and months. I don't know how we lived. There wasn't any to come; there wasn't any to get home. That isn't amusing when you're away, in a foreign town, without any friends. Mamma used to borrow, but people wouldn't always lend. You needn't be afraid – she won't borrow from you. We are rather better now. Something has been done in England; I don't understand what. It's only fivepence a year, but it has been settled; it comes regularly; it used to come only when we had written and begged and waited. But it made no difference; mamma was always up to her ears in books. They served her for food and drink. When she had nothing to eat she began a novel in ten volumes – the old-fashioned ones; they lasted longest. She knows every *cabinet de lecture* in every town; the little cheap, shabby ones, I mean, in the back streets, where they have odd volumes and only ask a sou, and the books are so old that they smell bad. She takes them to the cafés – the little cheap, shabby cafés, too – and she reads there all the evening. That's very well for her, but it doesn't feed me. I don't like a diet of dirty old novels. I sit there beside her, with nothing to do, not even a stocking to mend; she doesn't think that's *comme il faut*. I don't know what the people take me for. However, we have never been spoken to: any one can see mamma's a lady. As for me, I dare say I might be anything. If you're going to be an actress you must get used to being looked at. There were people in England who used to ask us to stay; some of them were our cousins – or mamma says they were. I have never been very clear about our cousins, and I don't think they were at all clear about us. Some of them are dead; the others don't ask us any more. You should hear mamma on the subject of our visits in England. It's very convenient when your cousins are dead, because that explains everything. Mamma has delightful phrases: "My family is almost extinct." Then your family may have been anything you like. Ours, of course, was magnificent. We did stay in a place once where there was a deer-park, and also private theatricals. I played in them; I was only fifteen years old, but I was very big and I thought I was in heaven. I will go anywhere you like; you needn't be afraid; we have been in places! I have learned a great deal that way; sitting beside mamma and watching people, their faces, their types, their movements. There's a great deal goes on in cafés: people come to them to talk things over, their private affairs, their complications; they have important meetings. Oh, I've observed scenes, between men and women – very quiet, terribly quiet, but tragic! Once I saw a woman do something that I'm going to do some day, when I'm great – if I can get the situation. I'll tell you what it is some day; I'll do it for you. Oh, it *is* the book of life!'

So Miriam discoursed, familiarly, disconnectedly, as the pair went their way down the Rue de Constantinople; and she continued to abound in anecdote and remark after they were seated face to face at a little marble table in an establishment which Sherringham selected carefully and he had caused her, at her request, to be accommodated with *sirop d'orgeat*. 'I know what it will come to: Madame Carré will want to keep me.' This was one of the announcements she presently made.

'To keep you?'

'For the French stage. She won't want to let you have me.' She said things of that kind, astounding in self-complacency, the assumption of quick success. She was in earnest, evidently prepared to work, but her imagination flew over preliminaries and probations, took no account of the steps in the process, especially the first tiresome ones, the test of patience. Sherringham had done nothing for her as yet, given no substantial pledge of interest; yet she was already talking as if his protection were assured and jealous. Certainly, however, she seemed to belong to him very much indeed, as she sat facing him in the Paris café, in her youth, her beauty and her talkative confidence. This degree of possession was highly agreeable to him, and he asked nothing more than to make it last and go further. The impulse to draw her out was irresistible, to encourage her to show herself to the end; for if he was really destined to take her career in hand he counted on some pleasant equivalent – such, for instance, as that she should at least amuse him.

'It's very singular; I know nothing like it,' he said – 'your equal mastery of two languages.'

'Say of half a dozen,' Miriam smiled.

'Oh, I don't believe in the others to the same degree. I don't imagine that, with all deference to your undeniable facility, you would be judged fit to address a German or an Italian audience in their own tongue. But you might a French, perfectly, and they are the most particular of all; for their idiom is supersensitive and they are incapable of enduring the *baragouinage* of foreigners, to which we listen with such complacency. In fact, your French is better than your English – it's more conventional; there are little queernesses and impurities in your English, as if you had lived abroad too much. Ah, you must work that.'

'I'll work it with you. I like the way you speak.'

'You must speak beautifully; you must do something for the standard.'

'For the standard?'

'There isn't any, after all; it has gone to the dogs.'

'Oh, I'll bring it back. I know what you mean.'

'No one knows, no one cares; the sense is gone – it isn't in the public,' Sherringham continued, ventilating a grievance he was rarely able to forget, the vision of which now suddenly made a mission full of sanctity for Miriam Rooth. 'Purity of speech, on our stage, doesn't exist. Every one speaks as he likes, and audiences never notice; it's the last thing they think of. The place is given up to abominable dialects and individual tricks, any vulgarity flourishes, and on the top of it all the Americans, with every conceivable crudity, come in to make confusion worse confounded. And when one laments it people stare; they don't know what one means.'

'Do you mean the grand manner, certain pompous pronunciations, the style of the Kembles?'

'I mean any style that *is* a style, that is a system, an art, that contributes a positive beauty to utterance. When I pay ten shillings to hear you speak, I want you to know how, *que diable!* Say that to people and they are mostly lost in stupor; only a few, the very intelligent ones, exclaim: 'Then do you want actors to be affected?'

'And *do* you?' asked Miriam, full of interest.

'My poor child, what else, under the sun, should they be? Isn't their whole art the affectation *par excellence*? The public won't stand that today, so one hears it said. If that be true, it simply means that the theatre, as I care for it, that is as a personal art, is at an end.'

'Never, never, never!' the girl cried, in a voice that made a dozen people look round.

'I sometimes think it – that the personal art *is* at an end, and that henceforth we shall have only the arts – capable, no doubt, of immense development in their way (indeed they have already reached it) – of the stage-carpenter and the costumer. In London the drama is already smothered in scenery; the interpretation scrambles off as it can. To get the old personal impression, which used to be everything, you must go to the poor countries, and most of all to Italy.'

'Oh, I've had it; it's very personal!' said Miriam, knowingly.

'You've seen the nudity of the stage, the poor painted, tattered screen behind, and in the empty space the histrionic figure, doing everything it knows how, in complete possession. The personality isn't our English personality, and it may not always carry us with it; but the direction is right, and it has the superiority that it's a human exhibition, not a mechanical one.'

'I can act just like an Italian,' said Miriam, eagerly.

'I would rather you acted like an Englishwoman, if an Englishwoman would only act.'

'Oh, I'll show you!'

'But you're not English,' said Sherringham, sociably, with his arms on the table.

'I beg your pardon; you should hear mamma about our "race".'

'You're a Jewess – I'm sure of that,' Sherringham went on.

She jumped at this, as he was destined to see, later, that she would jump at anything that would make her more interesting or striking; even at things which, grotesquely, contradicted or excluded each other. 'That's always possible, if one's clever. I'm very willing, because I want to be the English Rachel.'

'Then you must leave Madame Carré, as soon as you have got from her what she can give.'

'Oh, you needn't fear; you shan't lose me,' the girl replied, with gross, charming fatuity. 'My name is Jewish,' she went on, 'but it was that of my grandmother, my father's mother. She was a baroness, in Germany. That is she was the daughter of a baron.'

Sherringham accepted this statement with reservations, but he replied: 'Put all that together, and it makes you very sufficiently of Rachel's tribe.'

'I don't care, if I'm of her tribe artistically. I'm of the family of the artists; *je me fiche* of any other! I'm in the same style as that woman; I know it.'

'You speak as if you had seen her,' said Sherringham, amused at the way she talked of 'that woman'.

'Oh, I know all about her; I know all about all the great actors. But that won't prevent me from speaking divine English.'

'You must learn lots of verse; you must repeat it to me,' Sherringham went on. 'You must break yourself in till you can say anything. You must learn passages of Milton, passages of Wordsworth.'

'Did *they* write plays?'

'Oh, it isn't only a matter of plays! You can't speak a part properly till you can speak everything else, anything that comes up, especially in proportion as it's difficult. That gives you authority.'

'Oh, yes, I'm going in for authority. There's more chance in English,' the girl added, in the next breath. 'There are not so many others – the terrible competition. There are so many here

– not that I'm afraid,' she chattered on. 'But we've got America, and they haven't. America's a great place.'

'You talk like a theatrical agent. They're lucky not to have it as we have it. Some of them do go, and it ruins them.'

'Why, it fills their pockets!' Miriam cried.

'Yes, but see what they pay. It's the death of an actor to play to big populations that don't understand his language. It's nothing then but the *gros moyens*; all his delicacy perishes. However, they'll understand *you*.'

'Perhaps I shall be too affected,' said Miriam.

'You won't be more so than Garrick or Mrs Siddons or John Kemble or Edmund Kean. They understood Edmund Kean. All reflection is affectation, and all acting is reflection.'

'I don't know; mine is instinct,' Miriam replied.

'My dear young lady, you talk of "yours"; but don't be offended if I tell you that yours doesn't exist. Some day it will, if it comes off. Madame Carré's does, because she has reflected. The talent, the desire, the energy are an instinct; but by the time these things become a performance they are an instinct put in its place.'

'Madame Carré is very philosophic. I shall never be like her.'

'Of course you won't; you'll be original. But you'll have your own ideas.'

'I dare say I shall have a good many of yours,' said Miriam, smiling across the table.

They sat a moment looking at each other.

'Don't go in for coquetry; it's a waste of time.'

'Well, that's civil!' the girl cried.

'Oh, I don't mean for me; I mean for yourself. I want you to be so concentrated. I am bound to give you good advice. You don't strike me as flirtatious and that sort of thing, and that's in your favour.'

'In my favour!'

'It does save time.'

'Perhaps it saves too much. Don't you think the artist ought to have passions?'

Sherringham hesitated a moment: he thought an examination of this question premature. 'Flirtations are not passions,' he replied. 'No, you are simple – at least I suspect you are; for of course, with a woman, one would be clever to know.' She asked why he pronounced her simple, but he judged it best, and more consonant with fair play, to defer even a treatment of this branch of the question; so that, to change the subject, he said: 'Be sure you don't betray me to your friend Mr Nash.'

'Betray you? Do you mean about your recommending affectation?'

'Dear me, no; he recommends it himself. That is he practises it, and on a scale!'

'But he makes one hate it.'

'He proves what I mean,' said Sherringham: 'that the great comedian is the one who raises it to a science. If we paid ten shillings to listen to Mr Nash, we should think him very fine. But we want to know what it's supposed to be.'

'It's too odious, the way he talks about *us*!' Miriam cried, assentingly.

'About "us"?'

'Us poor actors.'

'It's the competition he dislikes,' said Sherringham, laughing.

'However, he is very good-natured; he lent mamma thirty pounds,' the girl added, honestly. Sherringham, at this information, was not able to repress a certain small twinge which his companion perceived and of which she appeared to mistake the meaning. 'Of course he'll get it back,' she went on, while Sherringham looked at her in silence for a minute. Fortune had not supplied him profusely with money, but his emotion was not caused by the apprehension that he too would probably have to put his hand in his pocket for Mrs Rooth. It was simply the instinctive recoil of a fastidious nature from the idea of familiar intimacy with people who lived from hand to mouth, and a sense that that intimacy would have to be defined if it was to go much further. He would wish to know what it was supposed to be, like Gabriel Nash's histrionics. After a moment Miriam mistook his thought still more completely, and in doing so gave him a flash of foreknowledge of the way it was in her to strike from time to time a note exasperatingly, almost consciously vulgar, which one would hate for the reason, among others, that by that time one would be in love with her. 'Well, then, he won't – if you don't believe it!' she exclaimed, with a laugh. He was saying to himself that the only possible form was that they should borrow only from him. 'You're a funny man: I make you blush,' Miriam persisted.

'I must reply with the *tu quoque*, though I have not that effect on you.'

'I don't understand,' said the girl.

'You're an extraordinary young lady.'

'You mean I'm horrid. Well, I dare say I am. But I'm better when you know me.'

Sherringham made no direct rejoinder to this, but after a moment he said: 'Your mother must repay that money. I'll give it to her.'

'You had better give it to him!' cried Miriam. 'If once *we* have it –' She interrupted herself, and with another and a softer tone, one of her professional transitions, she remarked: 'I suppose you have never known any one that's poor.'

'I'm poor myself. That is I'm very far from rich. But why receive favours –?' And here he, in turn, checked himself, with the sense that he was indeed taking a great deal on his back if he pretended already (he had not seen the pair three times) to regulate their intercourse with the rest of the world. But Miriam instantly carried out his thought and more than his thought.

'Favours from Mr Nash? Oh, he doesn't count!'

The way she dropped these words (they would have been admirable on the stage) made him laugh and say immediately: 'What I meant just now was that you are not to tell him, after all my swagger, that I consider that you and I are really required to save our theatre.'

'Oh, if we can save it, he shall know it!' Then Miriam added that she must positively get home; her mother would be in a state: she had really scarcely ever been out alone. He mightn't think it, but so it was. Her mother's ideas, those awfully proper ones, were not all talk. She *did* keep her! Sherringham accepted this – he had an adequate, and indeed an analytic vision of Mrs Rooth's conservatism; but he observed at the same time that his companion made no motion to rise. He made none, either; he only said –

'We are very frivolous, the way we chatter. What you want

to do, to get your foot in the stirrup, is supremely difficult. There is everything to overcome. You have neither an engagement nor the prospect of an engagement.'

'Oh, you'll get me one!' Miriam's manner expressed that this was so certain that it was not worth dilating upon; so instead of dilating she inquired abruptly, a second time: 'Why do you think I'm so simple?'

'I don't then. Didn't I tell you just now that you were extraordinary? That's the term moreover that you applied to yourself when you came to see me – when you said a girl had to be, to wish to go on the stage. It remains the right one, and your simplicity doesn't mitigate it. What's rare in you is that you have – as I suspect, at least – no nature of your own.' Miriam listened to this as if she were preparing to argue with it or not, only as it should strike her as being a pleasing picture; but as yet, naturally, she failed to understand. 'You are always playing something; there are no intervals. It's the absence of intervals, of a *fond* or background, that I don't comprehend. You're an embroidery without a canvas.'

'Yes, perhaps,' the girl replied, with her head on one side, as if she were looking at the pattern. 'But I'm very honest.'

'You can't be everything, a consummate actress and a flower of the field. You've got to choose.'

She looked at him a moment. 'I'm glad you think I'm so wonderful.'

'Your feigning may be honest, in the sense that your only feeling *is* your feigned one,' Sherringham went on. 'That's what I mean by the absence of a ground or of intervals. It's a kind of thing that's a labyrinth!'

'I know what I am,' said Miriam, sententiously.

But her companion continued, following his own train: 'Were you really so frightened, the first day you went to Madame Carré's?'

She stared a moment, and then with a flush, throwing back her head: 'Do you think I was pretending?'

'I think you always are. However, your vanity (if you had any!) would be natural.'

'I have plenty of that – I am not ashamed to own it.'

'You would be capable of pretending that you have. But excuse the audacity and the crudity of my speculations – it only proves my interest. What is it that you know you are?'

'Why, an artist. Isn't that a canvas?'

'Yes, an intellectual one, but not a moral.'

'Oh yes, it is, too. And I'm a good girl: won't that do?'

'It remains to be seen,' Sherringham laughed. 'A creature who is *all* an artist – I am curious to see that.'

'Surely it has been seen, in lots of painters, lots of musicians.'

'Yes, but those arts are not personal, like yours. I mean not so much so. There's something left for – what shall I call it? – for character.'

Miriam stared again, with her tragic light. 'And do you think I've got no character?' As he hesitated she pushed back her chair, rising rapidly.

He looked up at her an instant – she seemed so 'plastic'; and then, rising too, he answered: 'Delightful being, you've got a hundred!'

12

THE summer arrived and the dense air of the Paris theatres became in fact a still more complicated mixture; yet the occasions were not few on which Peter Sherringham, having placed a box, near the stage (most often a stuffy, dusky *baignoire*), at the disposal of Mrs Rooth and her daughter, found time to look in, as he said, to spend a part of the evening with them and point the moral of the performance. The pieces, the successes of the winter, had entered the automatic phase: they went on by the force of the impetus acquired, deriving little fresh life from the interpretation, and in ordinary conditions their strong points, as rendered by the actors, would have been as wearisome to Sherringham as an importunate repetition of a good story. But it was not long before he became aware that the conditions could not be regarded as ordinary. There was a new infusion in his consciousness – an element in his life which altered the relations of things. He was not easy till he had found the right name for it – a name the more satisfactory that it was simple, comprehensive and plausible. A new 'distraction', in the French sense, was what he flattered himself he had discovered; he could recognize that as freely as possible without being obliged to classify the agreeable resource as a new entanglement. He was neither too much nor too little diverted; he had all his usual attention to give to his work: he had only an employment for his odd hours, which, without being imperative, had over various others the advantage of a certain continuity.

And yet, I hasten to add, he was not so well pleased with it but that, among his friends, he maintained for the present a considerable reserve in regard to it. He had no irresistible impulse to tell people that he had disinterred a strange, handsome girl whom he was bringing up for the theatre. She had been seen by several of his associates at his rooms, but she was not soon to be seen there again. Sherringham's reserve might by the ill-natured have been termed dissimulation, inasmuch as when asked by the ladies of the Embassy what had become of the young person who amused them that day so cleverly, he gave it out that her whereabouts was uncertain and her destiny probably obscure; he let it be supposed in a word that his benevolence had scarcely survived an accidental, charitable occasion. As he went about his customary business, and perhaps even put a little more conscience into the transaction of it, there was nothing to suggest to his companions that he was engaged in a private speculation of a singular kind. It was perhaps his weakness that he carried the apprehension of ridicule too far; but his excuse may be said to be that he held it unpardonable for a man publicly enrolled in the service of his country to be ridiculous. It was of course not out of all order that such functionaries, their private situation permitting, should enjoy a personal acquaintance with stars of the dramatic, the lyric or even the choreographic stage: high diplomatists had indeed not rarely and not invisibly cultivated this privilege without its proving the sepulchre of their reputation. That a gentleman who was not a fool should consent a little to become one for the sake of a celebrated actress or singer – *cela s'était vu*, though it was not perhaps to be recommended. It was not a tendency that was fostered at headquarters, where even the most rising young men were not encouraged to believe they could never fall. Still, it might pass if it were kept in its place; and there were ancient worthies yet in the profession (not those, however, whom the tradition had helped to go furthest) who held that something of the sort was a graceful ornament of the diplomatic character. Sherringham was aware he was very 'rising'; but Miriam Rooth was not yet a celebrated actress. She was only a youthful artist, in conscientious process of formation, encumbered with a mother still more conscientious than herself. She was a young English lady, very earnest about artistic, about remunerative problems. He had accepted the position of a formative influence, and that was precisely what might provoke derision. He was a ministering angel – his patience and good-nature really entitled him to the epithet, and his rewards would doubtless some day define themselves; but meanwhile other promotions were in contingent prospect, for the failure of which these would not, even in their abundance, be a compensation. He kept an unembarrassed eye upon Downing Street; and while it may frankly be said for him that he was neither a pedant nor a prig, he remembered that the last impression he ought to wish to produce there was that of volatility.

He felt not particularly volatile, however, when he sat behind Miriam at the play and looked over her shoulder at the stage: her observation being so keen and her comments so unexpected in their vivacity that his curiosity was refreshed and his attention stretched beyond its wont. If the spectacle before the footlights had now lost much of its annual brilliancy, the fashion in which Miriam followed it came near being spectacle enough. Moreover, in most cases the attendance of the little party was at the Théâtre Français; and it has been sufficiently indicated that Sherringham, though the child of a sceptical age and the votary of a cynical science, was still candid enough to take the serious, the religious view of that establishment – the view of M. Sarcey and of the unregenerate provincial mind. 'In the trade that I follow we see things too much in the hard light of reason, of calculation,' he once remarked to his young *protégée*; 'but it's good for the mind to keep up a superstition or two: it leaves a margin, like having a second horse to your brougham for night-work. The arts, the amusements, the aesthetic part of life are night-work, if I may say so without suggesting the nefarious. At any rate you want your second horse – your superstition that stays at home when the sun is high – to go your rounds with. The Théâtre Français is my second horse.'

Miriam's appetite for this pleasure showed him vividly enough how rarely, in the past, it had been within her reach; and she pleased him at first by liking everything, seeing almost no difference and taking her deep draught undiluted. She leaned on the edge of the box with bright voracity, tasting to the core yet relishing the surface; watching each movement of each actor, attending to the way each thing was said or done as if it were the most important thing, and emitting from time to time applausive or restrictive sounds. It was a very pretty exhibition of enthusiasm, if enthusiasm be ever critical. Sherringham had his wonder about it, as it was a part of the attraction exerted by this young lady that she caused him to have his wonder about everything she did. Was it in fact an exhibition, a line taken for effect, so that at the comedy her own comedy was the most successful of all? That question danced attendance on the liberal intercourse of these young people and fortunately, as yet, did little to embitter Sherringham's share of it. His general sense that she was personating had its especial moments of suspense and perplexity and added variety and even occasionally a degree of excitement to their conversation. At the theatre, for the most part, she was really flushed with eagerness; and with the spectators who turned an admiring eye into the dim compartment of which she pervaded the front, she might have passed for a romantic, or at any rate an insatiable young woman from the country.

Mrs Rooth took a more placid view, but attended immensely to the story, in respect to which she manifested a patient good faith which had its surprises and its comicalities for Sherringham. She found no play too tedious, no entr'acte too long, no *baignoire* too hot, no tissue of incidents too complicated, no situation too unnatural and no sentiments too sublime. She gave Sherringham the measure of her power to sit and sit – an accomplishment to which she owed, in the struggle for existence, such superiority as she might be said to have achieved. She could outsit everyone, everything else; looking as if she had acquired the practice in repeated years of small frugality combined with large leisure – periods when she had nothing but time to spend and had learned to calculate, in any situation, how long she could stay. 'Staying' was so often a saving – a saving of candles, of fire and even (for it sometimes implied a vision of light refreshment) of food. Sherringham perceived soon enough that she was complete in her way, and if he had been addicted to studying the human mixture in its different combinations he would have found in her an interesting compendium of some of the infatuations that survive a hard discipline. He made indeed without difficulty the reflection that her life might have taught her the reality of things, at the same time that he could scarcely help thinking it clever of her to have so persistently declined the lesson. She appeared to have put it by with a deprecating, ladylike smile – a plea of being too soft and bland for experience.

She took the refined, sentimental, tender view of the universe, beginning with her own history and feelings. She believed in everything high and pure, disinterested and orthodox, and even at the Hôtel de la Garonne was unconscious of the shabby or the ugly side of the world. She never despaired: otherwise what would have been the use of being a Neville-Nugent? Only not to have been one – that would have been discouraging. She delighted in novels, poems, perversions, misrepresentations and evasions, and had a capacity for smooth, superfluous falsification which made Sherringham think her sometimes an amusing and sometimes a tedious inventor. But she was not dangerous even if you believed her; she was not even a warning if you didn't. It was harsh to call her a hypocrite, because you never could have resolved her back into her character: there was no reverse to her blazonry. She built in the air and was not less amiable than she pretended: only that was a pretence too. She moved altogether in a world of genteel fable and fancy, and Sherringham had to live in it with her, for Miriam's sake, in sociable, vulgar assent, in spite of his feeling that it was rather a low neighbourhood. He was at a loss how to take what she said – she talked sweetly and discursively of so many things – until he simply perceived that he could only take it always for untrue. When Miriam laughed at her he was rather disagreeably affected: 'dear mamma's fine stories' was a sufficiently cynical reference to the immemorial infirmity of a parent. But when the girl backed her up, as he phrased it to himself, he liked that even less.

Mrs Rooth was very fond of a moral and had never lost her taste for edification. She delighted in a beautiful character and was gratified to find so many represented in the contemporary French drama. She never failed to direct Miriam's attention to them and to remind her that there is nothing in life so precious as the ideal. Sherringham noted the difference between the mother and the daughter and thought it singularly marked – the way that one took everything for the sense, or behaved as if she did, caring above all for the subject and the romance, the triumph or defeat of virtue and the moral comfort of it all, and that the other was especially hungry for the manner and the art of it, the presentation and the vividness. Mrs Rooth abounded in impressive evocations, and yet he saw no link between her facile genius and that of which Miriam gave symptoms. The poor lady never could have been accused of successful deceit, whereas success in this line was exactly what her clever child went in for. She made even the true seem fictive, while Miriam's effort was to make the fictive true. Sherringham thought it an odd, unpromising stock (that of the Neville-Nugents) for a dramatic talent to have sprung from, till he reflected that the evolution was after all natural: the figurative impulse in the mother had become conscious, and therefore higher, through finding an aim, which was beauty, in the daughter. Likely enough the Hebraic Mr Rooth, with his love of old pots and Christian altar-cloths, had supplied, in the girl's composition, the aesthetic element, the sense of form. In their visits to the theatre there was nothing that Mrs Rooth more insisted upon than the unprofitableness of deceit, as shown by the most distinguished authors – the folly and degradation, the corrosive effect upon the spirit, of tortuous ways. Sherringham very soon gave up the futile task of piecing together her incongruous references to her early life and her family in England. He renounced even the doctrine that there was a residuum of truth in her claim of great relationships, for, existent or not, he cared equally little for her ramifications. The principle of this indifference was at bottom a certain desire to disconnect Miriam; for it was disagreeable not to be independent in dealing with her, and he could be fully so only if *she* were.

The early weeks of that summer (they went on indeed into

August) were destined to establish themselves in his memory as a season of pleasant things. The ambassador went away, and Sherringham had to wait for his own holiday, which he did, during the hot days, contentedly enough, in spacious halls, with a dim, bird-haunted garden. The official world, and most other worlds withdrew from Paris, and the Place de la Concorde, a larger, whiter desert than ever, became, by a reversal of custom, explorable with safety. The Champs Elysées were dusty and rural, with little creaking booths and exhibitions which made a noise like grasshoppers; the Arc de Triomphe threw its cool, sharp shadow for a mile; the Palais de l'Industrie glittered in the light of the long days; the cabmen, in their red waistcoats, dozed in their boxes; and Sherringham permitted himself a 'pot' hat and rarely met a friend. Thus was Miriam still more disconnected, and thus was it possible to deal with her still more independently. The theatres on the boulevard closed, for the most part, but the great temple of the Rue de Richelieu, with an aesthetic responsibility, continued imperturbably to dispense examples of style. Madame Carré was going to Vichy, but she had not yet taken flight, which was a great advantage for Miriam, who could now solicit her attention with the consciousness that she had no engagements *en ville*.

'I make her listen to me – I make her tell me,' said the ardent girl, who was always climbing the slope of the Rue de Constantinople, on the shady side, where in the July mornings there was a smell of violets from the moist flower-stands of fat, white-capped *bouquetières*, in the angles of doorways. Miriam liked the Paris of the summer mornings, the clever freshness of all the little trades and the open-air life, the cries, the talk from door to door, which reminded her of the south, where, in the multiplicity of her habitations, she had lived; and most of all the great amusement, or nearly, of her walk, the enviable baskets of the laundress, piled up with frilled and fluted whiteness – the certain luxury, she felt as she passed, with quick prevision of her own dawn of glory. The greatest amusement perhaps was to recognize the pretty sentiment of earliness, the particular congruity with the hour, in the studied, selected dress of the little tripping women who were taking the day, for important advantages, while it was tender. At any rate she always brought with her from her passage through the town good-humour enough (with the penny bunch of violets that she stuck in the front of her dress) for whatever awaited her at Madame Carré's. She told Sherringham that her dear mistress was terribly severe, giving her the most difficult, the most exhausting exercises – showing a kind of rage for breaking her in.

'So much the better,' Sherringham answered; but he asked no questions and was glad to let the preceptress and the pupil fight it out together. He wanted, for the moment, to know as little as possible about them: he had been overdosed with knowledge that second day he saw them together. He would send Madame Carré her money (she was really most obliging), and in the meantime he was conscious that Miriam could take care of herself. Sometimes he remarked to her that she needn't always talk 'shop' to him: there were times when he was very tired of shop – of hers. Moreover he frankly admitted that he was tired of his own, so that the restriction was not brutal. When she replied, staring: 'Why, I thought you considered it as such a beautiful, interesting art!' he had no rejoinder more philosophic than 'Well, I do; but there are moments when I'm sick of it, all the same.' At other times he said to her: 'Oh, yes, the results, the finished thing, the dish, perfectly seasoned and served: not the mess of preparation – at least not always – not the experiments that spoil the material.'

'I thought you thought just these questions of study, of the artistic education, as you have called it to me, so fascinating,' the girl persisted. Sometimes she was very lucid.

'Well, after all I'm not an actor myself,' Sherringham answered, laughing.

'You might be one if you were serious,' said Miriam. To this her friend replied that Mr Gabriel Nash ought to hear that; which made her exclaim, with a certain grimness, that she would settle *him* and his theories some day. Not to seem too inconsistent – for it was cruel to bewilder her when he had taken her up to enlighten – Sherringham repeated over that for a man like himself the interest of the whole thing depended on its being considered in a large, liberal way, with an intelligence that lifted it out of the question of the little tricks of the trade, gave it beauty and elevation. Miriam let him know that Madame Carré held that there were no *little* tricks; that everything had its importance as a means to a great end; and that if you were not willing to try to *approfondir* the reason why in a given situation you should scratch your nose with your left hand rather than with your right, you were not worthy to tread any stage that respected itself.

'That's very well; but if I must go into details read me a little Shelley,' said the young man, in the spirit of a high *raffiné*.

'You are worse than Madame Carré; you don't know what to invent: between you you'll kill me!' the girl declared. 'I think there's a secret league between you to spoil my voice, or at least to weaken my wind before I get it. But *à la guerre comme à la guerre!* How can I read Shelley, however, when I don't understand him?'

'That's just what I want to make you do. It's a part of your general training. You may do without that, of course – without culture and taste and perception; but in that case you'll be nothing but a vulgar *cabotine*, and nothing will be of any consequence.' Sherringham had a theory that the great lyric poets (he induced her to read and recite as well long passages of Wordsworth and Swinburne) would teach her many of the secrets of competent utterance, the mysteries of rhythm, the communicableness of style, the latent music of the language and the art of 'composing' copious speeches and of keeping her wind in hand. He held in perfect sincerity that there was an indirect enlightenment which would be of the highest importance to her and to which it was precisely, by good fortune, in his power to contribute. She would do better in proportion as she had more knowledge – even knowledge that might appear to have but a remote connection with her business. The actor's talent was essentially a gift, a thing by itself, implanted, instinctive, accidental, equally unconnected with intellect and with virtue – Sherringham was completely of that opinion; but it seemed to him no contradiction to consider at the same time that intellect (leaving virtue, for the moment, out of the question) might be brought into fruitful relation with it. It would be a larger thing if a better mind were projected upon it – without sacrificing the mind. So he lent Miriam books which she never read (she was on almost irreconcilable terms with the printed page), and in the long summer days, when he had leisure, took her to the Louvre to admire the great works of painting and sculpture. Here, as on all occasions, he was struck with the queer jumble of her taste, her mixture of intelligence and puerility. He saw that she never read what he gave her, though she sometimes would have liked him to suppose so; but in the presence of famous pictures and statues she had remarkable flashes of perception. She felt these things, she liked them, though it was always because she had an idea she could use them. The idea was often fantastic, but it showed what an eye she had to her business. 'I could look just like that, if I tried.' 'That's the dress I mean to wear when I do Portia.' Such were the observations that were apt to drop from her under the suggestion of antique marbles or when she stood before a Titian or a Bronzino.

When she uttered them, and many others besides, the effect was sometimes irritating to Sherringham, who had to reflect a little to remember that she was no more egotistical than the histrionic conscience demanded. He wondered if there were necessarily something vulgar in the histrionic conscience – something condemned to feel only the tricky personal question. Wasn't it better to be perfectly stupid than to have only one eye open and wear forever, in the great face of the world, the expression of a knowing wink? At the theatre, on the numerous July evenings when the Comédie Française played the repertory, with exponents determined the more sparse and provincial audience should thrill and gape with the tradition, her appreciation was tremendously technical and showed it was not for nothing she was now in and out of Madame Carré's innermost counsels. But there were moments when even her very acuteness seemed to him to drag the matter down, to see it in a small and superficial sense. What he flattered himself that he was trying to do for her (and through her for the stage of his time, since she was the instrument, and incontestably a fine one, that had come to his hand) was precisely to lift it up, make it rare, keep it in the region of distinction and breadth. However, she was doubtless right and he was wrong, he eventually reasoned: you could afford to be vague only if you hadn't a responsibility. He had fine ideas, but she was to do the acting, that is the application of them, and not he; and application was always of necessity a sort of vulgarization, a smaller thing than theory. If some day she should exhibit the great art that it was not purely fanciful to forecast for her, the subject would doubtless be sufficiently lifted up and it wouldn't matter that some of the onward steps should have been lame.

This was clear to him on several occasions when she repeated or acted something for him better than usual: then she quite carried him away, making him wish to ask no more questions but only let her disembroil herself in her own fashion. In these hours she gave him fitfully but forcibly that impression of beauty which was to be her justification. It was too soon for any general estimate of her progress; Madame Carré had at last given her an intelligent understanding, as well as a sore personal sense, of how bad she was. She had therefore begun on a new basis; she had returned to the alphabet and the drill. It was a phase of awkwardness, like the splashing of a young swimmer, but buoyancy would certainly come out of it. For the present there was for the most part no great alteration of the fact that when she did things according to her own idea they were not as yet, and seriously judged, worth the devil, as Madame Carré said; and when she did them according to that of her instructress they were too apt to be a gross parody of that lady's intention. None the less she gave glimpses, and her glimpses made him feel not only that she was not a fool (that was a small relief), but that *he* was not.

He made her stick to her English and read Shakespeare aloud to him. Mrs Rooth had recognized the importance of an apartment in which they should be able to receive so beneficent a visitor, and was now mistress of a small salon with a balcony and a rickety flower-stand (to say nothing of a view of many roofs and chimneys), a crooked waxed floor, an empire clock, an *armoire à glace* (highly convenient for Miriam's posturings), and several cupboard doors, covered over, allowing for treacherous gaps, with the faded magenta paper of the wall. The thing had been easily done, for Sherringham had said: 'Oh, we must have a sitting-room for our studies, you know. I'll settle it with the landlady.' Mrs Rooth had liked his 'we' (indeed she liked everything about him), and he saw in this way that she had no insuperable objection to being under a pecuniary obligation so long as it was distinctly understood to be temporary. That he should have his money back with interest as soon as Miriam was launched was a comfort so deeply implied that it only added to intimacy. The window stood open on the little balcony, and when the sun had left it Sherringham and Miriam could linger there, leaning on the rail and talking, above the great hum of Paris, with nothing but the neighbouring tiles and tall tubes to take account of. Mrs Rooth, in limp garments, much ungirdled, was on the sofa with a novel, making good her frequent assertion that she could put up with any life that would yield her these two articles. There were romantic works that Sherringham had never read, and as to which he had vaguely wondered to what class they were addressed – the earlier productions of M. Eugène Sue, the once-fashionable compositions of Madame Sophie Gay – with which Mrs Rooth was familiar and which she was ready to peruse once more if she could get nothing fresher. She had always a greasy volume tucked under her while her nose was bent upon the pages in hand. She scarcely looked up even when Miriam lifted her voice to show Sherringham what she could do. These tragic or pathetic notes all went out of the window and mingled with the undecipherable concert of Paris, so that no neighbour was disturbed by them. The girl shrieked and wailed when the occasion required it, and Mrs Rooth only turned her page, showing in this way a great aesthetic as well as a great personal trust.

She rather annoyed Sherringham by the serenity of her confidence (for a reason that he fully understood only later), save when Miriam caught an effect or a tone so well that she made him, in the pleasure of it, forget her parent was there. He continued to object to the girl's English, with the foreign patches which might pass in prose but were offensive in the recitation of verse, and he wanted to know why she could not speak like her mother. He had to do Mrs Rooth the justice of recognizing the charm of her voice and accent, which gave a certain richness even to the foolish things she said. They were of an excellent insular tradition, full both of natural and of cultivated sweetness, and they puzzled him when other indications seemed to betray her – to relegate her to the class of the simple dreary. They were like the reverberation of far-off drawing-rooms.

The connection between the development of Miriam's genius and the necessity of an occasional excursion to the country – the charming country that lies in so many directions beyond the Parisian *banlieue* – would not have been immediately apparent to a merely superficial observer; but a day, and then another, at Versailles, a day at Fontainebleau and a trip, particularly harmonious and happy, to Rambouillet, took their place in Sherringham's programme as a part of the legitimate indirect culture, an agency in the formation of taste. Intimations of the grand style, for instance, would proceed in abundance from the symmetrical palace and gardens of Louis XIV. Sherringham was very fond of Versailles, and went there more than once with the ladies of the Hôtel de la Garonne. They chose quiet hours, when the fountains were dry; and Mrs Rooth took an armful of novels and sat on a bench in the park, flanked by clipped hedges and old statues, while her young companions strolled away, walked to the Trianon, explored the long, straight vistas of the woods. Rambouillet was vague and pleasant and idle; they had an idea that they found suggestive associations there; and indeed there was an old white château which contained nothing else. They found, at any rate, luncheon and, in the landscape, a charming sense of summer and of little brushed French pictures.

I have said that in these days Sherringham wondered a good deal, and by the time his leave of absence was granted him this practice had engendered a particular speculation. He was surprised that he was not in love with Miriam Rooth, and he considered in moments of leisure the causes of his exemption. He had perceived from the first that she was a 'nature', and each time she met his eyes the more vividly it appeared to him that her beauty was rare. You had to get the view of her face, but when you did so it was a splendid mobile mask. And the possessor of this high advantage had frankness and courage and variety and the unusual and the unexpected. She had qualities that seldom went together – impulses and shynesses, audacities and lapses, something coarse, popular and strong, all intermingled with disdains and languors and nerves. And then, above all, she was there, she was accessible, she almost belonged to him. He reflected ingeniously that he owed his escape to a peculiar cause – the fact that they had together a positive outside object. Objective, as it were, was all their communion; not personal and selfish, but a matter of art and business and discussion. Discussion had saved him and would save him further; for they would always have something to quarrel about. Sherringham, who was not a diplomatist for nothing, who had his reasons for steering straight and wished neither to deprive the British public of a rising star nor to change his actual situation for that of a conjugal *impresario*, blessed the beneficence, the salubrity, the pure exorcism of art. At the same time, rather inconsistently and feeling that he had a completer vision than before of the odd animal the artist who happened to have been born a woman, he felt himself warned against a serious connection (he made a great point of the 'serious') with so slippery and ticklish a creature. The two ladies had only to stay in Paris, save their candle-ends and, as Madame Carré had enjoined, practise their scales: there were apparently no autumn visits to English country-houses in prospect for Mrs Rooth.

Sherringham parted with them on the understanding that in London he would look as thoroughly as possible into the question of an engagement for Miriam. The day before he began his holiday he went to see Madame Carré, who said to him: '*Vous devriez bien nous la laisser.*'

'She has got something, then?'

'She has got most things. She'll go far. It is the first time I ever was mistaken. But don't tell her so – I don't flatter her; she'll be too puffed up.'

'Is she very conceited?' Sherringham asked.

'*Mauvais sujet!*' said Madame Carré.

It was on the journey to London that he indulged in some of those questionings of his state which I have mentioned; but I must add that by the time he reached Charing Cross (he smoked a cigar, deferred till after the Channel, in a compartment by himself) it suddenly came over him that they were futile. Now that he had left the girl, a subversive, unpremeditated heart-beat told him – it made him hold his breath a minute in the carriage – that he had after all *not* escaped. He *was* in love with her: he had been in love with her from the first hour.

13

THE drive from Harsh to the Place, as it was called thereabouts, could be achieved by swift horses in less than ten minutes; and if Mrs Dallow's ponies were capital trotters the general high pitch of the occasion made it congruous that they should show their speed. The occasion was the polling-day, the hour after the battle. The ponies had worked, with all the rest, for the week before, passing and repassing the neat windows of the flat little town (Mrs Dallow had the complacent belief that there was none in the kingdom in which the flower-stands looked more respectable between the stiff muslin curtains), with their mistress behind them in her low, smart trap. Very often she was accompanied by the Liberal candidate, but even when she was not the equipage seemed scarcely less to represent his pleasant sociable confidence. It moved in a radiance of ribbons and handbills and hand-shakes and smiles; of quickened intercourse and sudden intimacy; of sympathy which assumed without presuming and gratitude which promised without soliciting. But, under Julia's guidance, the ponies pattered now, with no indication of a loss of freshness, along the firm, wide avenue which wound and curved, to make up in picturesque effect for not undulating, from the gates opening straight into the town to the Palladian mansion, high, square, gray and clean, which stood, among parterres and fountains, in the centre of the park. A generous steed had been sacrificed to bring the good news from Ghent to Aix, but no such extravagance was after all necessary for communicating with Lady Agnes.

She had remained at the house, not going to the Wheatsheaf, the Liberal inn, with the others; preferring to await in privacy, and indeed in solitude, the momentous result of the poll. She had come down to Harsh with the two girls in the course of the proceedings. Julia had not thought they would do much good, but she was expansive and indulgent now and she had liberally asked them. Lady Agnes had not a nice canvassing manner, effective as she might have been in the character of the high, benignant, affable mother – looking sweet participation but not interfering – of the young and handsome, the shining, convincing, wonderfully clever and certainly irresistible aspirant. Grace Dormer had zeal without art, and Lady Agnes, who during her husband's lifetime had seen their affairs follow the satisfactory principle of a tendency to defer to supreme merit, had never really learned the lesson that voting goes by favour. However, she could pray God if she couldn't flatter the cheesemonger, and Nick felt that she had stayed at home to pray for him. I must add that Julia Dallow was too happy now, flicking her whip in the bright summer air, to say anything so ungracious even to herself as that her companion had been returned in spite of his nearest female relatives. Besides, Biddy *had* been a rosy help: she had looked persuasively pretty, in white and pink, on platforms and in recurrent carriages, out of which she had tossed, blushing and making people remember her eyes, several words that were telling for their very simplicity.

Mrs Dallow was really too glad for any definite reflection, even for personal exultation, the vanity of recognizing her own large share of the work. Nick was in and he was beside her, tired, silent, vague, beflowered and beribboned, and he had been splendid from beginning to end, delightfully good-humoured and at the same time delightfully clever – still cleverer than she had supposed he could be. The sense that she had helped his cleverness and that she had been repaid by it, or by his gratitude (it came to the same thing), in a way she appreciated, was not triumphant and jealous; for the break of the long tension soothed her, it was as pleasant as an untied ligature. So nothing passed between them on their way to the house; there was no sound in the park but the happy rustle of summer (it seemed an applausive murmur) and the swift progress of the vehicle.

Lady Agnes already knew, for as soon as the result was declared Nick had dispatched a man on horseback to her, carrying the figures on a scrawled card. He had been far from getting away at once, having to respond to the hubbub of acclamation, to speak yet again, to thank his electors, individually and collectively, to chaff the Tories, to be carried, hither and yon, and above all to pretend that the interest of the business was now greater for him than ever. If he said never a word after he put himself in Julia's hands to go home, perhaps it was partly because the consciousness began to glimmer within him that that interest had on the contrary now suddenly diminished. He wanted to see his mother because he knew she wanted to see him, to fold him close in her arms. They had been open there for that purpose for the last half-hour, and her expectancy, now no longer an ache of suspense, was the reason of Julia's round pace. Yet this very expectancy somehow made Nick wince a little. Meeting his mother was like being elected over again.

The others had not come back yet – Lady Agnes was alone in the large bright drawing-room. When Nick went in with Mrs Dallow he saw her at the further end; she had evidently been walking to and fro, the whole length of it, and her tall, upright black figure seemed in possession of the fair vastness like an exclamation-point at the bottom of a blank page. The room, rich and simple, was a place of perfection as well as of splendour in delicate tints, with precious specimens of French furniture of the last century ranged against walls of pale brocade and here and there a small, almost priceless picture. George Dallow had made it, caring for these things and liking to talk about them (scarcely about anything else); so that it appeared to represent him still, what was best in his kindly, uniform nature – a friendly, competent, tiresome insistence upon purity and homogeneity. Nick Dormer could hear him yet, and could see him, too fat and with a congenital thickness in his speech, lounging there in loose clothes with his eternal cigarette. 'Now, my dear fellow, *that's* what I call form: I don't know what you call it' – that was the way he used to begin. The room was full of flowers in rare vases, but it looked like a place of which the beauty would have had a sweet odour even without them.

Lady Agnes had taken a white rose from one of the clusters and was holding it to her face, which was turned to the door as Nick crossed the threshold. The expression of her figure instantly told him (he saw the creased card that he had sent her lying on one of the beautiful bare tables) how she had been sailing up and down in a majesty of satisfaction. The inflation of her long, plain dress, the brightened dimness of her proud face were still in the air. In a moment he had kissed her and was being kissed, not in quick repetition, but in tender prolongation, with which the perfume of the white rose was mixed. But there was something else too – her sweet, smothered words in his ear:

'Oh, my boy, my boy – oh, your father, your father!' Neither the sense of pleasure nor that of pain, with Lady Agnes (and indeed with most of the persons with whom this history is concerned), was a liberation of chatter; so that for a minute all she said again was: 'I think of Sir Nicholas. I wish he were here'; addressing the words to Julia, who had wandered forward without looking at the mother and son.

'Poor Sir Nicholas!' said Mrs Dallow, vaguely.

'Did you make another speech?' Lady Agnes asked.

'I don't know; did I?' Nick inquired.

'I don't know!' Mrs Dallow replied, with her back turned, doing something to her hat before the glass.

'Oh, I can fancy the confusion, the bewilderment!' said Lady Agnes, in a tone rich in political reminiscence.

'It was really immense fun!' exclaimed Mrs Dallow.

'Dear Julia!' Lady Agnes went on. Then she added: 'It was you who made it sure.'

'There are a lot of people coming to dinner,' said Julia.

'Perhaps you'll have to speak again,' Lady Agnes smiled at her son.

'Thank you; I like the way you talk about it!' cried Nick. 'I'm like Iago: "from this time forth I never will speak word!"'

'Don't say that, Nick,' said his mother, gravely.

'Don't be afraid: he'll jabber like a magpie!' And Mrs Dallow went out of the room.

Nick had flung himself upon a sofa with an air of weariness, though not of completely vanished cheer; and Lady Agnes stood before him fingering her rose and looking down at him. His eyes looked away from hers: they seemed fixed on something she couldn't see. 'I hope you've thanked Julia,' Lady Agnes dropped.

'Why, of course, mother.'

'She has done as much as if you hadn't been sure.'

'I wasn't in the least sure – and she has done everything.'

'She has been too good – but *we*'ve done something. I hope you don't leave out your father,' Lady Agnes amplified, as Nick's glance appeared for a moment to question her 'we'.

'Never, never!' Nick uttered these words perhaps a little

mechanically, but the next minute he continued, as if he had suddenly been moved to think what he could say that would give his mother most pleasure: 'Of course his name has worked for me. Gone as he is, he is still a living force.' He felt a good deal of a hypocrite, but one didn't win a seat every day in the year. Probably indeed he should never win another.

'He hears you, he watches you, he rejoices in you,' Lady Agnes declared.

This idea was oppressive to Nick – that of the rejoicing almost as much as of the watching. He had made his concession, but, with a certain impulse to divert his mother from following up her advantage, he broke out 'Julia's a tremendously effective woman.'

'Of course she is!' answered Lady Agnes, knowingly.

'Her charming appearance is half the battle,' said Nick, explaining a little coldly what he meant. But he felt that his coldness was an inadequate protection to him when he heard his mother observe, with something of the same sapience –

'A woman is always effective when she likes a person.'

It discomposed him to be described as a person liked, and by a woman; and he asked abruptly: 'When are you going away?'

'The first moment that's civil – tomorrow morning. *You*'ll stay here, I hope.'

'Stay? What shall I stay for?'

'Why, you might stay to thank her.'

'I have everything to do.'

'I thought everything was done,' said Lady Agnes.

'Well, that's why,' her son replied, not very lucidly. 'I want to do other things – quite other things. I should like to take the next train.' And Nick looked at his watch.

'When there are people coming to dinner to meet you?'

'They'll meet *you* – that's better.'

'I'm sorry any one is coming,' Lady Agnes said, in a tone unencouraging to a deviation from the intensity of things. 'I wish we were alone – just as a family. It would please Julia today to feel that we *are* one. Do stay with her tomorrow.'

'How will that do, when she's alone?'

'She won't be alone, with Mrs Gresham.'

'Mrs Gresham doesn't count.'

'That's precisely why I want you to stop. And her cousin, almost her brother: what an idea that it won't do! Haven't you stayed here before, when there has been no one?'

'I have never stayed much, and there have always been people. At any rate, now it's different.'

'It's just because it is different. Besides, it isn't different, and it never was,' said Lady Agnes, more incoherent, in her earnestness, than it often happened to her to be. 'She always liked you, and she likes you now more than ever, if you call *that* different!' Nick got up at this and, without meeting her eyes, walked to one of the windows, where he stood with his back turned, looking out on the great greenness. She watched him a moment and she might well have been wishing, while he remained gazing there, as it appeared, that it would come to him with the same force as it had come to herself (very often before, but during these last days more than ever), that the level lands of Harsh, stretching away before the window, the French garden, with its symmetry, its screens and its statues, and a great many more things, of which these were the superficial token, were Julia's very own, to do with exactly as she liked. No word of appreciation or envy, however, dropped from the young man's lips, and his mother presently went on: 'What could be more natural than that after your triumphant contest you and she should have lots to settle and to talk about – no end of practical questions, no end of business? Aren't you her member, and can't her member pass a day with her, and she a great proprietor?'

Nick turned round at this, with an odd expression. '*Her* member – am I hers?'

Lady Agnes hesitated a moment; she felt that she had need of all her tact. 'Well, if the place is hers, and you represent the place –' she began. But she went no further, for Nick interrupted her with a laugh.

'What a droll thing to "represent", when one thinks of it! And what does *it* represent, poor torpid little borough, with its smell of meal and its curiously fat-faced inhabitants? Did you ever see such a collection of fat faces, turned up at the hustings? They looked like an enormous sofa, with the cheeks for the gathers and the eyes for the buttons.'

'Oh, well, the next time you shall have a great town,' Lady Agnes replied, smiling and feeling that she *was* tactful.

'It will only be a bigger sofa! I'm joking, of course,' Nick went on, 'and I ought to be ashamed of myself. They have done me the honour to elect me, and I shall never say a word that's not civil about them, poor dears. But even a new member may joke with his mother.'

'I wish you'd be serious with your mother,' said Lady Agnes, going nearer to him.

'The difficulty is that I'm two men; it's the strangest thing that ever was,' Nick pursued, bending his bright face upon her. 'I'm two quite distinct human beings, who have scarcely a point in common; not even the memory, on the part of one, of the achievements or the adventures of the other. One man wins the seat – but it's the other fellow who sits in it.'

'Oh, Nick, don't spoil your victory by your perversity!' Lady Agnes cried, clasping her hands to him.

'I went through it with great glee – I won't deny that: it excited me, it interested me, it amused me. When once I was in it I liked it. But now that I'm out of it again –'

'Out of it?' His mother stared. 'Isn't the whole point that you're in?'

'Ah, now I'm only in the House of Commons.'

For an instant Lady Agnes seemed not to understand and to be on the point of laying her finger quickly to her lips with a 'Hush!' as if the late Sir Nicholas might have heard the 'only'. Then, as if a comprehension of the young man's words promptly superseded that impulse, she replied with force: 'You will be in the Lords the day you determine to get there.'

This futile remark made Nick laugh afresh, and not only laugh but kiss her, which was always an intenser form of mystification for poor Lady Agnes, and apparently the one he liked best to practise; after which he said: 'The odd thing is, you know, that Harsh has no wants. At least it's not sharply, not eloquently conscious of them. We all talked them over together, and I promised to carry them in my heart of hearts. But upon my word I can't remember one of them. Julia says the wants of Harsh are simply the national wants – rather a pretty phrase for Julia. She means *she* does everything for the place; *she's* really their member, and this house in which we stand is their legislative chamber. Therefore the *lacunæ* that I have undertaken to fill up are the national wants. It will be rather a job to rectify some of them, won't it? I don't represent the appetites of Harsh – Harsh is gorged. I represent the ideas of my party. That's what Julia says.'

'Oh, never mind what Julia says!' Lady Agnes broke out, impatiently. This impatience made it singular that the very next words she uttered should be: 'My dearest son, I wish to heaven you'd marry her. It would be so fitting now!' she added.

'Why now?' asked Nick, frowning.

'She has shown you such sympathy, such devotion.'

'Is it for that she has shown it?'

'Ah, you might *feel* – I can't tell you!' said Lady Agnes, reproachfully.

Nick blushed at this, as if what he did feel was the reproach. 'Must I marry her because you like her?'

'I? Why, we are *all* as fond of her as we can be.'

'Dear mother, I hope that any woman I ever may marry will be a person agreeable not only to you, but also, since you make a point of it, to Grace and Biddy. But I must tell you this – that I shall marry no woman I am not unmistakably in love with.'

'And why are you not in love with Julia – charming, clever, generous as she is?' Lady Agnes laid her hands on him – she held him tight. 'My darling Nick, if you care anything in the world to make me happy, you'll stay over here tomorrow and be nice to her.'

'Be nice to her? Do you mean propose to her?'

'With a single word, with the glance of an eye, the movement of your little finger' – and Lady Agnes paused, looking intensely, imploringly up into Nick's face – 'In less time than it takes me to say what I say now, you may have it all.' As he made no answer, only returning her look, she added insistently, 'You know she's a fine creature – you know she is!'

'Dearest mother, what I seem to know better than anything else in the world is that I love my freedom. I set it far above everything.'

'Your freedom? What freedom is there in being poor? Talk

of that when Julia puts everything that she possesses at your feet!'

'I can't talk of it, mother – it's too terrible an idea. And I can't talk of *her*, nor of what I think of her. You must leave that to me. I do her perfect justice.'

'You don't, or you'd marry her tomorrow. You would feel that the opportunity is exquisitely rare, with everything in the world to make it perfect. Your father would have valued it for you beyond everything. Think a little what would have given *him* pleasure. That's what I meant when I spoke just now of us all. It wasn't of Grace and Biddy I was thinking – fancy! – it was of him. He's with you always; he takes with you, at your side, every step that you take yourself. He would bless devoutly your marriage to Julia; he would feel that it would be for you and for us all. I ask for no sacrifice, and he would ask for none. We only ask that you don't commit the crime –'

Nick Dormer stopped her with another kiss; he murmured: 'Mother, mother, mother!' as he bent over her.

He wished her not to go on, to let him off; but the deep deprecation in his voice did not prevent her saying: 'You know it – you know it perfectly. All, and more than all that I can tell you, you know.'

He drew her closer, kissed her again, held her there as he would have held a child in a paroxysm, soothing her silently till it should pass away. Her emotion had brought the tears to her eyes; she dried them as she disengaged herself. The next moment, however, she resumed, attacking him again:

'For a public man she would be the ideal companion. She's made for public life; she's made to shine, to be concerned in great things, to occupy a high position and to help him on. She would help you in everything as she has helped you in this. Together there is nothing you couldn't do. You can have the first house in England – yes, the first! What freedom *is* there in being poor? How can you do anything without money, and what money can you make for yourself – what money will ever come to you? That's the crime – to throw away such an instrument of power, such a blessed instrument of good.'

'It isn't everything to be rich, mother,' said Nick, looking at the floor in a certain patient way, with a provisional docility and his hands in his pockets. 'And it isn't so fearful to be poor.'

'It's vile – it's abject. Don't I know?'

'Are you in such acute want?' Nick asked, smiling.

'Ah, don't make me explain what you have only to look at to see!' his mother returned, as if with a richness of allusion to dark elements in her fate.

'Besides,' Nick went on, 'there's other money in the world than Julia's. I might come by some of that.'

'Do you mean Mr Carteret's?' The question made him laugh, as her feeble reference, five minutes before, to the House of Lords had done. But she pursued, too full of her idea to take account of such a poor substitute for an answer: 'Let me tell you one thing, for I have known Charles Carteret much longer than you, and I understand him better. There's nothing you could do that would do you more good with him than to marry Julia. I know the way he looks at things and I know exactly how that would strike him. It would please him, it would charm him; it would be the thing that would most prove to him that you're in earnest. You need to do something of that sort.'

'Haven't I come in for Harsh?' asked Nick.

'Oh, he's very canny. He likes to see people rich. *Then* he believes in them – then he's likely to believe more. He's kind to you because you're your father's son; but I'm sure your being poor takes just so much off.'

'He can remedy that so easily,' said Nick, smiling still. 'Is being kept by Julia what you call making an effort for myself?'

Lady Agnes hesitated; then: 'You needn't insult Julia!' she replied.

'Moreover, if I've *her* money, I shan't want his,' Nick hinted.

Again his mother waited an instant before answering; after which she produced: 'And pray wouldn't you wish to be independent?'

'You're delightful, dear mother – you're very delightful! I particularly like your conception of independence. Doesn't it occur to you that at a pinch I might improve my fortune by some other means than by making a mercenary marriage or by currying favour with a rich old gentleman? Doesn't it occur to you that I might work?'

'Work at politics? How does that make money, honourably?'

'I don't mean at politics.'

'What do you mean, then?' Lady Agnes demanded, looking at him as if she challenged him to phrase it if he dared. Her eye appeared to have a certain effect upon him, for he remained silent, and she continued: 'Are you elected or not?'

'It seems a dream,' said Nick.

'If you are, act accordingly and don't mix up things that are as wide asunder as the poles!' She spoke with sternness, and his silence might have been an admission that her sternness was wholesome to him. Possibly she was touched by it; at any rate, after a few moments, during which nothing more passed between them, she appealed to him in a gentler and more anxious key, which had this virtue to touch him, that he knew it was absolutely the first time in her life Lady Agnes had begged for anything. She had never been obliged to beg; she had got on without it and most things had come to her. He might judge therefore in what a light she regarded this boon for which, in her old age, she humbled herself to be a suitor. There was such a pride in her that he could feel what it cost her to go on her knees even to her son. He did judge how it was in his power to gratify her; and as he was generous and imaginative he was stirred and shaken as it came over him in a wave of figurative suggestion that he might make up to her for many things. He scarcely needed to hear her ask, with a pleading wail that was almost tragic: 'Don't you see how things have turned out for us; don't you know how unhappy I am – don't you know what a bitterness –?' She stopped for a moment, with a sob in her voice, and he recognized vividly this last tribulation, the unhealed wound of her bereavement and the way she had sunken from eminence to flatness. 'You know what Percival is and the comfort I have from him. You know the property and what he is doing with it and what comfort I get from *that*! Everything's dreary but what you can do for us. Everything's odious, down to living in a hole with one's girls who don't marry. Grace is impossible – I don't know what's the matter with her; no one will look at her, and she's so conceited with it – sometimes I feel as if I could beat her! And Biddy will never marry, and we are three dismal women in a filthy house. What are three dismal women, more or less, in London?'

So, with an unexpected rage of self-exposure, Lady Agnes talked of her disappointments and troubles, tore away the veil from her sadness and soreness. It almost frightened Nick to perceive how she hated her life, though at another time it might have amused him to note how she despised her gardenless house. Of course it was not a country-house, and Lady Agnes could not get used to that. Better than he could do – for it was the sort of thing into which, in any case, a woman enters more than a man – she felt what a lift into brighter air, what a regilding of his sisters' possibilities, his marriage to Julia would effect for them. He couldn't trace the difference, but his mother saw it all as a shining picture. She made the vision shine before him now, somehow, as she stood there like a poor woman crying for a kindness. What was filial in him, all the piety that he owed, especially to the revived spirit of his father, more than ever present on a day of such public pledges, was capable from one moment to the other of trembling into sympathetic response. He had the gift, so embarrassing when it is a question of consistent action, of seeing in an imaginative, interesting light anything that illustrated forcibly the life of another: such things effected a union with something in *his* life, and the recognition of them was ready to become a form of enthusiasm in which there was no consciousness of sacrifice – none scarcely of merit.

Rapidly, at present, this change of scene took place before his spiritual eye. He found himself believing, because his mother communicated the belief, that it was in his option to transform the social outlook of the three women who clung to him and who declared themselves dismal. This was not the highest kind of inspiration, but it was moving, and it associated itself with dim confusions of figures in the past – figures of authority and expectancy. Julia's wide kingdom opened out around him, making the future almost a dazzle of happy power. His mother and sisters floated in the rosy element with beaming faces, in transfigured safety. 'The first house in England,' she had called it; but it might be the first house in Europe, the first house in the world, by the fine air and the high humanities that should fill it. Everything that was beautiful in the place where he stood took on a more delicate charm; the house rose over his head like a museum of exquisite rewards, and the image of poor George

Dallow hovered there obsequious, as if to confess that he had only been the modest, tasteful forerunner, appointed to set it all in order and punctually retire. Lady Agnes's tone penetrated further into Nick's spirit than it had done yet, as she syllabled to him, supremely: 'Don't desert us – don't desert us.'

'Don't desert you?'

'Be great – be great,' said his mother. 'I'm old, I've lived, I've seen. Go in for a great material position. That will simplify everything else.'

'I will do what I can for you – anything, everything I can. Trust me – leave me alone,' said Nick Dormer.

'And you'll stay over – you'll spend the day with her?'

'I'll stay till she turns me out!'

His mother had hold of his hand again now; she raised it to her lips and kissed it. 'My dearest son, my only joy!' Then, 'I don't see how you can resist her,' she added.

'No more do I!'

Lady Agnes looked round the great room with a soft exhalation of gratitude and hope. 'If you're so fond of art, what art is equal to all this? The joy of living in the midst of it – of seeing the finest works every day! You'll have everything the world can give.'

'That's exactly what was just passing in my own mind. It's too much.'

'Don't be selfish!'

'Selfish?' Nick repeated.

'Don't be unselfish, then. You'll share it with us.'

'And with Julia a little, I hope,' said Nick.

'God bless you!' cried his mother, looking up at him. Her eyes were detained by the sudden perception of something in his own that was not clear to her; but before she had time to ask for an explanation of it Nick inquired, abruptly –

'Why do you talk so of poor Biddy? Why won't she marry?'

'You had better ask Peter Sherringham,' said Lady Agnes.

'What has he got to do with it?'

'How odd of you to ask, when it's so plain how she thinks of him that it's a matter of common chaff!'

'Yes, we've made it so, and she takes it like an angel. But Peter likes her.'

'Does he? Then it's the more shame to him to behave as he does. He had better leave his actresses alone. That's the love of art, too!' mocked Lady Agnes.

'Biddy's so charming – she'll marry some one else.'

'Never, if she loves him. But Julia will bring it about – Julia will help her,' said Lady Agnes, more cheerfully. 'That's what you'll do for us – that *she*'ll do everything!'

'Why then more than now?' Nick asked.

'Because we shall be yours.'

'You are mine already.'

'Yes, but *she* isn't. However, she's as good!' exulted Lady Agnes.

'She'll turn me out of the house,' said Nick.

'Come and tell me when she does! But there she is – go to her!' And she gave him a push towards one of the windows that stood open to the terrace. Mrs Dallow had become visible outside; she passed slowly along the terrace, with her long shadow. 'Go to her,' Lady Agnes repeated – 'she's waiting for you.'

Nick went out with the air of a man who was as ready to pass that way as any other, and at the same moment his two sisters, freshly restored from the excitements of the town, came into the room from another quarter.

'We go home tomorrow, but Nick will stay a day or two,' their mother said to them.

'Dear old Nick!' Grace ejaculated, looking at Lady Agnes.

'He's going to speak,' the latter went on. 'But don't mention it.'

'Don't mention it?' said Biddy, staring. 'Hasn't he spoken enough, poor fellow?'

'I mean to Julia,' Lady Agnes replied.

'Don't you understand, you goose?' Grace exclaimed to her sister.

14

THE next morning brought Nick Dormer many letters and telegrams, and his coffee was placed beside him in his room, where he remained until noon answering these communications. When

173

he came out he learned that his mother and sisters had left the house. This information was given him by Mrs. Gresham, whom he found dealing with her own voluminous budget at one of the tables in the library. She was a lady who received thirty letters a day, the subject-matter of which, as well as of her punctual answers, in a hand that would have been 'lady-like' in a manageress, was a puzzle to those who observed her.

She told Nick that Lady Agnes had not been willing to disturb him at his work to say good-bye, knowing she should see him in a day or two in town. Nick was amused at the way his mother had stolen off; as if she feared that further conversation might weaken the spell she believed herself to have wrought. The place was cleared, moreover, of its other visitors, so that, as Mrs Gresham said, the fun was at an end. This lady expressed the idea that the fun was after all rather heavy. At any rate now they could rest, Mrs Dallow and Nick and she, and she was glad Nick was going to stay for a little quiet. She liked Harsh best when it was not *en fête*: then one could see what a sympathetic old place it was. She hoped Nick was not dreadfully tired; she feared Julia was completely done up. Mrs Dallow, however, had transported her exhaustion to the grounds – she was wandering about somewhere. She thought more people would be coming to the house, people from the town, people from the country, and had gone out so as not to have to see them. She had not gone far – Nick could easily find her. Nick intimated that he himself was not eager for more people, whereupon Mrs Gresham said, rather archly smiling:

'And of course you hate *me* for being here!' He made some protest, and she added: 'But I'm almost a part of the house, you know – I'm one of the chairs or tables.' Nick declared that he had never seen a house so well furnished, and Mrs Gresham said: 'I believe there *are* to be some people to dinner: rather an interference, isn't it? Julia lives so in public. But it's all for you.' And after a moment she added: 'It's a wonderful constitution.' Nick at first failed to seize her allusion – he thought it a retarded political reference, a sudden tribute to the great unwritten instrument by which they were all governed. He was on the point of saying: 'The British? Wonderful!' when he perceived that the intention of his interlocutress was to praise Mrs Dallow's fine robustness. 'The surface so delicate, the action so easy, yet the frame of steel.'

Nick left Mrs Gresham to her correspondence and went out of the house; wondering as he walked whether she wanted him to do the same thing that his mother wanted, so that her words had been intended for a prick – whether even the two ladies had talked over their desire together. Mrs Gresham was a married woman who was usually taken for a widow; mainly because she was perpetually 'sent for' by her friends, and her friends never sent for Mr Gresham. She came in every case and had the air of being *répandue* at the expense of dingier belongings. Her figure was admired – that is it was sometimes mentioned – and she dressed as if it was expected of her to be smart, like a young woman in a shop or a servant much in view. She slipped in and out, accompanied at the piano, talking to the neglected visitors, walked in the rain and, after the arrival of the post, usually had conferences with her hostess, during which she stroked her chin and looked familiarly responsible. It was her peculiarity that people were always saying things to her in a lowered voice. She had all sorts of acquaintances and in small establishments she sometimes wrote the *menus*. Great ones, on the other hand, had no terrors for her: she had seen too many. No one had ever discovered whether any one else paid her.

If Lady Agnes, in a lowered tone, had discussed with her the propriety of a union between the mistress of Harsh and the hope of the Dormers our young man could take the circumstance for granted without irritation and even with cursory indulgence: for he was not unhappy now and his spirit was light and clear. The summer day was splendid and the world, as he looked at it from the terrace, offered no more worrying ambiguity than a vault of airy blue arching over a lap of solid green. The wide, still trees in the park appeared to be waiting for some daily inspection, and the rich fields, with their official frill of hedges, to rejoice in the light which approved them as named and numbered acres. The place looked happy to Nick, and he was struck with its having a charm to which he had perhaps not hitherto done justice; something of the impression that he had received, when he was younger, from showy 'views' of fine country-seats, as if they had been brighter and more established than life. There were a couple of peacocks on the terrace, and his eye was caught by the gleam of the swans on a distant lake, where there was also a little temple on an island; and these objects fell in with his humour, which at another time might have been ruffled by them as representing the Philistine in ornament.

It was certainly a proof of youth and health on his part that his spirits had risen as the tumult rose and that after he had taken his jump into the turbid waters of a contested election he had been able to tumble and splash, not only with a sense of awkwardness but with a considerable capacity for the frolic. Tepid as we saw him in Paris he had found his relation to his opportunity surprisingly altered by his little journey across the Channel. He saw things in a new perspective and he breathed an air that excited him unexpectedly. There was something in it that went to his head – an element that his mother and his sisters, his father from beyond the grave, Julia Dallow, the Liberal party and a hundred friends were both secretly and overtly occupied in pumping into it. If he was vague about success he liked the fray, and he had a general rule that when one was in a muddle there was refreshment in action. The embarrassment, that is the revival of scepticism, which might produce an inconsistency shameful to exhibit and yet very difficult to conceal, was safe enough to come later: indeed at the risk of making our young man appear a purely whimsical personage I may hint that some such sickly glow had even now begun to colour one quarter of his mental horizon.

I am afraid moreover that I have no better excuse for him than the one he had touched on in the momentous conversation with his mother which I have thought it useful to reproduce in full. He was conscious of a double nature; there were two men in him, quite separate, whose leading features had little in common and each of whom insisted on having an independent turn at life. Meanwhile if he was adequately aware that the bed of his moral existence would need a good deal of making over if he was to lie upon it without unseemly tossing, he was also alive to the propriety of not parading his inconsistencies, not letting his unrectified interests become a spectacle to the vulgar. He had none of that wish to appear complicated which is at the bottom of most forms of fatuity; he was perfectly willing to pass as simple; he only aspired to be continuous. If you were not really simple this presented difficulties; but he would have assented to the proposition that you must be as final as you can and that it contributes much to finality to consume the smoke of your inner fire. The fire was the great thing and not the chimney. He had no view of life which counted out the need of learning; it was teaching rather as to which he was conscious of no particular mission. He liked life, liked it immensely, and was willing to study the ways and means of it with a certain patience. He cherished the usual wise monitions, such as that one was not to make a fool of one's self and that one should not carry on one's technical experiments in public. It was because as yet he liked life in general better than it was clear to him that he liked any particular branch of it, that on the occasion of a constituency's holding out a cordial hand to him, while it extended another in a different direction, a certain bloom of boyhood that was on him had not resisted the idea of a match.

He rose to it as he had risen to matches at school, for his boyishness could take a pleasure in an inconsiderate show of agility. He could meet electors and conciliate bores and compliment women and answer questions and roll off speeches and chaff adversaries, because it was amusing and slightly dangerous, like playing football or ascending an Alp – pastimes for which nature had conferred on him an aptitude not so very different in kind from a gallant readiness on platforms. There were two voices which told him that all this was not really action at all, but only a pusillanimous imitation of it: one of them made itself fitfully audible in the depths of his own spirit and the other spoke, in the equivocal accents of a very crabbed hand, from a letter of four pages by Gabriel Nash. However, Nick acted as much as possible under the circumstances, and that was rectifying – it brought with it enjoyment and a working faith. He had not gone counter to the axiom that in a case of doubt one was to hold off; for that applied to choice, and he had not at present the slightest pretension to choosing. He knew he was lifted along, that what he was doing was not first-rate, that nothing was settled by it and that if there was essentially a problem in his life it would only grow tougher with keeping. But if doing one's sum tomorrow instead of today does not make the sum easier it at least makes today so.

Sometimes in the course of the following fortnight it seemed to him that he had gone in for Harsh because he was sure he should lose; sometimes he foresaw that he should win precisely to punish him for having tried and for his want of candour; and when presently he did win he was almost frightened at his success. Then it appeared to him that he had done something even worse than not choose – he had let others choose for him. The beauty of it was that they had chosen with only their own object in their eye: for what did they know about his strange alternative? He was rattled about so for a fortnight (Julia took care of that) that he had no time to think save when he tried to remember a quotation or an American story, and all his life became an overflow of verbiage. Thought retreated before increase of sound, which had to be pleasant and eloquent, and even superficially coherent, without its aid. Nick himself was surprised at the airs he could play; and often when the last thing at night he shut the door of his room he mentally exclaimed that he had had no idea he was such a mountebank.

I must add that if this reflection did not occupy him long, and if no meditation, after his return from Paris, held him for many moments, there was a reason better even than that he was tired or busy or excited by the agreeable combination of hits and hurrahs. That reason was simply Mrs Dallow, who had suddenly become a still larger fact in his consciousness than active politics. She *was*, indeed, active politics; that is, if the politics were his, how little soever, the activity was hers. She had ways of showing she was a clever woman that were better than saying clever things, which only prove at the most that one would be clever if one could. The accomplished fact itself was the demonstration that Mrs Dallow could; and when Nick came to his senses after the proclamation of the victor and the cessation of the noise her figure was, of all the queer phantasmagoria, the most substantial thing that survived. She had been always there, passing, repassing, before him, beside him, behind him. She had made the business infinitely prettier than it would have been without her, added music and flowers and ices, a charm, and converted it into a social game that had a strain of the heroic.

It was a garden-party with something at stake, or to celebrate something in advance, with the people let in. The concluded affair had bequeathed him not only a seat in the House of Commons, but a perception of what women may do in high embodiments, and an abyss of intimacy with one woman in particular.

She had wrapped him up in something, he didn't know what – a sense of facility, an overpowering fragrance – and they had moved together in an immense fraternity. There had been no love-making, no contact that was only personal, no vulgarity of flirtation: the hurry of the days and the sharpness with which they both tended to an outside object had made all that irrelevant. It was as if she had been too near for him to see her separate from himself; but none the less, when he now drew breath and looked back, what had happened met his eyes as a composed picture – a picture of which the subject was inveterately Julia and her ponies: Julia wonderfully fair and fine, holding her head more than ever in the manner characteristic of her, brilliant, benignant, waving her whip, cleaving the crowd, thanking people with her smile, carrying him beside her, carrying him to his doom. He had not supposed that in so few days he had driven about with her so much; but the image of it was there, in his consulted conscience, as well as in a personal glow not yet chilled: it looked large as it rose before him. The things his mother had said to him made a rich enough frame for it, and the whole impression, that night, had kept him much awake.

15

WHILE, after leaving Mrs Gresham, he was hesitating which way to go and was on the point of hailing a gardener to ask if Mrs Dallow had been seen, he noticed, as a spot of colour in an expanse of shrubbery, a far-away parasol moving in the direction of the lake. He took his course that way, across the park, and as the bearer of the parasol was strolling slowly it was not five minutes before he had joined her. He went to her soundlessly over the grass (he had been whistling at first, but as he got nearer he stopped), and it was not till he was close to her that she looked round. He had watched her moving as if she were turning things over in her mind, brushing the smooth walks and the clean turf with her dress, slowly making her parasol revolve on her shoulder and carrying in the hand which hung beside her a book which he perceived to be a monthly review.

'I came out to get away,' she remarked when he had begun to walk with her.

'Away from me?'

'Ah, that's impossible,' said Mrs Dallow. Then she added: 'The day is so nice.'

'Lovely weather,' Nick dropped. 'You want to get away from Mrs Gresham, I suppose.'

Mrs Dallow was silent a moment. 'From everything!'

'Well, I want to get away too.'

'It has been such a racket. Listen to the dear birds.'

'Yes, our noise isn't so good as theirs,' said Nick. 'I feel as if I had been married and had shoes and rice thrown after me,' he went on. 'But not to you, Julia – nothing so good as that.'

Mrs Dallow made no answer to this; she only turned her eyes on the ornamental water which stretched away at their right. In a moment she exclaimed: 'How nasty the lake looks!' and Nick recognized in the tone of the words a manifestation of that odd shyness – a perverse stiffness at a moment when she probably only wanted to be soft – which, taken in combination with her other qualities, was so far from being displeasing to him that it represented her nearest approach to extreme charm. *He* was not shy now, for he considered, this morning, that he saw things very straight and in a sense altogether superior and delightful. This enabled him to be generously sorry for his companion, if he were the reason of her being in any degree uncomfortable, and yet left him to enjoy the prettiness of some of the signs by which her discomfort was revealed. He would not insist on anything yet: so he observed that his cousin's standard in lakes was too high, and then talked a little about his mother and the girls, their having gone home, his not having seen them that morning, Lady Agnes's deep satisfaction in his victory and the fact that she would be obliged to 'do something' for the autumn – take a house or something.

'I'll lend her a house,' said Mrs Dallow.

'Oh, Julia, Julia!' Nick exclaimed.

But Mrs Dallow paid no attention to his exclamation; she only held up her review and said: 'See what I have brought with me to read – Mr Hoppus's article.'

'That's right; then *I* shan't have to. You'll tell me about it.' He uttered this without believing that she had meant or wished to read the article, which was entitled 'The Revision of the British Constitution', in spite of her having encumbered herself with the stiff, fresh magazine. He was conscious that she was not in want of such mental occupation as periodical literature could supply. They walked along and then he added: 'But *is* that what we are in for – reading Mr Hoppus? Is that the sort of thing our constituents expect? Or even worse, pretending to have read him when one hasn't? Oh, what a tangled web we weave!'

'People are talking about it. One has to know. It's the article of the month.'

Nick looked at his companion askance a moment. 'You say things every now and then for which I could kill you. "The article of the month", for instance: I could kill you for that.'

'Well, kill me!' Mrs Dallow exclaimed.

'Let me carry your book,' Nick rejoined, irrelevantly. The hand in which she held it was on the side of her on which he was walking, and he put out his own hand to take it. But for a couple of minutes she forbore to give it up, and they held it together, swinging it a little. Before she surrendered it he inquired where she was going.

'To the island,' she answered.

'Well, I'll go with you – I'll kill you there.'

'The things I say are the right things,' said Mrs Dallow.

'It's just the right things that are wrong. It's because you're so political,' Nick went on. 'It's your horrible ambition. The woman who has a salon should have read the article of the month. See how one dreadful thing leads to another.'

'There are some things that lead to nothing.'

'No doubt – no doubt. And how are you going to get over to your island?'

'I don't know.'

'Isn't there a boat?'

'I don't know.'

Nick had paused a moment, to look round for the boat, but

Mrs Dallow walked on without turning her head. 'Can you row?' her companion asked.

'Don't you know I can do everything?'

'Yes, to be sure. That's why I want to kill you. There's the boat.'

'Shall you drown me?'

'Oh, let me perish with you!' Nick answered with a sigh. The boat had been hidden from them by the bole of a great tree, which rose from the grass at the water's-edge. It was moored to a small place of embarkation and was large enough to hold as many persons as were likely to wish to visit at once the little temple in the middle of the lake, which Nick liked because it was absurd and Mrs Dallow had never had a particular esteem for. The lake, fed by a natural spring, was a liberal sheet of water, measured by the scale of park scenery; and though its principal merit was that, taken at a distance, it gave a gleam of abstraction to the concrete verdure, doing the office of an open eye in a dull face, it could also be approached without derision on a sweet summer morning, when it made a lapping sound and reflected candidly various things that were probably finer than itself – the sky, the great trees, the flight of birds.

A man of taste, a hundred years before, coming back from Rome, had caused a small ornamental structure to be erected, on artificial foundations, on its bosom, and had endeavoured to make this architectural pleasantry as nearly as possible a reminiscence of the small ruined rotunda which stands on the bank of the Tiber and is declared by *ciceroni* to have been dedicated to Vesta. It was circular, it was roofed with old tiles, it was surrounded by white columns and it was considerably dilapidated. George Dallow had taken an interest in it (it reminded him not in the least of Rome, but of other things that he liked), and had amused himself with restoring it.

'Give me your hand; sit there, and I'll ferry you,' Nick Dormer said.

Mrs Dallow complied, placing herself opposite to him in the boat; but as he took up the paddles she declared that she preferred to remain on the water – there was too much malice prepense in the temple. He asked her what she meant by that, and she said it was ridiculous to withdraw to an island a few feet square on purpose to meditate. She had nothing to meditate about which required so much attitude.

'On the contrary, it would be just to change the *pose*. It's what we have been doing for a week that's attitude; and to be for half an hour where nobody's looking and one hasn't to keep it up is just what I wanted to put in an idle, irresponsible day for. I am not keeping it up now – I suppose you've noticed,' Nick went on, as they floated and he scarcely dipped the oars.

'I don't understand you,' said Mrs Dallow, leaning back in the boat.

Nick gave no further explanation than to ask in a minute: 'Have you people to dinner tonight?'

'I believe there are three or four, but I'll put them off if you like.'

'Must you *always* live in public, Julia?' Nick continued.

She looked at him a moment, and he could see that she coloured slightly. 'We'll go home – I'll put them off.'

'Ah no, don't go home; it's too jolly here. Let them come – let them come, poor wretches!'

'How little you know me, when, ever so many times, I have lived here for months without a creature.'

'Except Mrs Gresham, I suppose.'

'I have had to have the house going, I admit.'

'You are perfect, you are admirable, and I don't criticize you.'

'I don't understand you!' she tossed back.

'That only adds to the generosity of what you have done for me,' Nick returned, beginning to pull faster. He bent over the oars and sent the boat forward, keeping this up for ten minutes, during which they both remained silent. His companion, in her place, motionless, reclining (the seat in the stern was very comfortable), looked only at the water, the sky, the trees. At last Nick headed for the little temple, saying first however: 'Shan't we visit the ruin?'

'If you like. I don't mind seeing how they keep it.'

They reached the white steps which led up to it. Nick held the boat and Mrs Dallow got out. He fastened the boat and they went up the steps together, passing through the open door.

'They keep it very well,' Nick said, looking round. 'It's a capital place to give up everything.'

'It might do for you to explain what you mean,' said Julia, sitting down.

'I mean to pretend for half an hour that I don't represent the burgesses of Harsh. It's charming – it's very delicate work. Surely it has been retouched.'

The interior of the pavilion, lighted by windows which the circle of columns was supposed, outside and at a distance, to conceal, had a vaulted ceiling and was occupied by a few pieces of last-century furniture, spare and faded, of which the colours matched with the decoration of the walls. These and the ceiling, tinted and not exempt from indications of damp, were covered with fine mouldings and medallions. It was a very elegant little tea-house.

Mrs Dallow sat on the edge of a sofa, rolling her parasol and remarking: 'You ought to read Mr Hoppus's article to me.'

'Why, is *this* your salon?' asked Nick, smiling.

'Why are you always talking of that? It's an invention of your own.'

'But isn't it the idea you care most about?'

Suddenly, nervously, Mrs Dallow put up her parasol and sat under it, as if she were not quite sensible of what she was doing. 'How much you know me! I don't care about anything – that you will ever guess.'

Nick Dormer wandered about the room, looking at various things it contained – the odd volumes on the tables, the bits of quaint china on the shelves. 'They keep it very well; you've got charming things.'

'They're supposed to come over every day and look after them.'

'They must come over in force.'

'Oh, no one knows.'

'It's spick and span. How well you have everything done!'

'I think you've some reason to say so,' said Mrs Dallow. Her parasol was down and she was again rolling it tight.

'But you're right about my not knowing you. Why were you so ready to do so much for me?'

He stopped in front of her and she looked up at him. Her eyes rested on his a minute; then she broke out: 'Why do you hate me so?'

'Was it because you like me personally?' Nick asked. 'You may think that an odd, or even an odious question; but isn't it natural, my wanting to know?'

'Oh, if you don't know!' Mrs Dallow exclaimed.

'It's a question of being sure.'

'Well, then, if you're not sure –'

'Was it done for me as a friend, as a man?'

'You're not a man; you're a child,' said his hostess, with a face that was cold, though she had been smiling the moment before.

'After all, I was a good candidate,' Nick went on.

'What do I care for candidates?'

'You're the most delightful woman, Julia,' said Nick, sitting down beside her, 'and I can't imagine what you mean by my hating you.'

'If you haven't discovered that I like you, you might as well.'

'Might as well discover it?'

Mrs Dallow was grave; he had never seen her so pale and never so beautiful. She had stopped rolling her parasol now: her hands were folded in her lap and her eyes were bent on them. Nick sat looking at them too, a trifle awkwardly. 'Might as well have hated me,' said Mrs Dallow.

'We have got on so beautifully together, all these days: why shouldn't we get on as well forever and ever?' Mrs Dallow made no answer, and suddenly Nick said to her: 'Ah, Julia, I don't know what you have done to me, but you've done it. You've done it by strange ways, but it will serve. Yes, I hate you,' he added, in a different tone, with his face nearer to hers.

'Dear Nick – dear Nick –' she began. But she stopped, for she suddenly felt that he was altogether nearer, nearer than he had ever been to her before, that his arm was round her, that he was in possession of her. She closed her eyes but she heard him ask: 'Why shouldn't it be forever, forever?' in a voice that had, for her ear, such a vibration as no voice had ever had.

'You've done it – you've done it,' Nick repeated.

'What do you want of me?' she demanded.

'To stay with me, this way, always.'

'Ah, not this way,' she answered, softly, but as if in pain, and making an effort, with a certain force, to detach herself.

'This way, then – or this!' He took such insistent advantage of her that he had quickly kissed her. She rose as quickly, but he held her yet, and while he did so he said to her in the same tender tone: 'If you'll marry me, why shouldn't it be so simple, so good?' He drew her closer again, too close for her to answer. But her struggle ceased and she rested upon him for a minute, she buried her face on his breast.

'You're hard, and it's cruel!' she then exclaimed, breaking away.

'Hard – cruel?'

'You do it with so little!' And with this, unexpectedly to Nick, Mrs Dallow burst straight into tears. Before he could stop her she was at the door of the pavilion, as if she wished to quit it immediately. There, however, he stopped her, bending over her while she sobbed, unspeakably gentle with her.

'So little? It's with everything – with everything I have.'

'I have done it, you say? What do you accuse me of doing?' Her tears were already over.

'Of making me yours; of being so precious, Julia, so exactly what a man wants, as it seems to me. I didn't know you could,' he went on, smiling down at her. 'I didn't – no, I didn't.'

'It's what I say – that you've always hated me.'

'I'll make it up to you.'

She leaned on the doorway with her head against the lintel. 'You don't even deny it.'

'Contradict you *now*? I'll admit it, though it's rubbish, on purpose to live it down.'

'It doesn't matter,' she said, slowly; 'for however much you might have liked me, you would never have done so half as much as I have cared for you.'

'Oh, I'm so poor!' Nick murmured, cheerfully.

She looked at him, smiling, and slowly shook her head. Then she declared: 'You never can.'

'I like that! Haven't I asked you to marry me? When did you ever ask me?'

'Every day of my life! As I say, it's hard – for a proud woman.'

'Yes, you're too proud even to answer me.'

'We must think of it, we must talk of it.'

'Think of it? I've thought of it ever so much.'

'I mean together. There are things to be said.'

'The principal thing is to give me your word.'

Mrs Dallow looked at him in silence; then she exclaimed: 'I wish I didn't adore you!' She went straight down the steps.

'You don't, if you leave me now. Why do you go? It's so charming here, and we are so delightfully alone.'

'Detach the boat; we'll go on the water,' said Mrs Dallow.

Nick was at the top of the steps, looking down at her. 'Ah, stay a little – *do* stay!' he pleaded.

'I'll get in myself, I'll put off,' she answered.

At this Nick came down and he bent a little to undo the rope. He was close to her, and as he raised his head he felt it caught; she had seized it in her hands and she pressed her lips to the first place they encountered. The next instant she was in the boat.

This time he dipped the oars very slowly indeed; and while, for a period that was longer than it seemed to them, they floated vaguely, they mainly sat and glowed at each other, as if everything had been settled. There were reasons enough why Nick should be happy; but it is a singular fact that the leading one was the sense of having escaped from a great mistake. The final result of his mother's appeal to him the day before had been the idea that he must act with unimpeachable honour. He was capable of taking it as an assurance that Julia had placed him under an obligation which a gentleman could regard only in one way. If *she* had understood it so, putting the vision, or at any rate the appreciation, of a closer tie into everything she had done for him, the case was conspicuously simple and his course unmistakably plain. That is why he had been gay when he came out of the house to look for her: he could be gay when his course was plain. He could be all the gayer, naturally, I must add, that in turning things over, as he had done half the night, what he had turned up oftenest was the recognition that Julia now had a new personal power over him. It was not for nothing that she had thrown herself personally into his life. She had by her act made him live twice as much, and such a service, if a man had accepted and deeply tasted it, was certainly a thing to put him on his honour. Nick gladly recognized that there was nothing he could do in preference that would not be spoiled for him by any deflection from that point. His mother had made him uncomfortable by intimating to him that Julia was in love with him (he didn't like, in general, to be told such things); but the responsibility seemed easier to carry and he was less shy about it when once he was away from other eyes, with only Julia's own to express that truth and with indifferent nature all around. Besides, what discovery had he made this morning but that he also was in love?

'You must be a very great man,' she said to him, in the middle of the lake. 'I don't know what you mean about my salon; but I *am* ambitious.'

'We must look at life in a large, bold way,' Nick replied, resting his oars.

'That's what I mean. If I didn't think you could I wouldn't look at you.'

'I could what?'

'Do everything you ought – everything I imagine, I dream of. You *are* clever: you can never make me believe the contrary, after your speech on Tuesday. Don't speak to me! I've seen, I've heard and I know what's in you. I shall hold you to it. You are everything that you pretend not to be.'

Nick sat looking at the water while she talked. 'Will it always be so amusing?' he asked.

'Will what always be?'

'Why, my career.'

'Shan't I make it so?'

'It will be yours; it won't be mine,' said Nick.

'Ah, don't say that: don't make me out that sort of woman! If they should say it's me, I'd drown myself.'

'If they should say what's you?'

'Why, you're getting on. If they should say I push you, that I do things for you.'

'Well, won't you do them? It's just what I count on.'

'Don't be dreadful,' said Mrs Dallow. 'It would be loathsome if I were said to be cleverer than you. That's not the sort of man I want to marry.'

'Oh, I shall make you work, my dear!'

'Ah, that!' exclaimed Mrs Dallow, in a tone that might come back to a man in after years.

'You will do the great thing, you will make my life delightful,' Nick declared, as if he fully perceived the sweetness of it. 'I dare say that will keep me in heart.'

'In heart? Why shouldn't you be in heart?' Julia's eyes, lingering on him, searching him, seemed to question him still more than her lips.

'Oh, it will be all right!' cried Nick.

'You'll like success, as well as any one else. Don't tell me – you're not so ethereal!'

'Yes, I shall like success.'

'So shall I! And of course I am glad that you'll be able to do things,' Mrs Dallow went on. 'I'm glad you'll have things. I'm glad I'm not poor.'

'Ah, don't speak of that,' Nick murmured. 'Only be nice to my mother: we shall make her supremely happy.'

'I'm glad I like your people,' Mrs Dallow dropped. 'Leave them to me!'

'You're generous – you're noble,' stammered Nick.

'Your mother must live at Broadwood; she must have it for life. It's not at all bad.'

'Ah, Julia,' her companion replied, 'it's well I love you!'

'Why shouldn't you?' laughed Julia; and after this there was nothing said between them till the boat touched the shore. When she had got out Mrs Dallow remarked that it was time for luncheon; but they took no action in consequence, strolling in a direction which was not that of the house. There was a vista that drew them on, a grassy path skirting the foundations of scattered beeches and leading to a stile from which the charmed wanderer might drop into another division of Mrs Dallow's property. This lady said something about their going as far as the stile; then the next instant she exclaimed: 'How stupid of you – you've forgotten Mr Hoppus!'

'We left him in the temple of Vesta. Darling, I had other things to think of there.'

'I'll send for him,' said Mrs Dallow.

'Lord, can you think of him now?' Nick asked.

'Of course I can – more than ever.'

'Shall we go back for him?' Nick inquired, pausing.

Mrs Dallow made no answer; she continued to walk, saying

they would go as far as the stile. 'Of course I know you're fearfully vague,' she presently resumed.

'I wasn't vague at all. But you were in such a hurry to get away.'

'It doesn't signify. I have another one at home.'

'Another summer-house?' suggested Nick.

'A copy of Mr Hoppus.'

'Mercy, how you go in for him! Fancy having two!'

'He sent me the number of the magazine; and the other is the one that comes every month.'

'Every month – I see,' said Nick, in a manner justifying considerably Mrs Dallow's charge of vagueness. They had reached the stile and he leaned over it, looking at a great mild meadow and at the browsing beasts in the distance.

'Did you suppose they come every day?' asked Mrs Dallow.

'Dear no, thank God!' They remained there a little; he continued to look at the animals and before long he added: 'Delightful English pastoral scene. Why do they say it won't paint?'

'Who says it won't?'

'I don't know – some of them. It will in France; but somehow it won't here.'

'What are you talking about?' Mrs Dallow demanded.

Nick appeared unable to satisfy her on this point; at any rate instead of answering her directly he said: 'Is Broadwood very charming?'

'Have you never been there? It shows how you've treated me. We used to go there in August. George had ideas about it,' added Mrs Dallow. She had never affected not to speak of her late husband, especially with Nick, whose kinsman in a manner he had been and who had liked him better than some others did.

'George had ideas about a great many things.'

Julia Dallow appeared to be conscious that it would be rather odd on such an occasion to take this up. It was even odd in Nick to have said it. 'Broadwood is just right,' she rejoined at last. 'It's neither too small nor too big, and it takes care of itself. There's nothing to be done: you can't spend a penny.'

'And don't you want to use it?'

'We can go and stay with *them*,' said Mrs Dallow.

'They'll think I bring them an angel.' And Nick covered her hand, which was resting on the stile, with his own large one.

'As they regard you yourself as an angel they will take it as natural of you to associate with your kind.'

'Oh, *my* kind!' murmured Nick, looking at the cows.

Mrs Dallow turned away from him as if she were starting homeward, and he began to retrace his steps with her. Suddenly she said: 'What did you mean that night in Paris?'

'That night?'

'When you came to the hotel with me, after we had all dined at that place with Peter.'

'What did I mean?'

'About your caring so much for the fine arts. You seemed to want to frighten me.'

'Why should you have been frightened? I can't imagine what I had in my head: not now.'

'You *are* vague,' said Julia, with a little flush.

'Not about the great thing.'

'The great thing?'

'That I owe you everything an honest man has to offer. How can I care about the fine arts now?'

Mrs Dallow stopped, looking at him. 'Is it because you think you *owe* it –' and she paused, still with the heightened colour in her cheek; then she went on – 'that you have spoken to me as you did there?' She tossed her head towards the lake.

'I think I spoke to you because I couldn't help it.'

'You *are* vague.' And Mrs Dallow walked on again.

'You affect me differently from any other woman.'

'Oh, other women! Why shouldn't you care about the fine arts now?' she added.

'There will be no time. All my days and my years will be none too much to do what you expect of me.'

'I don't expect you to give up anything. I only expect you to do more.'

'To do more I must do less. I have no talent.'

'No talent?'

'I mean for painting.'

Mrs Dallow stopped again. 'That's odious! You *have* – you must.'

191

Nick burst out laughing. 'You're altogether delightful. But how little you know about it – about the honourable practice of any art!'

'What do you call practice? You'll have all our things – you'll live in the midst of them.'

'Certainly I shall enjoy looking at them, being so near them.'

'Don't say I've taken you away then.'

'Taken me away?'

'From the love of art. I like them myself now, poor George's treasures. I didn't, of old, so much, because it seemed to me he made too much of them – he was always talking.'

'Well, I won't talk,' said Nick.

'You may do as you like – they're yours.'

'Give them to the nation,' Nick went on.

'I like that! When we have done with them.'

'We shall have done with them when your Vandykes and Moronis have cured me of the delusion that I may be of *their* family. Surely that won't take long.'

'You shall paint *me*,' said Julia.

'Never, never, never!' Nick uttered these words in a tone that made his companion stare; and he appeared slightly embarrassed at this result of his emphasis. To relieve himself he said, as they had come back to the place beside the lake where the boat was moored: 'Shan't we really go and fetch Mr Hoppus?'

She hesitated. 'You may go; I won't, please.'

'That's not what I want.'

'Oblige me by going. I'll wait here.' With which Mrs Dallow sat down on the bench attached to the little landing.

Nick, at this, got into the boat and put off; he smiled at her as she sat there watching him. He made his short journey, disembarked and went into the pavilion; but when he came out with the object of his errand he saw that Mrs Dallow had quitted her station – she had returned to the house without him. He rowed back quickly, sprang ashore and followed her with long steps. Apparently she had gone fast; she had almost reached the door when he overtook her.

'Why did you basely desert me?' he asked, stopping her there.

'I don't know. Because I'm so happy.'

'May I tell mother?'
'You may tell her she shall have Broadwood.'

16

NICK lost no time in going down to see Mr Carteret, to whom he had written immediately after the election and who had answered him in twelve revised pages of historical parallel. He used often to envy Mr Carteret's leisure, a sense of which came to him now afresh, in the summer evening, as he walked up the hill towards the quiet house where enjoyment, for him, had ever been mingled with a vague oppression. He was a little boy again, under Mr Carteret's roof – a little boy on whom it had been duly impressed that in the wide, plain, peaceful rooms he was not to 'touch'. When he paid a visit to his father's old friend there were in fact many things – many topics – from which he instinctively kept his hands. Even Mr Chayter, the immemorial blank butler, who was so like his master that he might have been a twin brother, helped to remind him that he must be good. Mr Carteret seemed to Nick a very grave person, but he had the sense that Chayter thought him rather frivolous.

Our young man always came on foot from the station, leaving his portmanteau to be carried: the direct way was steep and he liked the slow approach, which gave him a chance to look about the place and smell the new-mown hay. At this season the air was full of it – the fields were so near that it was in the small, empty streets. Nick would never have thought of rattling up to Mr Carteret's door. It had an old brass plate with his name, as if he had been the principal surgeon. The house was in the high part, and the neat roofs of other houses, lower down the hill, made an immediate prospect for it, scarcely counting however, for the green country was just below these, familiar and inter-penetrating, in the shape of small but thick-tufted gardens. There was something growing in all the intervals, and the only disorder of the place was that there were sometimes oats on the pavements. A crooked lane, very clean, with cobblestones, opened opposite to Mr Carteret's house and wandered towards the old abbey: for the abbey was the secondary fact of Beauclere, after Mr Carteret. Mr Carteret sometimes went away and the abbey never did; yet somehow it was most of the essence of the place that it possessed the proprietor of the squarest of the square red houses, with the finest of the arched hall-windows, in three divisions, over the widest of the last-century doorways. You saw the great abbey from the doorstep, beyond the gardens of course, and in the stillness you could hear the flutter of the birds that circled round its huge, short towers. The towers had never been finished, save as time finishes things, by perpetuating their incompleteness. There is something right in old monuments that have been wrong for centuries: some such moral as that was usually in Nick's mind as an emanation of Beauclere, when he looked at the magnificent line of the roof, riding the sky and unsurpassed for length.

When the door with the brass plate was opened and Mr Chayter appeared in the middle distance (he always advanced just to the same spot, like a prime minister receiving an ambassador), Nick saw anew that he would be wonderfully like Mr Carteret if he had had an expression. He did not permit himself this freedom; never giving a sign of recognition, often as the young man had been at the house. He was most attentive to the visitor's wants, but apparently feared that if he allowed a familiarity it might go too far. There was always the same question to be asked – had Mr Carteret finished his nap? He usually had not finished it, and this left Nick what he liked – time to smoke a cigarette in the garden or even, before dinner, to take a turn about the place. He observed now, every time he came, that Mr Carteret's nap lasted a little longer. There was each year a little more strength to be gathered for the ceremony of dinner: this was the principal symptom – almost the only one – that the clear-cheeked old gentleman gave of not being so fresh as of yore. He was still wonderful for his age. Today he was particularly careful: Chayter went so far as to mention to Nick that four gentlemen were expected to dinner – an effusiveness perhaps partly explained by the circumstance that Lord Bottomley was one of them.

The prospect of Lord Bottomley was somehow not stirring; it only made the young man say to himself with a quick, thin sigh: 'This time I *am* in for it!' And he immediately had the unpolitical sense again that there was nothing so pleasant as the way the quiet bachelor house had its best rooms on the big garden, which seemed to advance into them through their wide windows and ruralize their dullness.

'I expect it will be a latish eight, sir,' said Mr Chayter, superintending in the library the production of tea on a large scale. Everything at Mr Carteret's appeared to Nick to be on a larger scale than anywhere else – the tea-cups, the knives and forks, the door-handles, the chair-backs, the legs of mutton, the candles and the lumps of coal: they represented and apparently exhausted the master's sense of pleasing effect, for the house was not otherwise decorated. Nick thought it really hideous, but he was capable at the same time of extracting a degree of amusement from anything that was strongly characteristic, and Mr Carteret's interior expressed a whole view of life. Our young man was generous enough to find a hundred instructive intimations in it even at the time it came over him (as it always did at Beauclere) that this was the view he himself was expected to take. Nowhere were the boiled eggs at breakfast so big or in such big receptacles; his own shoes, arranged in his room, looked to him vaster there than at home. He went out into the garden and remembered what enormous strawberries they should have for dinner. In the house there was a great deal of Landseer, of oilcloth, of woodwork painted and 'grained'.

Finding that he should have time before the evening meal, or before Mr Carteret would be able to see him, he quitted the house and took a stroll towards the abbey. It covered acres of ground on the summit of the hill, and there were aspects in which its vast bulk reminded him of the ark left high and dry upon Ararat. At least it was the image of a great wreck, of the indestructible vessel of a faith, washed up there by a storm centuries before. The injury of time added to this appearance – the infirmities around which, as he knew, the battle of restoration had begun to be fought. The cry had been raised to save the splendid pile, and the counter-cry by the purists, the sentimentalists, whatever they were, to save it from being saved. They were all exchanging compliments in the morning papers.

Nick sauntered round the church – it took a good while; he

leaned against low things and looked up at it while he smoked another cigarette. It struck him as a great pity it should be touched: so much of the past was buried there that it was like desecrating, like digging up a grave. And the years seemed to be letting it down so gently: why jostle the elbow of slow-fingering time? The fading afternoon was exquisitely pure; the place was empty; he heard nothing but the cries of several children, which sounded sweet, who were playing on the flatness of the very old tombs. He knew that this would inevitably be one of the topics at dinner, the restoration of the abbey; it would give rise to a considerable deal of orderly debate. Lord Bottomley, oddly enough, would probably oppose the expensive project, but on grounds that would be characteristic of him even if the attitude were not. Nick's nerves, on this spot, always knew what it was to be soothed; but he shifted his position with a slight impatience as the vision came over him of Lord Bottomley's treating a question of aesthetics. It was enough to make one want to take the other side, the idea of having the same taste as his lordship: one would have it for such different reasons.

Dear Mr Carteret would be deliberate and fair all round, and would, like his noble friend, exhibit much more architectural knowledge than he, Nick, possessed: which would not make it a whit less droll to our young man that an artistic idea, so little really assimilated, should be broached at that table and in that air. It would remain so outside of their minds and their minds would remain so outside of it. It would be dropped at last however, after half an hour's gentle worrying, and the conversation would incline itself to public affairs. Mr Carteret would find his natural level – the production of anecdote in regard to the formation of early ministries. He knew more than any one else about the personages of whom certain cabinets would have consisted if they had not consisted of others. His favourite exercise was to illustrate how different everything might have been from what it was, and how the reason of the difference had always been somebody's inability to 'see his way' to accept the view of somebody else – a view usually, at the time, discussed, in strict confidence, with Mr Carteret, who surrounded his actual violation of that confidence, thirty years later, with many precautions against scandal. In this retrospective vein, at the head of his table, the old gentleman always enjoyed an audience or at any rate commanded a silence, often profound. Every one left it to some one else to ask another question; and when by chance some one else did so every one was struck with admiration at any one's being able to say anything. Nick knew the moment when he himself would take a glass of a particular port and, surreptitiously looking at his watch, perceive it was ten o'clock. It might as well be 1830.

All this would be a part of the suggestion of leisure that invariably descended upon him at Beauclere – the image of a sloping shore where the tide of time broke with a ripple too faint to be a warning. But there was another admonition that was almost equally sure to descend upon his spirit in a summer hour, in a stroll about the grand abbey; to sink into it as the light lingered on the rough red walls and the local accent of the children sounded soft in the churchyard. It was simply the sense of England – a sort of apprehended revelation of his country. The dim annals of the place appeared to be in the air (foundations bafflingly early, a great monastic life, wars of the Roses, with battles and blood in the streets, and then the long quietude of the respectable centuries, all cornfields and magistrates and vicars), and these things were connected with an emotion that arose from the green country, the rich land so infinitely lived in, and laid on him a hand that was too ghostly to press and yet somehow too urgent to be light. It produced a throb that he could not have spoken of, it was so deep, and that was half imagination and half responsibility. These impressions melted together and made a general appeal, of which, with his new honours as a legislator, he was the sentient subject. If he had a love for this particular scene of life, might it not have a love for him and expect something of him? What fate could be so high as to grow old in a national affection. What a grand kind of reciprocity, making mere soreness of all the balms of indifference!

The great church was still open, and he turned into it and wandered a little in the twilight, which had gathered earlier there. The whole structure, with its immensity of height and distance, seemed to rest on tremendous facts – facts of achievement and endurance – and the huge Norman pillars to loom through the dimness like the ghosts of heroes. Nick was more struck with its human than with its divine significance, and he felt the oppression of his conscience as he walked slowly about. It was in his mind that nothing in life was really clear, all things were mingled and charged, and that patriotism might be an uplifting passion even if it had to allow for Lord Bottomley and for Mr Carteret's blindness on certain sides. Presently he perceived it was nearly half-past seven, and as he went back to his old friend's he could not have told you whether he was in a state of gladness or of gloom.

'Mr Carteret will be in the drawing-room at a quarter to eight, sir,' Chayter said; and Nick, as he went to his chamber, asked himself what was the use of being a member of Parliament if one was still sensitive to an intimation on the part of such a functionary that one ought really to have begun to dress. Chayter's words meant that Mr Carteret would expect to have a little comfortable conversation with him before dinner. Nick's usual rapidity in dressing was however quite adequate to the occasion, and his host had not appeared when he went down. There were flowers in the unfeminine saloon, which contained several paintings, in addition to the engravings of pictures of animals; but nothing could prevent its reminding Nick of a comfortable committee-room.

Mr Carteret presently came in, with his gold-headed stick, a laugh like a series of little warning coughs and the air of embarrassment that our young man always perceived in him at first. He was nearly eighty, but he was still shy – he laughed a great deal, faintly and vaguely, at nothing, as if to make up for the seriousness with which he took some jokes. He always began by looking away from his interlocutor, and it was only little by little that his eyes came round; after which their limpid and benevolent blue made you wonder why they should ever be circumspect. He was clean shaven and had a long upper lip. When he had seated himself he talked of 'majorities' and showed a disposition to converse on the general subject of the fluctuation of Liberal gains. He had an extraordinary memory for facts of this sort and could mention the figures relating to elections in innumerable places in particular years. To many of these facts he attached great importance, in his simple, kindly, presupposing way; returning five minutes later and correcting himself if he had said that some one, in 1857, had had 6014 instead of 6004.

Nick always felt a great hypocrite as he listened to him, in spite of the old man's courtesy – a thing so charming in itself that it would have been grossness to speak of him as a bore. The difficulty was that he took for granted all kinds of positive assent, and Nick, in his company, found himself immersed in an atmosphere of tacit pledges which constituted the very medium of intercourse and yet made him draw his breath a little in pain when, for a moment, he measured them. There would have been no hypocrisy at all if he could have regarded Mr Carteret as a mere sweet spectacle, the last or almost the last illustration of a departing tradition of manners. But he represented something more than manners; he represented what he believed to be morals and ideas – ideas as regards which he took your personal deference (not discovering how natural that was) for participation. Nick liked to think that his father, though ten years younger, had found it congruous to make his best friend of the owner of so nice a nature: it gave a softness to his feeling for that memory to be reminded that Sir Nicholas had been of the same general type – a type so pure, so disinterested, so anxious about the public good. Just so it endeared Mr Carteret to him to perceive that he considered his father had done a definite work, prematurely interrupted, which had been an absolute benefit to the people of England. The oddity was however that though both Mr Carteret's aspect and his appreciation were still so fresh, this relation of his to his late distinguished friend made the latter appear to Nick even more irrecoverably dead. The good old man had almost a vocabulary of his own, made up of old-fashioned political phrases and quite untainted with the new terms, mostly borrowed from America; indeed, his language and his tone made those of almost any one who might be talking with him appear by contrast rather American. He was, at least nowadays, never severe or denunciatory; but sometimes in telling an anecdote he dropped such an expression as 'the rascal said to me', or such an epithet as 'the vulgar dog'.

Nick was always struck with the rare simplicity (it came out in his countenance) of one who had lived so long and seen so much of affairs that draw forth the passions and perversities of men. It often made him say to himself that Mr Carteret must have been very remarkable to achieve with his means so many things requiring cleverness. It was as if experience, though coming to him in abundance, had dealt with him with such clean hands as to leave no stain and had never provoked him to any general reflection. He had never proceeded in any ironic way from the particular to the general; certainly he had never made a reflection upon anything so unparliamentary as Life. He would have questioned the taste of such an excrescence, and if he had encountered it on the part of another would have regarded it as a kind of French toy, with the uses of which he was unacquainted. Life, for him, was a purely practical function, not a question of phrasing. It must be added that he had, to Nick's perception, his variations – his back windows opening into grounds more private. That was visible from the way his eye grew cold and his whole polite face rather austere when he listened to something that he didn't agree with or perhaps even understand; as if his modesty did not in strictness forbid the suspicion that a thing he didn't understand would have a probability against it. At such times there was something a little deadly in the silence in which he simply waited, with a lapse in his face, without helping his interlocutor out. Nick would have been very sorry to attempt to communicate to him a matter which he probably would not understand. This cut off of course a multitude of subjects.

The evening passed exactly as Nick had foreseen, even to the rather early dispersal of the guests, two of whom were 'local' men, earnest and distinct, though not particularly distinguished. The third was a young, slim, uninitiated gentleman whom Lord Bottomley brought with him and concerning whom Nick was informed beforehand that he was engaged to be married to the Honourable Jane, his lordship's second daughter. There were recurrent allusions to Nick's victory, as to which he had the fear that he might appear to exhibit less interest in it than the company did. He took energetic precautions against this and felt repeatedly a little spent with them, for the subject always came up once more. Yet it was not as his but as theirs that they liked the triumph. Mr Carteret took leave of him for the night directly after the other guests had gone, using at this moment the words that he had often used before –

'You may sit up to any hour you like. I only ask that you don't read in bed.'

17

NICK's little visit was to terminate immediately after luncheon the following day: much as the old man enjoyed his being there he would not have dreamed of asking for more of his time now that it had such great public uses. He liked infinitely better that his young friend should be occupied with parliamentary work than only occupied in talking about it with him. Talk about it however was the next best thing, as on the morrow after breakfast Mr Carteret showed Nick that he considered. They sat in the garden, the morning being warm, and the old man had a table beside him, covered with the letters and newspapers that the post had brought. He was proud of his correspondence, which was altogether on public affairs, and proud in a manner of the fact that he now dictated almost everything. That had more in it of the statesman in retirement, a character indeed not consciously assumed by Mr Carteret, but always tacitly attributed to him by Nick, who took it rather from the pictorial point of view: remembering on each occasion only afterwards that though he was in retirement he had not exactly been a statesman. A young man, a very sharp, handy young man, came every morning at ten o'clock and wrote for him till lunchtime. The young man had a holiday today, in honour of Nick's visit – a fact the mention of which led Nick to make some not particularly sincere speech about *his* being ready to write anything if Mr Carteret were at all pressed.

'Ah, but your own budget: what will become of that?' the old gentleman objected, glancing at Nick's pockets as if he was rather surprised not to see them stuffed out with documents in split envelopes. His visitor had to confess that he had not directed his letters to meet him at Beauclere: he should find them in town that afternoon. This led to a little homily from Mr Carteret which made him feel rather guilty; there was such an implication of neglected duty in the way the old man said: 'You won't do them justice – you won't do them justice.' He talked for ten minutes, in his rich, simple, urbane way, about the fatal consequences of getting behind. It was his favourite doctrine that one should always be a little before; and his own eminently regular respiration seemed to illustrate the idea. A man was certainly before who had so much in his rear.

This led to the bestowal of a good deal of general advice as to the mistakes to avoid at the beginning of a parliamentary career; as to which Mr Carteret spoke with the experience of one who had sat for fifty years in the House of Commons. Nick was amused, but also mystified and even a little irritated by his talk: it was founded on the idea of observation and yet Nick was unable to regard Mr Carteret as an observer. 'He doesn't observe *me*,' he said to himself; 'if he did he would see, he wouldn't think –' And the end of this private cogitation was a vague impatience of all the things his venerable host took for granted. He didn't see any of the things that Nick saw. Some of these latter were the light touches of the summer morning scattered through the sweet old garden. The time passed there a good deal as if it were sitting still, with a plaid under its feet, while Mr Carteret distilled a little more of the wisdom that he had drawn from his fifty years. This immense term had something fabulous and monstrous for Nick, who wondered whether it were the sort of thing his companion supposed *he* had gone in for. It was not strange Mr Carteret should be different; he might originally have been more – to himself Nick was not obliged to phrase it: what our young man meant was more of what it was perceptible to him that his host was not. Should even he, Nick, be like that at the end of fifty years? What Mr Carteret was so good as to expect for him was that he should be much more distinguished; and wouldn't this exactly mean much more like that? Of course Nick heard some things that he had heard before; as for instance the circumstances that had originally led the old man to settle at Beauclere. He had been returned for that locality (it was his second seat), in years far remote, and had come to live there because he then had a conscientious conviction (modified indeed by later experience) that a member should be constantly resident. He spoke of this now, smiling rosily, as he might have spoken of some wild aberration of his youth; yet he called Nick's attention to the fact that he still so far clung to his conviction as to hold (though of what might be urged on the other side he was perfectly aware) that a representative should at least be as resident as possible. This gave Nick an opening for saying something that had been on and off his lips all the morning.

'According to that I ought to take up my abode at Harsh.'

'In the measure of the convenient I should not be sorry to see you do it.'

'It ought to be rather convenient,' Nick replied, smiling. 'I've got a piece of news for you which I've kept, as one keeps that sort of thing (for it's very good), till the last.' He waited a little, to see if Mr Carteret would guess, and at first he thought nothing would come of this. But after resting his young-looking eyes on him for a moment the old man said –

'I should indeed be very happy to hear that you have arranged to take a wife.'

'Mrs Dallow has been so good as to say that she will marry me,' Nick went on.

'That is very suitable. I should think it would answer.'

'It's very jolly,' said Nick. It was well that Mr Carteret was not what his guest called observant, or he might have thought there was less gaiety in the sound of this sentence than in the sense.

'Your dear father would have liked it.'

'So my mother says.'

'And *she* must be delighted.'

'Mrs Dallow, do you mean?' Nick asked.

'I was thinking of your mother. But I don't exclude the charming lady. I remember her as a little girl. I must have seen her at Windrush. Now I understand the zeal and amiability with which she threw herself into your canvass.'

'It was her they elected,' said Nick.

'I don't know that I have ever been an enthusiast for political women, but there is no doubt that, in approaching the mass of electors, a graceful, affable manner, the manner of the real English lady, is a force not to be despised.'

'Mrs Dallow is a real English lady, and at the same time she's a very political woman,' Nick remarked.

'Isn't it rather in the family? I remember once going to see her mother in town and finding the leaders of both parties sitting with her.'

'My principal friend, of the others, is her brother Peter. I don't think he troubles himself much about that sort of thing.'

'What does he trouble himself about?' Mr Carteret inquired, with a certain gravity.

'He's in the diplomatic service; he's a secretary in Paris.'

'That may be serious,' said the old man.

'He takes a great interest in the theatre; I suppose you'll say that may be serious too,' Nick added, laughing.

'Oh!' exclaimed Mr Carteret, looking as if he scarcely understood. Then he continued: 'Well, it can't hurt you.'

'It can't hurt me?'

'If Mrs Dallow takes an interest in your interests.'

'When a man's in my situation he feels as if nothing could hurt him.'

'I'm very glad you're happy,' said Mr Carteret. He rested his mild eyes on our young man, who had a sense of seeing in them for a moment the faint ghost of an old story, the dim revival of a sentiment that had become the memory of a memory. This glimmer of wonder and envy, the revelation of a life intensely celibate, was for an instant infinitely touching. Nick had always had a theory, suggested by a vague allusion from his father, who had been discreet, that their benevolent friend had had in his youth an unhappy love-affair which had led him to forswear for ever the commerce of woman. What remained in him of conscious renunciation gave a throb as he looked at his bright companion, who proposed to take the matter so much the other way. 'It's good to marry, and I think it's right. I've not done right, I know it. If she's a good woman it's the best thing,' Mr Carteret went on. 'It's what I've been hoping for you. Sometimes I've thought of speaking to you.'

'She's a very good woman,' said Nick.

'And I hope she's not poor.' Mr Carteret spoke with exactly the same blandness.

'No, indeed, she's rich. Her husband, whom I knew and liked, left her a large fortune.'

'And on what terms does she enjoy it?'

'I haven't the least idea,' said Nick.

Mr Carteret was silent a moment. 'I see. It doesn't concern you. It needn't concern you,' he added in a moment.

Nick thought of his mother, at this, but he remarked: 'I dare say she can do what she likes with her money.'

'So can I, my dear young friend,' said Mr Carteret.

Nick tried not to look conscious, for he felt a significance in the old man's face. He turned his own everywhere but towards it, thinking again of his mother. 'That must be very pleasant, if one has any.'

'I wish you had a little more.'

'I don't particularly care,' said Nick.

'Your marriage will assist you; you can't help that,' Mr Carteret went on. 'But I should like you to be under obligations not quite so heavy.'

'Oh, I'm so obliged to her for caring for me!'

'That the rest doesn't count? Certainly it's nice of her to like you. But why shouldn't she? Other people do.'

'Some of them make me feel as if I abused it,' said Nick, looking at his host. 'That is, they don't make me, but I feel it,' he added, correcting himself.

'I have no son,' said Mr Carteret. 'Shan't you be very kind to her?' he pursued. 'You'll gratify her ambition.'

'Oh, she thinks me cleverer than I am.'

'That's because she's in love,' hinted the old gentleman, as if this were very subtle. 'However, you must be as clever as we think you. If you don't prove so –' And he paused, with his folded hands.

'Well, if I don't?' asked Nick.

'Oh, it won't do – it won't do,' said Mr Carteret, in a tone his companion was destined to remember afterwards. 'I say I have no son,' he continued; 'but if I had had one he should have risen high.'

'It's well for me such a person doesn't exist. I shouldn't easily have found a wife.'

'He should have gone to the altar with a little money in his pocket.'

'That would have been the least of his advantages, sir.'

'When are you to be married?' Mr Carteret asked.

'Ah, that's the question. Mrs Dallow won't say.'

'Well, you may consider that when it comes off I will make you a settlement.'

'I feel your kindness more than I can say,' Nick replied; 'but that will probably be the moment when I shall be least conscious of wanting anything.'

'You'll appreciate it later – you'll appreciate it very soon. I shall like you to appreciate it,' Mr Carteret went on, as if he had a just vision of the way a young man of a proper spirit should feel. Then he added: 'Your father would have liked you to appreciate it.'

'Poor father!' Nick exclaimed vaguely, rather embarrassed, reflecting on the oddity of a position in which the ground for holding up his head as the husband of a rich woman would be that he had accepted a present of money from another source. It was plain that he was not fated to go in for independence; the most that he could treat himself to would be dependence that was truly grateful. 'How much you expect of me!' he pursued, with a grave face.

'It's only what your father did. He so often spoke of you, I remember, at the last, just after you had been with him alone – you know I saw him then. He was greatly moved by his interview with you, and so was I by what he told me of it. He said he should live on in you – he should work in you. It has always given me a very peculiar feeling, if I may use the expression, about you.'

'The feelings are indeed peculiar, dear Mr Carteret, which take so munificent a form. But you do – oh, you do – expect too much.'

'I expect you to repay me!' said the old man gaily. 'As for the form, I have it in my mind.'

'The form of repayment?'

'No, no – of settlement.'

'Ah, don't talk of it now,' said Nick, 'for, you see, nothing else is settled. No one has been told except my mother. She has only consented to my telling you.'

'Lady Agnes, do you mean?'

'Ah, no; dear mother would like to publish it on the housetops. She's so glad – she wants us to have it over tomorrow. But

Julia wishes to wait. Therefore kindly mention it for the present to no one.'

'My dear boy, at this rate there is nothing to mention. What does Julia want to wait for?'

'Till I like her better – that's what she says.'

'It's the way to make you like her worse. Hasn't she your affection?'

'So much so that her delay makes me exceedingly unhappy.'

Mr Carteret looked at his young friend as if he didn't strike him as very unhappy; but he demanded: 'Then what more does she want?' Nick laughed out at this, but he perceived his host had not meant it as an epigram; while the latter went on: 'I don't understand. You're engaged or you're not engaged.'

'She is, but I am not. That's what she says about it. The trouble is she doesn't believe in me.'

'Doesn't she love you then?'

'That's what I ask her. Her answer is that she loves me only too well. She's so afraid of being a burden to me that she gives me my freedom till I've taken another year to think.'

'I like the way you talk about other years!' Mr Carteret exclaimed. 'You had better do it while I'm here to bless you.'

'She thinks I proposed to her because she got me in for Harsh,' said Nick.

'Well, I'm sure it would be a very pretty return.'

'Ah, she doesn't believe in me,' Nick murmured.

'Then I don't believe in her.'

'Don't say that – don't say that. She's a very rare creature. But she's proud, shy, suspicious.'

'Suspicious of what?'

'Of everything. She thinks I'm not persistent.'

'Persistent?'

'She can't believe I shall arrrive at true eminence.'

'A good wife should believe what her husband believes,' said Mr Carteret.

'Ah, unfortunately I don't believe it either.'

Mr Carteret looked serious. 'Your dear father did.'

'I think of that – I think of that,' Nick replied. 'Certainly it will help me. If I say we're engaged,' he went on, 'it's because I consider it so. She gives me my liberty, but I don't take it.'

'Does she expect you to take back your word?'

'That's what I ask her. *She* never will. Therefore we're as good as tied.'

'I don't like it,' said Mr Carteret, after a moment. 'I don't like ambiguous, uncertain situations. They please me much better when they are definite and clear.' The retreat of expression had been sounded in his face – the aspect it wore when he wished not to be encouraging. But after an instant he added in a tone softer than this: 'Don't disappoint me, my dear boy.'

'Disappoint you?'

'I've told you what I want to do for you. See that the conditions come about promptly in which I *may* do it. Are you sure that you do everything to satisfy Mrs Dallow?' Mr Carteret continued.

'I think I'm very nice to her,' Nick protested. 'But she's so ambitious. Frankly speaking, it's a pity – for her – that she likes me.'

'She can't help that.'

'Possibly. But isn't it a reason for taking me as I am? What she wants to do is to take me as I may be a year hence.'

'I don't understand if, as you say, even then she won't take back her word,' said Mr Carteret.

'If she doesn't marry me I think she'll never marry again at all.'

'What then does she gain by delay?'

'Simply this, as I make it out – that she'll feel she has been very magnanimous. She won't have to reproach herself with not having given me a chance to change.'

'To change? What does she think you liable to do?'

Nick was silent a minute. 'I don't know!' he said, not at all candidly.

'Everything has altered: young people in my day looked at these questions more naturally,' Mr Carteret declared. 'A woman in love has no need to be magnanimous. If she is, she isn't in love,' he added shrewdly.

'Oh, Mrs Dallow's safe – she's safe,' Nick smiled.

'If it were a question between you and another gentleman one might comprehend. But what does it mean, between you and nothing?'

'I'm much obliged to you, sir,' Nick returned. 'The trouble is that she doesn't know what she has got hold of.'

'Ah, if you can't make it clear to her!'

'I'm such a humbug,' said the young man. His companion stared, and he continued: 'I deceive people without in the least intending it.'

'What on earth do you mean? Are you deceiving me?'

'I don't know – it depends on what you think.'

'I think you're flighty,' said Mr Carteret, with the nearest approach to sternness that Nick had ever observed in him. 'I never thought so before.'

'Forgive me; it's all right. I'm not frivolous; that I affirm I'm not.'

'You *have* deceived me if you are.'

'It's all right,' Nick stammered, with a blush.

'Remember your name – carry it high.'

'I will – as high as possible.'

'You've no excuse. Don't tell me, after your speeches at Harsh!' Nick was on the point of declaring again that he was a humbug, so vivid was his inner sense of what *he* thought of his factitious public utterances, which had the cursed property of creating dreadful responsibilities and importunate credulities for him. If *he* was 'clever', what fools many other people were! He repressed his impulse, and Mr Carteret pursued: 'If, as you express it, Mrs Dallow doesn't know what she has got hold of, won't it clear the matter up a little if you inform her that the day before your marriage is definitely settled to take place you will come into something comfortable?'

A quick vision of what Mr Carteret would be likely to regard as something comfortable flitted before Nick, but it did not prevent him from replying: 'Oh, I'm afraid that won't do any good. It would make her like you better, but it wouldn't make her like me. I'm afraid she won't care for any benefit that comes to me from another hand than hers. Her affection is a very jealous sentiment.'

'It's a very peculiar one!' sighed Mr Carteret. 'Mine's a jealous sentiment too. However, if she takes it that way don't tell her.'

'I'll let you know as soon as she comes round,' said Nick.

'And you'll tell your mother,' said Mr Carteret. 'I shall like her to know.'

'It will be delightful news to her. But she's keen enough already.'

'I know that. I may mention now that she has written to me,' the old man added.

'So I suspected.'

'We have corresponded on the subject,' Mr Carteret continued to confess. 'My view of the advantageous character of such an alliance has entirely coincided with hers.'

'It was very good-natured of you to leave me to speak first,' said Nick.

'I should have been disappointed if you hadn't. I don't like all you have told me. But don't disappoint me now.'

'Dear Mr Carteret!' Nick exclaimed.

'I won't disappoint *you*,' the old man went on, looking at his big, old-fashioned watch.

18

AT first Peter Sherringham thought of asking to be transferred to another post and went so far, in London, as to take what he believed to be good advice on the subject. The advice perhaps struck him as the better for consisting of a strong recommendation to do nothing so foolish. Two or three reasons were mentioned to him why such a request would not, in the particular circumstances, raise him in the esteem of his superiors, and he promptly recognized their force. It next appeared to him that it might help him (not with his superiors, but with himself) to apply for an extension of leave; but on further reflection he remained convinced that though there are some dangers before which it is perfectly consistent with honour to flee, it was better for everyone concerned that he should fight this especial battle on the spot. During his holiday his plan of campaign gave him plenty of occupation. He refurbished his arms, rubbed up his strategy, laid out his lines of defence.

There was only one thing in life that his mind had been very much made up to, but on this question he had never wavered:

he would get on to the utmost in his profession. It was a point on which it was perfectly lawful to be unamiable to others – to be vigilant, eager, suspicious, selfish. He had not in fact been unamiable to others, for his affairs had not required it: he had got on well enough without hardening his heart. Fortune had been kind to him and he had passed so many competitors on the way that he could forswear jealousy and be generous. But he had always flattered himself that his hand would not falter on the day he should find it necessary to drop bitterness into his cup. This day would be sure to dawn, for no career was all clear water to the end; and then the sacrifice would find him ready. His mind was familiar with the thought of a sacrifice: it is true that nothing could be plain in advance about the occasion, the object, the victim. All that was tolerably definite was that the propitiatory offering would have to be some cherished enjoyment. Very likely indeed this enjoyment would be associated with the charms of another person – a probability pregnant with the idea that such charms would have to be dashed out of sight. At any rate it never had occurred to Sherringham that he himself might be the sacrifice. You had to pay to get on; but at least you borrowed from others to do it. When you couldn't borrow you didn't get on: for what was the situation in life in which you met the whole requisition yourself?

Least of all had it occurred to our friend that the wrench might come through his interest in that branch of art on which Nick Dormer had rallied him. The beauty of a love of the theatre was precisely that it was a passion exercised on the easiest terms. This was not the region of responsibility. It had the discredit of being sniffed at by the austere; but if it was not, as they said, a serious field, was not the compensation just that you could not be seriously entangled in it? Sherringham's great advantage, as he regarded the matter, was that he had always kept his taste for the drama quite in its place. His facetious cousin was free to pretend that it sprawled through his life; but this was nonsense, as any unprejudiced observer of that life would unhesitatingly attest. There had not been the least sprawling, and his fancy for the art of Garrick had never worn the proportions of an eccentricity. It had never drawn down from above anything approaching a reprimand, a remonstrance, a remark. Sherringham was positively proud of his discretion; for he was a little proud of what he did know about the stage. Trifling for trifling there were plenty of his fellows who had in their lives private infatuations much sillier and less confessable. Had he not known men who collected old invitation-cards (hungry for those of the last century), and others who had a secret passion for shuffleboard? His little weaknesses were intellectual – they were a part of the life of the mind. All the same on the day they showed a symptom of interfering they should be plucked off with a turn of the wrist.

Sherringham scented interference now, and interference in rather an invidious form. It might be a bore, from the point of view of the profession, to find one's self, as a critic of the stage, in love with a *coquine*; but it was a much greater bore to find one's self in love with a young woman whose character remained to be estimated. Miriam Rooth was neither fish nor flesh: one had with her neither the guarantees of one's own class nor the immunities of hers. What was hers, if one came to that? A certain puzzlement about this very point was part of the fascination which she had ended by throwing over him. Poor Sherringham's scheme for getting on had contained no proviso against falling in love, but it had embodied an important clause on the subject of surprises. It was always a surprise to fall in love, especially if one were looking out for it; so this contingency had not been worth official paper. But it became a man who respected the service he had undertaken for the State to be on his guard against predicaments from which the only issue was the rigour of matrimony. An ambitious diplomatist would probably be wise to marry, but only with his eyes very much open. That was the fatal surprise – to be led to the altar in a dream. Sherringham's view of the proprieties attached to such a step was high and strict; and if he held that a man in his position was, especially as the position improved, essentially a representative of the greatness of his country, he considered that the wife of such a personage would exercise in her degree (for instance, at a foreign court) a function no less symbolic. She would always be, in short, a very important quantity, and the scene was strewn with illustrations of this general truth. She might be such a help and she might be such a blight that common prudence required that one should test her in advance. Sherringham had seen women in the career who were stupid or vulgar make a mess of things – it was enough to wring your heart. Then he had his positive idea of the perfect ambassadress, the full-blown lily of the future; and with this idea Miriam Rooth presented no analogy whatever.

The girl had described herself with characteristic directness as 'all right'; and so she might be, so she assuredly was: only all right for what? He had divined that she was not sentimental – that whatever capacity she might have for responding to a devotion or for desiring it was at any rate not in the direction of vague philandering. With him certainly she had no disposition to philander. Sherringham was almost afraid to think of this, lest it should beget in him a rage convertible mainly into caring for her more. Rage or no rage, it would be charming to be in love with her if there were no complications; but the complications were in advance just what was clearest in the business. He was perhaps cold-blooded to think of them; but it must be remembered that they were the particular thing which his training had equipped him for dealing with. He was at all events not too cold-blooded to have, for the two months of his holiday, very little inner vision of anything more abstract than Miriam's face. The desire to see it again was as pressing as thirst; but he tried to teach himself the endurance of the traveller in the desert. He kept the Channel between them, but his spirit moved every day an inch nearer to her, until (and it was not long) there were no more inches left. The last thing he expected the future ambassadress to have been was a *fille de théâtre*. The answer to this objection was of course that Miriam was not yet so much of one but that he could easily head her off. Then came worrying retorts to that, chief among which was the sense that to his artistic conscience heading her off would be simple shallowness. The poor girl had a right to her chance, and he should not really alter anything by taking it away from her; for was she not the artist to the tips of her tresses (the ambassadress never in the world), and would she not take it out in something else if one were to make her deviate? So certain was that irrepressible deviltry to insist ever on its own.

Besides, *could* one make her deviate? If she had no disposition to philander, what was his warrant for supposing that she could be corrupted into respectability? How could the career (*his* career) speak to a nature which had glimpses, as vivid as they were crude, of such a different range, and for which success meant quite another sauce to the dish? Would the brilliancy of marrying Peter Sherringham be such a bribe to relinquishment? How could he think so without fatuity – how could he regard himself as a high prize? Relinquishment of the opportunity to exercise a rare talent was not, in the nature of things, an easy effort to a young lady who was conceited as well as ambitious. Besides, she might eat her cake and have it – might make her fortune both on the stage and in the world. Successful actresses had ended by marrying dukes, and was not that better than remaining obscure and marrying a commoner? There were moments when Sherringham tried to think that Miriam's talent was not a force to reckon with; there was so little to show for it as yet that the caprice of believing in it would perhaps suddenly leave her. But his suspicion that it was real was too uneasy to make such an experiment peaceful, and he came back moreover to his deepest impression – that of her being of the turn of mind for which the only consistency is talent. Had not Madame Carré said at the last that she could 'do anything'? It was true that if Madame Carré had been mistaken in the first place she might also be mistaken in the second. But in this latter case she would be mistaken with him, and such an error would be too much like a truth.

I ought possibly to hesitate to say how much Sherringham felt the discomfort, for him, of the advantage that Miriam had of him – the advantage of her presenting herself in a light which rendered any passion that he might entertain an implication of duty as well as of pleasure. Why there should be this implication was more than he could say; sometimes he declared to himself that he was superstitious for seeing it. He didn't know, he could scarcely conceive of another case of the same general type in which he would have seen it. In foreign countries there were very few ladies of Miss Rooth's intended profession who would not have regarded it as too strong an order that to console them for not being admitted into drawing-rooms they should have no offset but the exercise of a virtue in which no one would believe.

Because in foreign countries actresses were not admitted into drawing-rooms: that was a pure English drollery, ministering equally little to histrionics and to the tone of these resorts. Did the sanctity which to his imagination made it a burden to have to reckon with Miriam come from her being English? Sherringham could remember cases in which that privilege operated as little as possible as a restriction. It came a great deal from Mrs Rooth, in whom he apprehended depths of calculation as to what she might achieve for her daughter by 'working' the idea of a blameless life. Her romantic turn of mind would not in the least prevent her from regarding that idea as a substantial capital, to be laid out to the best worldly advantage. Miriam's essential irreverence was capable, on a pretext, of making mincemeat of it – that he was sure of; for the only capital she recognized was the talent which some day managers and agents would outbid each other in paying for. She was a good-natured creature; she was fond of her mother, would do anything to oblige (that might work in all sorts of ways), and would probably like the loose slippers of blamelessness quite as well as the high standards of the opposite camp.

Sherringham, I may add, had no desire that she should indulge a different preference: it was foreign to him to compute the probabilities of a young lady's misbehaving for his advantage (that seemed to him definitely base), and he would have thought himself a blackguard if, professing a tenderness for Miriam, he had not wished the thing that was best for her. The thing that was best for her would no doubt be to become the wife of the man to whose suit she should incline her ear. That this would be the best thing for the gentleman in question was however a very different matter, and Sherringham's final conviction was that it would never do for him to act the part of that hypothetic personage. He asked for no removal and no extension of leave, and he proved to himself how well he knew what he was about by never addressing a line, during his absence, to the Hôtel de la Garonne. He would simply go straight and inflict as little injury upon Peter Sherringham as upon any one else. He remained away to the last hour of his privilege and continued to act lucidly in having nothing to do with the mother and daughter for several days after his return to Paris.

It was when this discipline came to an end, one afternoon after a week had passed, that he felt most of the force of the reference that has just been made to Mrs Rooth's private reckonings. He found her at home alone, writing a letter under the lamp, and as soon as he came in she cried out that he was the very person to whom the letter was addressed. She could bear it no longer; she had permitted herself to reproach him with his terrible silence – to ask why he had quite forsaken them. It was an illustration of the way in which her visitor had come to regard her that he rather disbelieved than believed this description of the crumpled papers lying on the table. He was not sure even that he believed that Miriam had just gone out. He told her mother how busy he had been all the while he was away and how much time in particular he had had to give, in London, to seeing on her daughter's behalf the people connected with the theatres.

'Ah, if you pity me, tell me that you've got her an engagement!' Mrs Rooth cried, clasping her hands.

'I took a great deal of trouble; I wrote ever so many notes, sought introductions, talked with people – such impossible people some of them. In short I knocked at every door, I went into the question exhaustively.' And he enumerated the things he had done, imparted some of the knowledge he had gathered. The difficulties were serious, and even with the influence he could command (such as it was) there was very little to be achieved in face of them. Still he had gained ground: there were two or three fellows, men with small theatres, who had listened to him better than the others, and there was one in particular whom he had a hope he might really have interested. From him he had extracted certain merciful assurances: he would see Miriam, he would listen to her, he would do for her what he could. The trouble was that no one would lift a finger for a girl unless she were known, and yet that she never could become known until innumerable fingers were lifted. You couldn't go into the water unless you could swim, and you couldn't swim until you had been in the water.

'But new performers appear; they get theatres, they get audiences, they get notices in the newspapers,' Mrs Rooth objected.

'I know of these things only what Miriam tells me. It's no knowledge that I was born to.'

'It's perfectly true; it's all done with money.'

'And how do they come by money?' Mrs Rooth asked, candidly.

'When they're women people give it to them.'

'Well, what people, now?'

'People who believe in them.'

'As you believe in Miriam?'

Sherringham was silent a moment. 'No, rather differently. A poor man doesn't believe anything in the same way that a rich man does.'

'Ah, don't call yourself poor!' groaned Mrs Rooth.

'What good would it do me to be rich?'

'Why, you could take a theatre; you could do it all yourself.'

'And what good would that do me?'

'Why don't you delight in her genius?' demanded Mrs Rooth.

'I delight in her mother. You think me more disinterested than I am,' Sherringham added, with a certain soreness of irritation.

'I know why you didn't write!' Mrs Rooth declared, archly.

'You must go to London,' Peter said, without heeding this remark.

'Ah, if we could only get there it would be a relief. I should draw a long breath. There at least I know where I am, and what people are. But here one lives on hollow ground!'

'The sooner you get away the better,' Sherringham went on.

'I know why you say that.'

'It's just what I'm explaining.'

'I couldn't have held out if I hadn't been so sure of Miriam,' said Mrs Rooth.

'Well, you needn't hold out any longer.'

'Don't *you* trust her?' asked Sherringham's hostess.

'Trust her?'

'You don't trust yourself. That's why you were silent, why we might have thought you were dead, why we might have perished ourselves.'

'I don't think I understand you; I don't know what you're talking about,' Sherringham said. 'But it doesn't matter.'

217

'Doesn't it? Let yourself go: why should you struggle?' the old woman inquired.

Her unexpected insistence annoyed her visitor, and he was silent again, looking at her, on the point of telling her that he didn't like her tone. But he had his tongue under such control that he was able presently to say, instead of this – and it was a relief for him to give audible voice to the reflection: 'It's a great mistake, either way, for a man to be in love with an actress. Either it means nothing serious, and what's the use of that? or it means everything, and that's still more delusive.'

'Delusive?'

'Idle, unprofitable.'

'Surely a pure affection is its own reward,' Mrs Rooth rejoined, with soft reasonableness.

'In such a case how can it be pure?'

'I thought you were talking of an English gentleman,' said Mrs Rooth.

'Call the poor fellow whatever you like: a man with his life to lead, his way to make, his work, his duties, his career to attend to. If it means nothing, as I say, the thing it means least of all is marriage.'

'Oh, my own Miriam!' murmured Mrs Rooth.

'On the other hand fancy the complication if such a man marries a woman who's on the stage.'

Mrs Rooth looked as if she were trying to follow. 'Miriam isn't on the stage yet.'

'Go to London and she soon will be.'

'Yes, and then you'll have your excuse.'

'My excuse?'

'For deserting us altogether.'

Sherringham broke into laughter at this, the tone was so droll. Then he rejoined: 'Show me some good acting and I won't desert you.'

'Good acting? Ah, what is the best acting compared with the position of an English lady? If you'll take her as she is you may have her,' Mrs Rooth suddenly added.

'As she is, with all her ambitions unassuaged?'

'To marry *you* – might not that be an ambition?'

218

'A very paltry one. Don't answer for her, don't attempt that,' said Sherringham. 'You can do much better.'

'Do you think *you* can?' smiled Mrs Rooth.

'I don't want to; I only want to let it alone. She's an artist; you must give her her head,' Peter went on. 'You must always give an artist his head.'

'But I have known great ladies who were artists. In English society there is always a field.'

'Don't talk to me of English society! Thank heaven, in the first place, I don't live in it. Do you want her to give up her genius?'

'I thought you didn't care for it.'

'She'd say "No, I thank you, dear mamma."'

'My gifted child!' Mrs Rooth murmured.

'Have you ever proposed it to her?'

'Proposed it?'

'That she should give up trying.'

Mrs Rooth hesitated, looking down. 'Not for the reason you mean. We don't talk about love,' she simpered.

'Then it's so much less time wasted. Don't stretch out your hand to the worse when it may some day grasp the better,' Sherringham pursued. Mrs Rooth raised her eyes at him as if she recognized the force there might be in that, and he added: 'Let her blaze out, let her look about her. Then you may talk to me if you like.'

'It's very puzzling,' the old woman remarked, artlessly.

Sherringham laughed again; then he said: 'Now don't tell me I'm not a good friend.'

'You are indeed – you're a very noble gentleman. That's just why a quiet life with you –'

'It wouldn't be quiet for me!' Sherringham broke in. 'And that's not what Miriam was made for.'

'Don't say that, for my precious one!' Mrs Rooth quavered.

'Go to London – go to London,' her visitor repeated.

Thoughtfully, after an instant, she extended her hand and took from the table the letter on the composition of which he had found her engaged. Then with a quick movement she tore it up. 'That's what Mr Dashwood says.'

219

'Mr Dashwood?'

'I forgot you don't know him. He's the brother of that lady we met the day you were so good as to receive us; the one who was so kind to us – Mrs Lovick.'

'I never heard of him.'

'Don't you remember that she spoke of him and Mr Lovick didn't seem very kind about him? She told us that if he were to meet us – and she was so good as to insinuate that it would be a pleasure to him to do so – he might give us, as she said, a tip.'

Sherringham indulged in a visible effort to recollect. 'Yes, he comes back to me. He's an actor.'

'He's a gentleman too,' said Mrs Rooth.

'And you've met him and he *has* given you a tip?'

'As I say, he wants us to go to London.'

'I see, but even I can tell you that.'

'Oh, yes,' said Mrs Rooth; 'but *he* says he can help us.'

'Keep hold of him then, if he's in the business.'

'He's a perfect gentleman,' said Mrs Rooth. 'He's immensely struck with Miriam.'

'Better and better. Keep hold of him.'

'Well, I'm glad you don't object,' Mrs Rooth smiled.

'Why should I object?'

'You don't consider us as *all* your own?'

'My own? Why, I regard you as the public's – the world's.'

Mrs Rooth gave a little shudder. 'There's a sort of chill in that. It's grand, but it's cold. However I needn't hesitate then to tell you that it's with Mr Dashwood that Miriam has gone out.'

'Why hesitate, gracious heaven?' But in the next breath Sherringham asked: 'Where has she gone?'

'You don't like it!' laughed Mrs Rooth.

'Why should it be a thing to be enthusiastic about?'

'Well, he's charming, and *I* trust him.'

'So do I,' said Sherringham.

'They've gone to see Madame Carré.'

'She has come back then?'

'She was expected back last week. Miriam wants to show her how she has improved.'

'And *has* she improved?'

'How can I tell – with my mother's heart?' asked Mrs Rooth. 'I don't judge; I only wait and pray. But Mr Dashwood thinks she's wonderful.'

'That's a blessing. And when did he turn up?'

'About a fortnight ago. We met Mrs Lovick at the English church, and she was so good as to recognize us and speak to us. She said she had been away with her children, or she would have come to see us. She had just returned to Paris.'

'Yes, I've not yet seen her,' said Sherringham. 'I see Lovick, but he doesn't talk of his brother-in-law.'

'I didn't, that day, like his tone about him,' Mrs Rooth observed. 'We walked a little way with Mrs Lovick and she asked Miriam about her prospects and if she were working. Miriam said she had no prospects.'

'That was not very nice to *me*,' Sherringham interrupted.

'But when you had left us in black darkness, where *were* our prospects?'

'I see; it's all right. Go on.'

'Then Mrs Lovick said her brother was to be in Paris a few days and she would tell him to come and see us. He arrived, she told him and he came. *Voilà!*' said Mrs Rooth.

'So that now (so far as *he* is concerned) Miss Rooth has prospects?'

'He isn't a manager unfortunately.'

'Where does he act?'

'He isn't acting just now; he has been abroad. He has been to Italy, I believe, and he is just stopping here on his way to London.'

'I see; he *is* a perfect gentleman,' said Sherringham.

'Ah, you're jealous of him.'

'No, but you're trying to make me so. The more competitors there are for the glory of bringing her out, the better for her.'

'Mr Dashwood wants to take a theatre,' said Mrs Rooth.

'Then perhaps he's our man.'

'Oh, if you'd help him?' cried Mrs Rooth.

'Help him?'

'Help him to help us.'

'We'll all work together; it will be very jolly,' said Sherringham gaily. 'It's a sacred cause, the love of art, and we shall be a happy band. Dashwood's his name?' he added in a moment. 'Mrs Lovick wasn't a Dashwood.'

'It's his *nom de théâtre* – Basil Dashwood. Do you like it?' Mrs Rooth inquired.

'You say that as Miriam might do: her talent is catching.'

'She's always practising – always saying things over and over, to seize the tone. I have her voice in my ears. He wants *her* not to have any.'

'Not to have any?'

'Any *nom de théâtre*. He wants her to use her own; he likes it so much. He says it will do so well – you can't better it.'

'He's a capital adviser,' said Sherringham, getting up. 'I'll come back tomorrow.'

'I won't ask you to wait till they return, they may be so long,' Mrs Rooth replied.

'Will he come back with her?' Sherringham inquired, smoothing his hat.

'I hope so, at this hour. With my child in the streets I tremble. We don't live in cabs, as you may easily suppose.'

'Did they go on foot?' Sherringham continued.

'Oh yes; they started in high spirits.'

'And is Mr Basil Dashwood acquainted with Madame Carré?'

'Oh no, but he longed to be introduced to her; he implored Miriam to take him. Naturally she wishes to oblige him. She's very nice to him – if he can do anything.'

'Quite right; that's the way.'

'And she also wanted him to see what she can do for the great critic,' Mrs Rooth added.

'The great critic?'

'I mean that terrible old woman in the red wig.'

'That's what I should like to see too,' said Sherringham.

'Oh, she has gone ahead; she is pleased with herself. "Work, work, work," said Madame Carré. Well, she has worked, worked, worked. That's what Mr Dashwood is pleased with even more than with other things.'

'What do you mean by other things?'

'Oh, her genius and her fine appearance.'

'He approves of her fine appearance? I ask because you think he knows what will take.'

'I know why you ask,' said Mrs Rooth. 'He says it will be worth hundreds of thousands to her.'

'That's the sort of thing I like to hear,' Sherringham rejoined. 'I'll come in tomorrow,' he repeated.

'And shall you mind if Mr Dashwood's here?'

'Does he come every day?'

'Oh, they're always at it.'

'Always at it?'

'Why, she acts to him – every sort of thing – and he says if it will do.'

'How many days has he been here then?'

Mrs Rooth reflected. 'Oh, I don't know. Since he turned up they've passed so quickly.'

'So far from "minding" it I'm eager to see him,' Sherringham declared; 'and I can imagine nothing better than what you describe – if he isn't an ass.'

'Dear me, if he isn't clever you must tell us: we can't afford to be deceived!' Mrs Rooth exclaimed, innocently and plaintively. 'What do we know – how can we judge?' she added.

Sherringham hesitated, with his hand on the latch. 'Oh, I'll tell you what I think of him!'

19

WHEN he got into the street he looked about him for a cab, but he was obliged to walk some distance before encountering one. In this little interval he saw no reason to modify the determination he had formed in descending the steep staircase of the Hôtel de la Garonne; indeed the desire which prompted it only quickened his pace. He had an hour to spare and he too would go to see Madame Carré. If Miriam and her companion had proceeded to the Rue de Constantinople on foot he would probably reach the house as soon as they. It was all quite logical: he was eager to see Miriam – that was natural enough; and he had admitted to Mrs Rooth that he was keen on the subject of Mrs Lovick's theatrical brother, in whom such effective aid might perhaps reside. To catch Miriam really revealing herself to the old actress (since that was her errand), with the jump she believed herself to have taken, would be a very happy stroke, the thought of which made her benefactor impatient. He presently found his cab and, as he bounded in, bade the coachman drive fast. He learned from Madame Carré's portress that her illustrious *locataire* was at home and that a lady and a gentleman had gone up some time before.

In the little antechamber, after he was admitted, he heard a high voice issue from the salon, and, stopping a moment to listen, perceived that Miriam was already launched in a recitation. He was able to make out the words, all the more that before he could prevent the movement the maid-servant who had let him in had already opened the door of the room (one of the wings of it, there being, as in most French doors, two pieces), before which, within, a heavy curtain was suspended. Miriam was in the act of rolling out some speech from the English poetic drama –

> 'For I am sick and capable of fears,
> Oppressed with wrongs and therefore full of fears.'

He recognized one of the great tirades of Shakespeare's Constance and saw she had just begun the magnificent scene at the beginning of the third act of *King John*, in which the passionate, injured mother and widow sweeps in wild organ-tones up and down the scale of her irony and wrath. The curtain concealed him and he lurked there for three minutes after he had motioned to the *femme de chambre* to retire on tiptoe. The trio in the salon, absorbed in the performance, had apparently not heard his entrance or the opening of the door, which was covered by the girl's splendid declamation. Sherringham listened intently, he was so arrested by the spirit with which she attacked her formidable verses. He had needed to hear her utter but half a dozen of them to comprehend the long stride she had taken in his absence; they told him that she had leaped into possession of her means. He remained where he was till she arrived at –

> 'Then speak again; not all thy former tale,
> But this one word, whether thy tale be true.'

This apostrophe, being briefly responded to in another voice, gave him time quickly to raise the curtain and show himself, passing into the room with a 'Go on, go on!' and a gesture earnestly deprecating a stop.

Miriam, in the full swing of her part, paused but for an instant and let herself ring out again, while Peter sank into the nearest chair and she fixed him with her illumined eyes, or rather with those of the raving Constance. Madame Carré, buried in a chair, kissed her hand to him, and a young man who stood near the girl giving her the cue stared at him over the top of a little book. 'Admirable – magnificent; go on,' Sherringham repeated – 'go on to the end of the scene – do it all!' Miriam flushed a little, but he immediately discovered that she had no personal emotion in seeing him again; the cold passion of art had perched on her banner and she listened to herself with an ear as vigilant as if she had been a Paganini drawing a fiddle-bow. This effect deepened as she went on, rising and rising to the great occasion, moving with extraordinary ease and in the largest, clearest style on the dizzy ridge of her idea. That she had an idea was visible enough and that the whole thing was very different from all that Sherringham had hitherto heard her attempt. It belonged quite to another class of effort; she seemed now like the finished statue lifted from the ground to its pedestal. It was as if the sun of her talent had risen above the hills and she knew that she was moving, that she would always move, in its guiding light. This conviction was the one artless thing that glimmered like a young joy through the tragic mask of Constance, and Sherringham's heart beat faster as he caught it in her face. It only made her appear more intelligent; and yet there had been a time when he had thought her stupid! Intelligent was the whole spirit in which she carried the scene, making him cry to himself from point to point: 'How she feels it – how she sees it – how she creates it!'

He looked at moments at Madame Carré and perceived that she had an open book in her lap, apparently a French prose version, brought by her visitors, of the play; but she never either glanced at him or at the volume; she only sat screwing into the girl her hard bright eyes, polished by experience like fine old brasses. The young man uttering the lines of the other speakers was attentive in another degree; he followed Miriam in his own copy of the play, to be sure not to miss the cue; but he was elated and expressive, was evidently even surprised; he coloured and smiled, and when he extended his hand to assist Constance to rise, after Miriam, acting out her text, had seated herself grandly on 'the huge, firm earth,' he bowed over her as obsequiously as if she had been his veritable sovereign. He was a very good-looking young man, tall, well-proportioned, straight-featured and fair, of whom manifestly the first thing to be said on any occasion was that he looked remarkably like a gentleman. He carried this appearance, which proved inveterate and importunate, to a point that was almost a negation of its spirit; that is, it might have been a question whether it could be in good taste to wear any character, even that particular one, so much on one's sleeve. It was literally on his sleeve that this young man partly wore his own; for it resided considerably in his attire and especially in a certain close-fitting dark blue frock-coat (a miracle of a fit), which moulded his young form just enough and not too much and constituted (as Sherringham was destined to perceive later) his perpetual uniform or badge. It was not till later that Sherringham began to feel exasperated by Basil Dashwood's 'type' (the young stranger was of course Basil Dashwood), and even by his blue frock-coat, the recurrent, unvarying, imperturbable 'good form' of his aspect. This unprofessional air ended by striking the observer as the profession that he had adopted, and was indeed (so far as had as yet been indicated) his theatrical capital, his main qualification for the stage.

The powerful, ample manner in which Miriam handled her scene produced its full impression, the art with which she surmounted its difficulties, the liberality with which she met its great demand upon the voice, and the variety of expression that she threw into a torrent of objurgation. It was a real composition, studded with passages that called a suppressed 'Brava!' to the lips and seeming to show that a talent capable of such an exhibition was capable of anything.

> 'But thou art fair, and at thy birth, dear boy,
> Nature and Fortune join'd to make thee great:
> Of Nature's gifts thou mayst with lilies boast,
> And with the half-blown rose.'

As Miriam turned to her imagined child with this exquisite apostrophe (she addressed Mr Dashwood as if he were playing

Arthur, and he lowered his book, dropped his head and his eyes and looked handsome and ingenuous), she opened at a stroke to Sherringham's vision a prospect that they would yet see her express tenderness better even than anything else. Her voice was enchanting in these lines, and the beauty of her performance was that while she uttered the full fury of the part she missed none of its poetry.

'Where did she get hold of that – where did she get hold of that?' Sherringham wondered while his whole sense vibrated. 'She hadn't got hold of it when I went away.' And the assurance flowed over him again that she had found the key to her box of treasures. In the summer, during their weeks of frequent meeting, she had only fumbled with the lock. One October day, while he was away, the key had slipped in, had fitted, or her finger at last had touched the right spring, and the capricious casket had flown open.

It was during the present solemnity that Sherringham, excited by the way she came out and with a hundred startled ideas about her wheeling through his mind, was for the first time and most vividly visited by a perception that ended by becoming frequent with him – that of the perfect presence of mind, unconfused, unhurried by emotion, that any artistic performance requires and that all, whatever the instrument, require in exactly the same degree: the application, in other words, clear and calculated, crystal-firm as it were, of the idea conceived in the glow of experience, of suffering, of joy. Sherringham afterwards often talked of this with Miriam, who however was not able to present him with a neat theory of the subject. She had no knowledge that it was publicly discussed; she was just practically on the side of those who hold that at the moment of production the artist cannot have his wits too much about him. When Peter told her there were people who maintained that in such a crisis he must lose himself in the flurry she stared with surprise and then broke out: 'Ah, the idiots!' She eventually became in her judgements, in impatience and the expression of contempt, very free and absolutely irreverent.

'What a splendid scolding!' Sherringham exclaimed when, on the entrance of the Pope's legate, her companion closed the book upon the scene. Peter pressed his lips to Madame Carré's finger-tips; the old actress got up and held out her arms to Miriam. The girl never took her eyes off Sherringham while she passed into Madame Carré's embrace and remained there. They were full of their usual sombre fire, and it was always the case that they expressed too much anything that they expressed at all; but they were not defiant nor even triumphant now – they were only deeply explicative; they seemed to say: 'That's the sort of thing I meant; that's what I had in mind when I asked you to try to do something for me.' Madame Carré folded her pupil to her bosom, holding her there as the old marquise in a *comédie de mœurs* might, in the last scene, have held her god-daughter the *ingénue*.

'Have you got me an engagement?' Miriam asked of Sherringham. 'Yes, he has done something splendid for me,' she went on to Madame Carré, resting her hand caressingly on one of the actress's while the old woman discoursed with Mr Dashwood, who was telling her in very pretty French that he was tremendously excited about Miss Rooth. Madame Carré looked at him as if she wondered how he appeared when he was calm and how, as a dramatic artist, he expressed that condition.

'Yes, yes, something splendid, for a beginning,' Sherringham answered, radiantly, recklessly; feeling now only that he would say anything, do anything, to please her. He spent on the spot, in imagination, his last penny.

'It's such a pity you couldn't follow it; you would have liked it so much better,' Mr Dashwood observed to his hostess.

'Couldn't follow it? Do you take me for *une sotte*?' the celebrated artist cried. 'I suspect I followed it *de plus près que vous, monsieur*!'

'Ah, you see the language is so awfully fine,' Basil Dashwood replied, looking at his shoes.

'The language? Why, she rails like a fish-wife. Is that what you call language? Ours is another business.'

'If you understood – if you understood you would see the greatness of it,' Miriam declared. And then, in another tone, 'Such delicious expressions!'

'*On dit que c'est très fort*. But who can tell if you really say it?' Madame Carré demanded.

'Ah, *par exemple*, I can!' Sherringham exclaimed.

'Oh, you – you're a Frenchman.'

'Couldn't he tell if he were not?' asked Basil Dashwood.

The old woman shrugged her shoulders. 'He wouldn't know.'

'That's flattering to me.'

'Oh, you – don't you pretend to complain,' Madame Carré said. 'I prefer *our* imprecations – those of Camille,' she went on. 'They have the beauty *des plus belles choses*.'

'I can say them too,' Miriam broke in.

'*Insolente!*' smiled Madame Carré. 'Camille doesn't squat down on the floor in the middle of them.'

> 'For grief is proud and makes his owner stoop.
> To me and to the state of my great grief
> Let kings assemble,'

Miriam quickly declaimed. 'Ah, if you don't feel the way she makes a throne of it!'

'It's really tremendously fine, *chère madame*,' Sherringham said. 'There's nothing like it.'

'*Vous êtes insupportables*,' the old woman answered. 'Stay with us. I'll teach you Phèdre.'

'Ah, Phaedra – Phaedra!' Basil Dashwood vaguely ejaculated, looking more gentlemanly than ever.

'You have learned all I have taught you, but where the devil have you learned what I haven't taught you?' Madame Carré went on.

'I've worked – I have; you'd call it work – all through the bright, late summer, all through the hot, dull, empty days. I've battered down the door – I did hear it crash one day. But I'm not so very good yet: I'm only in the right direction.'

'*Malicieuse!*' murmured Madame Carré.

'Oh, I can beat that,' the girl went on.

'Didn't you wake up one morning and find you had grown a pair of wings?' Sherringham asked. 'Because that's what the difference amounts to – you really soar. Moreover you're an angel,' he added, charmed with her unexpectedness, the good-nature of her forbearance to reproach him for not having written to her. And it seemed to him privately that she *was* angelic when in answer to this she said ever so blandly:

'You know you read *King John* with me before you went

away. I thought over constantly what you said. I didn't understand it much at the time – I was so stupid. But it all came to me later.'

'I wish you could see yourself,' Sherringham answered.

'My dear fellow, I do. What do you take me for? I didn't miss a vibration of my voice, a fold of my robe.'

'I didn't see you looking,' Sherringham returned.

'No one ever will. Do you think I would show it?'

'*Ars celare artem*,' Basil Dashwood jocosely dropped.

'You must first have the art to hide,' said Sherringham, wondering a little why Miriam didn't introduce her young friend to him. She was, however, both then and later, perfectly neglectful of such cares, never thinking or heeding how other people got on together. When she found they didn't get on she laughed at them: that was the nearest she came to arranging for them. Sherringham observed, from the moment she felt her strength, the immense increase of her good-humoured inattention to detail – all detail save that of her work, to which she was ready to sacrifice holocausts of feelings when the feelings were other people's. This conferred on her a kind of profanity, an absence of ceremony in her social relations, which was both amusing, because it suggested that she would take what she gave, and formidable, because it was inconvenient and you might not care to give what she would take.

'If you haven't got any art it's not quite the same as if you didn't hide it, is it?' asked Basil Dashwood.

'That's right – say one of your clever things!' murmured Miriam, sweetly, to the young man.

'You're always acting,' he answered, in English, with a laugh, while Sherringham remained struck with his expressing just what he himself had felt weeks before.

'And when you have shown them your fish-wife, to your public *de là-bas*, what will you do next?' asked Madame Carré.

'I'll do Juliet – I'll do Cleopatra.'

'Rather a big bill, isn't it?' Mr Dashwood volunteered to Sherringham, in a friendly, discriminating manner.

'Constance and Juliet – take care you don't mix them,' said Sherringham.

'I want to be various. You once told me I had a hundred characters,' Miriam replied.

'Ah, *vous-en-êtes là*?' cried the old actress. 'You may have a hundred characters, but you have only three plays. I'm told that's all there are in English.'

Miriam appealed to Sherringham. 'What arrangements have you made? What do the people want?'

'The people at the theatre?'

'I'm afraid they don't want King John, and I don't believe they hunger for Antony and Cleopatra,' Basil Dashwood suggested. 'Ships and sieges and armies and pyramids, you know: we mustn't be too heavy.'

'Oh, I hate scenery!' sighed Miriam.

'*Elle est superbe*,' said Madame Carré. 'You must put those pieces on the stage: how will you do it?'

'Oh, we know how to get up a play in London, Madame Carré,' Basil Dashwood responded, genially. 'They put money on it, you know.'

'On it? But what do they put *in* it? Who will interpret them? Who will manage a style like that – the style of which the verses she just repeated are a specimen? Whom have you got that one has ever heard of?'

'Oh, you'll hear of a good deal when once she gets started,' Basil Dashwood contended, cheerfully.

Madame Carré looked at him a moment; then, 'You'll become very bad,' she said to Miriam. 'I'm glad I shan't see it.'

'People will do things for me – I'll make them,' the girl declared. 'I'll stir them up so that they'll have ideas.'

'What people, pray?'

'Ah, terrible woman!' Sherringham moaned, theatrically.

'We translate your pieces – there will be plenty of parts,' Basil Dashwood said.

'Why then go out of the door to come in at the window? – especially if you smash it! An English arrangement of a French piece is a pretty woman with her back turned.'

'Do you really want to keep her?' Sherringham asked of Madame Carré, as if he were thinking for a moment that this after all might be possible.

She bent her strange eyes on him. 'No, you are all too queer together; we couldn't be bothered with you, and you're not worth it.'

'I'm glad it's together; we can console each other.'

'If you only would; but you don't seem to! In short, I don't understand you – I give you up. But it doesn't matter,' said the old woman, wearily, 'for the theatre is dead and even you, *ma toute-belle*, won't bring it to life. Everything is going from bad to worse, and I don't care what becomes of you. You wouldn't understand us here and they won't understand you there, and everything is impossible, and no one is a whit the wiser, and it's not of the least consequence. Only when you raise your arms lift them just a little higher,' Madame Carré added.

'My mother will be happier *chez nous*,' said Miriam, throwing her arms straight up with a noble tragic movement.

'You won't be in the least in the right path till your mother's in despair.'

'Well, perhaps we can bring that about even in London,' Sherringham suggested, laughing.

'Dear Mrs Rooth – she's great fun,' Mr Dashwood dropped.

Miriam transferred the gloomy beauty of her gaze to him, as if she were practising. '*You* won't upset her, at any rate.' Then she stood with her fatal mask before Madame Carré. 'I want to do the modern too. I want to do *le drame*, with realistic effects.'

'And do you want to look like the portico of the Madeleine when it's draped for a funeral?' her instructress mocked. 'Never, never. I don't believe you're various: that's not the way I see you. You're pure tragedy, with *de grands effets de voix*, in the great style, or you're nothing.'

'Be beautiful – be only that,' Sherringham advised. 'Be only what you can be so well – something that one may turn to for a glimpse of perfection, to lift one out of all the vulgarities of the day.'

Thus apostrophized, the girl broke out with one of the speeches of Racine's Phaedra, hushing her companions on the instant. 'You'll be the English Rachel,' said Basil Dashwood when she stopped.

'Acting in French!' Madame Carré exclaimed. 'I don't believe in an English Rachel.'

'I shall have to work it out, what I shall be,' Miriam responded with a rich, pensive effect.

'You're in wonderfully good form today,' Sherringham said to her; his appreciation revealing a personal subjection which he was unable to conceal from his companions, much as he wished it.

'I really mean to do everything.'

'Very well; after all, Garrick did.'

'Well, I shall be the Garrick of my sex.'

'There's a very clever author doing something for me; I should like you to see it,' said Basil Dashwood, addressing himself equally to Miriam and to her diplomatic friend.

'Ah, if you have very clever authors!' Madame Carré spun the sound to the finest satiric thread.

'I shall be very happy to see it,' said Sherringham.

This response was so benevolent that Basil Dashwood presently began: 'May I ask you at what theatre you have made arrangements?'

Sherringham looked at him a moment. 'Come and see me at the Embassy and I'll tell you.' Then he added: 'I know your sister, Mrs Lovick.'

'So I supposed: that's why I took the liberty of asking such a question.'

'It's no liberty; but Mr Sherringham doesn't appear to be able to tell you,' said Miriam.

'Well, you know it's a very curious world, all those theatrical people over there,' Sherringham said.

'Ah, don't say anything against them, when I'm one of them,' Basil Dashwood laughed.

'I might plead the absence of information, as Miss Rooth has neglected to make us acquainted.'

Miriam smiled: 'I know you both so little.' But she presented them, with a great stately air, to each other, and the two men shook hands while Madame Carré observed them.

'*Tiens!* you gentlemen meet here for the first time? You do right to become friends – that's the best thing. Live together in peace and mutual confidence. *C'est de beaucoup le plus sage.*'

'Certainly, for yoke-fellows,' said Sherringham.

He began the next moment to repeat to his new acquaintance

some of the things he had been told in London; but their hostess stopped him off, waving the talk away with charming overdone stage-horror and the young hands of the heroines of Marivaux. 'Ah, wait till you go, for that! Do you suppose I care for news of your mountebanks' booths?'

20

As many people know, there are not, in the famous Théâtre Français, more than a dozen good seats accessible to ladies. The stalls are forbidden them, the boxes are a quarter of a mile from the stage and the balcony is a delusion save for a few chairs at either end of its vast horseshoe. But there are two excellent *baignoires d'avant-scène*, which indeed are by no means always to be had. It was however into one of them that, immediately after his return to Paris, Sherringham ushered Mrs Rooth and her daughter, with the further escort of Basil Dashwood. He had chosen the evening of the reappearance of the celebrated Mademoiselle Voisin (she had been enjoying a *congé* of three months), an actress whom Miriam had seen several times before and for whose method she professed a high though somewhat critical esteem. It was only for the return of this charming performer that Sherringham had been waiting to respond to Miriam's most ardent wish – that of spending an hour in the *foyer des artistes* of the great theatre. She was the person whom he knew best in the house of Molière; he could count upon her to do them the honours some night when she was in the 'bill', and make the occasion sociable. Miriam had been impatient for it – she was so convinced that her eyes would be opened in the holy of holies; but wishing particularly, as he did, to participate in her impression he had made her promise that she would not taste of this experience without him – not let Madame Carré, for instance, take her in his absence. There were questions the girl wished to put to Mademoiselle Voisin – questions which, having admired her from the balcony, she felt she was exactly the person to answer. She was more 'in it' now, after all, than Madame Carré, in spite of her slenderer talent: she was younger, fresher, more modern and (Miriam found the word) less academic. Sherringham perfectly foresaw the day when his young friend would make indulgent allowances for poor Madame Carré, patronizing her as an old woman of good intentions.

The play, tonight, was six months old, a large, serious, successful comedy, by the most distinguished of authors, with a thesis, a chorus, embodied in one character, a *scène à faire* and a part full of opportunities for Mademoiselle Voisin. There were things to be said about this artist, strictures to be dropped as to the general quality of her art, and Miriam leaned back now, making her comments as if they cost her less; but the actress had knowledge and distinction and pathos, and our young lady repeated several times: 'How quiet she is, how wonderfully quiet! Scarcely anything moves but her face and her voice. *Le geste rare*, but really expressive when it comes. I like that economy; it's the only way to make the gesture significant.'

'I don't admire the way she holds her arms,' Basil Dashwood said: 'like a *demoiselle de magasin* trying on a jacket.'

'Well, she holds them, at any rate. I dare say it's more than you do with yours.'

'Oh, yes, she holds them; there's no mistake about that. "I hold them, I hope, *hein*?" she seems to say to all the house.' The young English professional laughed good-humouredly, and Sherringham was struck with the pleasant familiarity he had established with their brave companion. He was knowing and ready, and he said, in the first entr'acte (they were waiting for the second to go behind), amusing, perceptive things. 'They teach them to be ladylike, and Voisin is always trying to show that. "See how I walk, see how I sit, see how quiet I am and how I have *le geste rare*. Now can you say I ain't a lady?" She does it all as if she had a class.'

'Well, tonight I'm her class,' said Miriam.

'Oh, I don't mean of actresses, but of *femmes du monde*. She shows them how to act in society.'

'You had better take a few lessons,' Miriam retorted.

'You should see Voisin in society,' Sherringham interposed.

'Does she go into it?' Mrs Rooth demanded, with interest.

Sherringham hesitated. 'She receives a great many people.'

'Why shouldn't she, when they're nice?' Mrs Rooth continued.

'When the people are nice?' Miriam asked.

'Now don't tell me *she's* not what one would wish,' said Mrs Rooth to Sherringham.

'It depends upon what that is,' he answered, smiling.

'What I should wish if she were my daughter,' the old woman rejoined, blandly.

'Ah, wish your daughter to act as well as that, and you'll do the handsome thing for her!'

'Well, she *seems* to feel what she says,' Mrs Rooth murmured, piously.

'She has some stiff things to say. I mean about her past,' Basil Dashwood remarked. 'The past – the dreadful past – on the stage!'

'Wait till the end, to see how she comes out. We must all be merciful!' sighed Mrs Rooth.

'We've seen it before; you know what happens,' Miriam observed to her mother.

'I've seen so many, I get them mixed.'

'Yes, they're all in queer predicaments. Poor old mother – what we show you!' laughed the girl.

'Ah, it will be what *you* show me: something noble and wise!'

'I want to do this; it's a magnificent part,' said Miriam.

'You couldn't put it on in London; they wouldn't swallow it,' Basil Dashwood declared.

'Aren't there things they do there, to get over the difficulties?'

'You can't get over what *she* did,' the young man replied.

'Yes, we must pay, we must expiate!' Mrs Rooth moaned, as the curtain rose again.

When the second act was over our friends passed out of their *baignoire* into those corridors of tribulation where the bristling *ouvreuse*, like a pawnbroker driving a roaring trade, mounts guard upon piles of heterogeneous clothing, and, gaining the top of the fine staircase which forms the state entrance and connects the statued vestibule of the basement with the grand tier of boxes, opened an ambiguous door, composed of little mirrors, and found themselves in the society of the initiated. The janitors were courteous folk who greeted Sherringham as an acquaintance, and he had no difficulty in marshalling his little troop towards the foyer. They traversed a low, curving lobby, hung with pictures and furnished with velvet-covered benches, where several unrecognized persons, of both sexes, looked at them without hostility, and arrived at an opening on the right from which by a short flight of steps there was a descent to one of the wings of the stage. Here Miriam paused in silent excitement, like a young warrior arrested by a glimpse of the battle-field. Her vision was carried off through a lane of light to the point of vantage from which the actor held the house; but there was a hushed guard over the place, and curiosity could only glance and pass.

Then she came with her companions to a sort of parlour with a polished floor, not large and rather vacant, where her attention flew delightedly to a coat-tree, in a corner, from which three or four dresses were suspended – dresses that she immediately perceived to be costumes in that night's play – accompanied by a saucer of something and a much-worn powder-puff casually left upon a sofa. This was a familiar note in a general impression (it had begun at the threshold) of high decorum – a sense of majesty in the place. Miriam rushed at the powder-puff (there was no one in the room), snatched it up and gazed at it with droll veneration; then stood rapt a moment before the charming petticoats ('That's Dunoyer's first under-skirt,' she said to her mother), while Sherringham explained that in this apartment an actress traditionally changed her gown, when the transaction was simple enough, to save the long ascent to her *loge*. He felt like a cicerone showing a church to a party of provincials; and indeed there was a grave hospitality in the air, mingled with something academic and important, the tone of an institution, a temple, which made them all, out of respect and delicacy, hold their breath a little and tread the shining floors with discretion.

These precautions increased (Mrs Rooth crept in like a friendly but undomesticated cat) after they entered the foyer itself, a square spacious saloon, covered with pictures and relics and draped in official green velvet, where the *genius loci* holds a reception every night in the year. The effect was freshly charming to Sherringham; he was fond of the place, always saw it again with pleasure, enjoyed its honourable look and the way, among the portraits and scrolls, the records of a splendid history, the green velvet and the waxed floors, the *genius loci* seemed to be 'at home' in the quiet lamplight. At the end of the room, in an ample chimney, blazed a fire of logs. Miriam said nothing; they looked about, noting that most of the portraits and pictures were 'old-fashioned', and Basil Dashwood expressed disappointment at the absence of all the people they wanted most to see. Three or four gentlemen in evening dress circulated slowly, looking, like themselves, at the pictures, and another gentleman stood before a lady, with whom he was in conversation, seated against the wall. The foyer, in these conditions, resembled a ball-room cleared for the dance, before the guests or the music had arrived.

'Oh, it's enough to see *this*; it makes my heart beat,' said Miriam. 'It's full of the vanished past, it makes me cry. I feel them here, the great artists I shall never see. Think of Rachel (look at her grand portrait there!) and how she stood on these very boards and trailed over them the robes of Hermione and Phèdre!' The girl broke out theatrically, as on the spot was right, not a bit afraid of her voice as soon as it rolled through the room; appealing to her companions as they stood under the chandelier and making the other persons present, who had already given her some attention, turn round to stare at so unusual a specimen of the English miss. She laughed musically when she noticed this, and her mother, scandalized, begged her to lower her tone. 'It's all right. I produce an effect,' said Miriam: 'it shan't be said that I too haven't had my little success in the *maison de Molière*.' And Sherringham repeated that it was all right – the place was familiar with mirth and passion, there was often wonderful talk there, and it was only the setting that was still and solemn. It happened that this evening – there was no knowing in advance – the scene was not characteristically brilliant; but to confirm his assertion, at the moment he spoke, Mademoiselle Dunoyer, who was also in the play, came into the room attended by a pair of gentlemen.

She was the celebrated, the perpetual, the necessary *ingénue*, who with all her talent could not have represented a woman of her actual age. She had the gliding, hopping movement of a small bird, the same air of having nothing to do with time, and the clear, sure, piercing note, a miracle of exact vocalization. She chaffed her companions, she chaffed the room; she seemed to be a very clever little girl trying to personate a more innocent big one. She scattered her amiability (showing Miriam how much the children of Molière took their ease), and it quickly placed her in the friendliest communication with Peter Sherringham, who already enjoyed her acquaintance and who now extended it to his companions and in particular to the young lady *sur le point d'entrer au théâtre*.

'You deserve a happier lot,' said the actress, looking up at Miriam brightly, as if to a great height, and taking her in; upon which Sherringham left them together a little and led Mrs Rooth and young Dashwood to consider further some of the pictures.

'Most delightful, most curious,' the old woman murmured, about everything; while Basil Dashwood exclaimed, in the presence of most of the portraits: 'But their ugliness – their ugliness: did you ever see such a collection of hideous people? And those who were supposed to be good-looking – the beauties of the past – they are worse than the others. Ah, you may say what you will, *nous sommes mieux que ça!*' Sherringham suspected him of irritation, of not liking the theatre of the great rival nation to be thrust down his throat. They returned to Miriam and Mademoiselle Dunoyer, and Sherringham asked the actress a question about one of the portraits, to which there was no name attached. She replied, like a child who had only played about the room, that she was *toute honteuse* not to be able to tell him the original: she had forgotten, she had never asked – '*Vous allez me trouver bien légère!*' She appealed to the other persons present, who formed a gallery for her, and laughed in delightful ripples at their suggestions, which she covered with ridicule. She bestirred herself; she declared she would ascertain, she should not be happy till she did, and swam out of the room, with the prettiest paddles, to obtain the information, leaving behind her a perfume of delicate kindness and gaiety. She seemed above all things obliging, and Sherringham said that she was almost as natural off the stage as on. She didn't come back.

21

WHETHER Sherringham had prearranged it is more than I can say, but Mademoiselle Voisin delayed so long to show herself that Mrs Rooth, who wished to see the rest of the play, though she had sat it out on another occasion, expressed a returning relish for her corner of the *baignoire* and gave her conductor the best pretext he could have desired for asking Basil Dashwood to be so good as to escort her back. When the young actor, of whose personal preference Sherringham was quite aware, had led Mrs Rooth away with an absence of moroseness which showed that his striking analogy with a gentleman was not kept for the footlights, the two others sat on a divan in the part of the room furthest from the entrance, so that it gave them a degree of privacy, and Miriam watched the coming and going of their fellow-visitors and the indefinite people, attached to the theatre, hanging about, while her companion gave a name to some of the figures, Parisian celebrities.

'Fancy poor Dashwood, cooped up there with mamma!' the girl exclaimed, whimsically.

'You're awfully cruel to him; but that's of course,' said Sherringham.

'It seems to me I'm as kind as you: you sent him off.'

'That was for your mother: she was tired.'

'Oh, gammon! And why, if I *were* cruel, should it be of course?'

'Because you must destroy and torment and consume – that's your nature. But you can't help your type, can you?'

'My type?' the girl repeated.

'It's bad, perverse, dangerous. It's essentially insolent.'

'And pray what is yours, when you talk like that? Would you say such things if you didn't know the depths of my good-nature?'

'Your good-nature all comes back to that,' said Sherringham. 'It's an abyss of ruin – for others. You have no respect. I'm speaking of the artistic character, in the direction and in the plenitude in which you have it. It's unscrupulous, nervous, capricious, wanton.'

'I don't know about respect: one can be good,' Miriam reasoned.

'It doesn't matter, so long as one is powerful,' answered Sherringham. 'We can't have everything, and surely we ought to understand that we must pay for things. A splendid organization for a special end, like yours, is so rare and rich and fine that we oughtn't to grudge it its conditions.'

'What do you call its conditions?' Miriam demanded, turning and looking at him.

'Oh, the need to take its ease, to take up space, to make itself at home in the world, to square its elbows and knock others about. That's large and free; it's the good-nature you speak of. You must forage and ravage and leave a track behind you; you must live upon the country you occupy. And you give such delight that, after all, you are welcome – you are infinitely welcome!'

'I don't know what you mean. I only care for the idea,' Miriam said.

'That's exactly what I pretend; and we must all help you to it. You use us, you push us about, you break us up. We are your tables and chairs, the simple furniture of your life.'

'Whom do you mean by "we"?'

Sherringham gave a laugh. 'Oh, don't be afraid – there will be plenty of others.'

Miriam made no rejoinder to this, but after a moment she broke out again: 'Poor Dashwood, immured with mamma – he's like a lame chair that one has put into the corner.'

'Don't break him up before he has served. I really believe that something will come out of him,' her companion went on. 'However, you'll break me up first,' he added, 'and him probably never at all.'

'And why shall I honour you so much more?'

'Because I'm a better article, and you'll feel that.'

'You have the superiority of modesty – I see.'

'I'm better than a young mountebank – I've vanity enough to say that.'

She turned upon him with a flush in her cheek and a splendid dramatic face. 'How you hate us! Yes, at bottom, below your little taste, you *hate* us!' she repeated.

He coloured too, met her eyes, looked into them a minute, seemed to accept the imputation, and then said quickly: 'Give it up; come away with me.'

'Come away with you?'

'Leave this place: give it up.'

'You brought me here, you insisted it should be only you, and now you must stay,' she declared, with a head-shake and a laugh. 'You should know what you want, dear Mr Sherringham.'

'I do – I know now. Come away, before she comes.'

'Before she comes?'

'She's success – this wonderful Voisin – she's triumph, she's full accomplishment: the hard, brilliant realization of what I want to avert for you.' Miriam looked at him in silence, the angry light still in her face, and he repeated: 'Give it up – give it up.'

Her eyes softened after a moment; she smiled and then she said: 'Yes, you're better than poor Dashwood.'

'Give it up and we'll live for ourselves, *in* ourselves, in something that can have a sanctity.'

'All the same, you do hate us,' the girl went on.

'I don't want to be conceited, but I mean that I'm sufficiently fine and complicated to tempt you. I'm an expensive modern watch, with a wonderful escapement – therefore you'll smash me if you can.'

'Never – never!' said the girl, getting up. 'You tell me the hour too well.' She quitted her companion and stood looking at Gérôme's fine portrait of the pale Rachel, invested with the antique attributes of tragedy. The rise of the curtain had drawn away most of the company. Sherringham, from his bench, watched Miriam a little, turning his eye from her to the vivid image of the dead actress and thinking that his companion suffered little by the juxtaposition. Presently he came over and joined her again, and she said to him: 'I wonder if that is what your cousin had in his mind.'

'My cousin?'

'What was his name? Mr Dormer; that first day at Madame Carré's. He offered to paint my portrait.'

'I remember. I put him up to it.'

'Was he thinking of this?'

'I don't think he has ever seen it. I dare say I was.'

'Well, when we go to London he must do it,' said Miriam.

'Oh, there's no hurry,' Sherringham replied.

'Don't you want my picture?' asked the girl, with one of her successful touches.

'I'm not sure I want it from you. I don't know quite what he'd make of you.'

'He looked so clever – I liked him. I saw him again at your party.'

'He's a dear fellow; but what is one to say of a painter who goes for his inspiration to the House of Commons?'

'To the House of Commons?'

'He has lately got himself elected.'

'Dear me, what a pity! I wanted to sit for him; but perhaps he won't have me, as I'm not a member of Parliament.'

'It's my sister, rather, who has got him in.'

'Your sister who was at your house that day? What has she to do with it?'

'Why, she's his cousin, just as I am. And in addition,' Sherringham went on, 'she's to be married to him.'

'Married – really? So he paints *her*, I suppose?'

'Not much, probably. His talent in that line isn't what she esteems in him most.'

'It isn't great, then?'

'I haven't the least idea.'

'And in the political line?'

'I scarcely can tell. He's very clever.'

'He does paint then?'

'I dare say.'

Miriam looked at Gérôme's picture again. 'Fancy his going into the House of Commons! And your sister put him there?'

'She worked, she canvassed.'

'Ah, you're a queer family!' the girl exclaimed, turning round at the sound of a step.

'We're lost – here's Mademoiselle Voisin,' said Sherringham.

This celebrity presented herself smiling and addressing Miriam. 'I acted for *you* tonight – I did my best.'

'What a pleasure to speak to you, to thank you!' the girl murmured, admiringly. She was startled and dazzled.

243

'I couldn't come to you before, but now I've got a rest – for half an hour,' the actress went on. Gracious and passive, as if she were a little tired, she let Sherringham, without looking at him, take her hand and raise it to his lips. 'I'm sorry I make you lose the others – they are so good in this act,' she added.

'We have seen them before, and there's nothing so good as you,' Miriam replied.

'I like my part,' said Mademoiselle Voisin, gently, smiling still at our young lady with clear, charming eyes. 'One is always better in that case.'

'She's so bad sometimes, you know!' Sherringham jested, to Miriam; leading the actress to glance at him kindly and vaguely, with a little silence which, with her, you could not call embarrassment, but which was still less affectation.

'And it's so interesting to be here – *so* interesting!' Miriam declared.

'Ah, you like our old house? Yes, we are very proud of it.' And Mademoiselle Voisin smiled again at Sherringham, good-humouredly, as if to say: 'Well, here I am, and what do you want of me? Don't ask me to invent it myself, but if you'll tell me I'll do it.' Miriam admired the note of discreet interrogation in her voice – the slight suggestion of surprise at their 'old house' being liked. The actress was already an astonishment to her, from her seeming still more perfect on a nearer view; which was not, she had an idea, what actresses usually did. This was very encouraging to her: it widened the programme of a young lady about to embrace the scenic career. To have so much to show before the footlights and yet to have so much left when you came off – that was really wonderful. Mademoiselle Voisin's eyes, as one looked into them, were still more agreeable than the distant spectator would have supposed; and there was in her appearance an extreme finish which instantly suggested to Miriam that she herself, in comparison, was big and rough and coarse.

'You're lovely tonight – you're particularly lovely,' said Sherringham, very frankly, translating Miriam's own impression and at the same time giving her an illustration of the way that, in Paris at least, gentlemen expressed themselves to the stars of the drama. She thought she knew her companion very well and had been witness of the degree to which, under these circumstances, his familiarity could increase; but his address to the slim, distinguished, harmonious woman before them had a different quality, the note of a special usage. If Miriam had had any apprehension that such directness might be taken as excessive, it was removed by the manner in which Mademoiselle Voisin returned –

'Oh, one is always well enough when one is made up; one is always exactly the same.' That served as an example of the good taste with which a star of the drama could receive homage that was wanting in originality. Miriam determined on the spot that this should be the way *she* would receive it. The grace of her new acquaintance was the greater as the becoming bloom which she alluded to as artificial was the result of a science so consummate that it had none of the grossness of a mask. The perception of all this was exciting to our young aspirant, and her excitement relieved itself in the inquiry, which struck her as rude as soon as she had uttered it –

'You acted for *me*? How did you know? What am I to you?'

'Monsieur Sherringham has told me about you. He says we are nothing beside you; that you are to be the great star of the future. I'm proud that you've seen me.'

'That of course is what I tell every one,' Sherringham said, a trifle awkwardly, to Miriam.

'I can believe it when I see you. *Je vous ai bien observée,*' the actress continued, in her sweet, conciliatory tone.

Miriam looked from one of her interlocutors to the other, as if there was a joy to her in this report of Sherringham's remarks, accompanied however, and partly mitigated, by a quicker vision of what might have passed between a secretary of embassy and a creature so exquisite as Mademoiselle Voisin. 'Ah, you're wonderful people – a most interesting impression!' she sighed.

'I was looking for you; he had prepared me. We are such old friends!' said the actress, in a tone courteously exempt from intention: upon which Sherringham again took her hand and raised it to his lips, with a tenderness which her whole appearance seemed to bespeak for her, a sort of practical consideration and carefulness of touch, as if she were an object precious and frail, an instrument for producing rare sounds, to be handled, like a legendary violin, with a recognition of its value.

'Your dressing-room is so pretty – show her your dressing-room,' said Sherringham.

'Willingly, if she'll come up. *Vous savez, c'est une montée.*'

'It's a shame to inflict it on *you*,' Miriam objected.

'*Comment donc?* If it will interest you in the least!' They exchanged civilities, almost caresses, trying which could have the nicest manner to the other. It was the actress's manner that struck Miriam most; it denoted such a training, so much taste, expressed such a ripe conception of urbanity.

'No wonder she acts well when she has that tact – feels, perceives, is so remarkable, *mon Dieu, mon Dieu!*' Miriam said to herself as they followed their conductress into another corridor and up a wide, plain staircase. The staircase was spacious and long, and this part of the establishment was sombre and still, with the gravity of a college or a convent. They reached another passage, lined with little doors, on each of which the name of a comedian was painted; and here the aspect became still more monastic, like that of a row of solitary cells. Mademoiselle Voisin led the way to her own door, obligingly, as if she wished to be hospitable, dropping little subdued, friendly attempts at explanation on the way. At her threshold the monasticism stopped. Miriam found herself in a wonderfully upholstered nook, a nest of lamplight and delicate cretonne. Save for its pair of long glasses it looked like a tiny boudoir, with a water-colour drawing of value in each panel of stretched stuff, its crackling fire, its charming order. It was intensely bright and extremely hot, singularly pretty and exempt from litter. Nothing was lying about, but a small draped doorway led into an inner sanctuary. To Miriam it seemed royal; it immediately made the art of the comedian the most distinguished thing in the world. It was just such a place as they *should* have, in the intervals, if they were expected to be great artists. It was a result of the same evolution as Mademoiselle Voisin herself – not that our young lady found this particular term to her hand, to express her idea. But her mind was flooded with an impression of style, of refinement, of the long continuity of a tradition. The actress said, '*Voilà, c'est tout!*' as if it were little enough and there were even something clumsy in her having brought them so far for nothing and in their all sitting there waiting and looking at each other till it was time for her to change her dress. But to Miriam it was occupation enough to note what she did and said: these things and her whole person and carriage struck her as exquisite in their adaptation to the particular occasion. She had had an idea that foreign actresses were rather of the *cabotin* order; but her hostess suggested to her much more a princess than a cabotine. She would do things as she liked, and straight off: Miriam couldn't fancy her in the gropings and humiliations of rehearsal. Everything in her had been sifted and formed, her tone was perfect, her amiability complete, and she might have been the charming young wife of a secretary of state receiving a pair of strangers of distinction. Miriam observed all her movements. And then, as Sherringham had said, she was particularly lovely.

Suddenly she told Sherringham that she must put him *à la porte* – she wanted to change her dress. He retired and returned to the foyer, where Miriam was to rejoin him after remaining the few minutes more with Mademoiselle Voisin and coming down with her. He waited for his companion, walking up and down and making up his mind; and when she presently came in he said to her:

'Please don't go back for the rest of the play. Stay here.' They now had the foyer virtually to themselves.

'I want to stay here. I like it better.' She moved back to the chimney-piece, from above which the cold portrait of Rachel looked down, and as he accompanied her he said:

'I meant what I said just now.'

'What you said to Voisin?'

'No, no; to you. Give it up and live with *me*.'

'Give it up?' And she turned her stage face upon him.

'Give it up and I'll marry you tomorrow.'

'This is a happy time to ask it!' she mocked. 'And this is a good place.'

'Very good indeed, and that's why I speak: it's a place to make one choose – it puts all before one.'

'To make *you* choose, you mean. I'm much obliged, but that's not my choice,' laughed Miriam.

'You shall be anything you like – except this.'

247

'Except what I most want to be? I *am* much obliged.'

'Don't you care for me? Haven't you any gratitude?' Sherringham asked.

'Gratitude for kindly removing the blessed cup from my lips? I want to be what *she* is – I want it more than ever.'

'Ah, what she is!' he replied impatiently.

'Do you mean I can't? We'll see if I can't. Tell me more about her – tell me everything.'

'Haven't you seen for yourself, and can't you judge?'

'She's strange, she's mysterious,' Miriam declared, looking at the fire. 'She showed us nothing – nothing of her real self.'

'So much the better, all things considered.'

'Are there all sorts of other things in her life? That's what I believe,' Miriam went on, raising her eyes to him.

'I can't tell you what there is in the life of such a woman.'

'Imagine – when she's so perfect!' the girl exclaimed, thoughtfully. 'Ah, she kept me off – she kept me off! Her charming manner is in itself a kind of contempt. It's an abyss – it's the wall of China. She has a hard polish, an inimitable surface, like some wonderful porcelain that costs more than you'd think.'

'Do you want to become like that?' Sherringham asked.

'If I could I should be enchanted. One can always try.'

'You must act better than she,' said Sherringham.

'Better? I thought you wanted me to give it up.'

'Ah, I don't know what I want, and you torment me and turn me inside out! What I want is you yourself.'

'Oh, don't worry,' said Miriam, kindly. Then she added that Mademoiselle Voisin had asked her to come to see her; to which Sherringham replied, with a certain dryness, that she would probably not find that necessary. This made Miriam stare, and she asked, 'do you mean it won't do, on account of mamma's prejudices?'

'Say, this time, on account of mine.'

'Do you mean because she has lovers?'

'Her lovers are none of our business.'

'None of mine, I see. So you have been one of them?'

'No such luck.'

'What a pity! I should have liked to see that. One must see everything, to be able to do everything.' And as he inquired what she had wished to see she replied: 'The way a woman like that receives one of the old ones.'

Sherringham gave a groan at this, which was at the same time partly a laugh, and, turning away and dropping upon a bench, ejaculated: 'You'll do – you'll do!'

He sat there some minutes with his elbows on his knees and his face in his hands. Miriam remained looking at the portrait of Rachel; after which she demanded: 'Doesn't such a woman as that receive – receive every one?'

'Every one who goes to see her, no doubt.'

'And who goes?'

'Lots of men – clever men, eminent men.'

'Ah, what a charming life! Then doesn't she go out?'

'Not what we Philistines mean by that – not into society, never. She never enters a lady's drawing-room.'

'How strange, when one's as distinguished as that; except that she must escape a lot of stupidities and *corvées*. Then where does she learn such manners?'

'She teaches manners, *à ses heures*: she doesn't need to learn them.'

'Oh, she has given me ideas! But in London actresses go into society,' Miriam continued.

'Oh, in London *nous mêlons les genres*!'

'And shan't *I* go – I mean if I want?'

'You'll have every facility to bore yourself. Don't doubt of it.'

'And doesn't she feel excluded?' Miriam asked.

'Excluded from what? She has the fullest life.'

'The fullest?'

'An intense artistic life. The cleverest men in Paris talk over her work with her; the principal authors of plays discuss with her subjects and characters and questions of treatment. She lives in the world of art.'

'Ah, the world of art – how I envy her! And you offer me Dashwood!'

Sherringham rose in his emotion. 'I offer you –?'

Miriam burst out laughing. 'You look so droll! You offer me yourself then, instead of all these things.'

'My child, I also am a very clever man,' he said, smiling, though conscious that for a moment he had stood gaping.

'You are – you are; I delight in you. No ladies at all – no *femmes comme il faut*?' Miriam began again.

'Ah, what do *they* matter? Your business is the artistic life!' he broke out with inconsequence and with a little irritation at hearing her sound that trivial note again.

'You're a dear – your charming good sense comes back to you! What do you want of me then?'

'I want you for myself – not for others; and now, in time, before anything's done.'

'Why then did you bring me here? Everything's done; I feel it tonight.'

'I know the way you should look at it – if you do look at it at all,' Sherringham conceded.

'That's so easy! I thought you liked the stage so,' Miriam said, artfully.

'Don't you want me to be a great swell?'

'And don't you want *me* to be?'

'You will be – you'll share my glory.'

'So will you share mine.'

'The husband of an actress? Yes, I see that!' Sherringham cried, with a frank ring of disgust.

'It's a silly position, no doubt. But if you're too good for it why talk about it? Don't you think I'm important?' Miriam inquired. Her companion stood looking at her, and she suddenly said in a different tone: 'Ah, why should we quarrel, when you have been so kind, so generous? Can't we always be friends – the solidest friends?'

Her voice sank to the sweetest cadence and her eyes were grateful and good as they rested on him. She sometimes said things with such perfection that they seemed dishonest, but in this case Sherringham was stirred to an expressive response. Just as he was making it, however, he was moved to utter other words – 'Take care, here's Dashwood!' Mrs Rooth's companion was in the doorway. He had come back to say that they really must relieve him.

22

MRS DALLOW came up to London soon after the meeting of Parliament; she made no secret of the fact that she was fond of the place, and naturally in present conditions it would not have become less attractive to her. But she prepared to withdraw from it again for the Easter vacation, not to return to Harsh, but to pay a couple of country visits. She did not however leave town with the crowd – she never did anything with the crowd – but waited till the Monday after Parliament rose; facing with composure, in Great Stanhope Street, the horrors, as she had been taught to consider them, of a Sunday out of the session. She had done what she could to mitigate them by asking a handful of 'stray men' to dine with her that evening. Several members of this disconsolate class sought comfort in Great Stanhope Street in the afternoon, and them for the most part she also invited to come back at eight o'clock. There were therefore almost too many people at dinner – there were even a couple of wives. Nick Dormer came to dinner, but he was not present in the afternoon. Each of the persons who were had said on coming in: 'So you've not gone – I'm awfully glad.' Mrs Dallow had replied, 'No, I've not gone,' but she had in no case added that she was glad, nor had she offered an explanation. She never offered explanations: she always assumed that no one could invent them so well as those who had the florid taste to desire them.

And in this case she was right, for it is probable that few of her visitors failed to say to themselves that her not having gone would have had something to do with Dormer. That could pass for an explanation with many of Mrs Dallow's visitors, who as a general thing were not morbidly analytic; especially with those who met Nick as a matter of course at the dinner. His being present at this lady's entertainments, being in her house whenever, as the phrase was, a candle was lighted, was taken as a sign that there was something rather particular between them. Nick had said to her more than once that people would wonder why they didn't marry; but he was wrong in this, inasmuch as there were many of their friends to whom it would not have occurred that his position could be improved by it. That they were cousins was a fact not so evident to others as to themselves, in consequence of which they appeared remarkably intimate. The person seeing clearest in the matter was Mrs Gresham, who lived so much in the world that being alone had become her idea of true sociability. She knew very well that if she had been privately engaged to a young man as amiable as Nick Dormer she would have managed that publicity should not play such a part in their intercourse; and she had her secret scorn for the stupidity of people whose conception of Nick's relation to Julia Dallow rested on the fact that he was always included in her parties. 'If he never was there they might talk,' she said to herself. But Mrs Gresham was supersubtle. To her it would have appeared natural that Julia should celebrate the parliamentary recess by going down to Harsh and securing Nick's company there for a fortnight; she recognized Mrs Dallow's actual plans as a comparatively poor substitute – the project of spending the holidays in other people's houses, to which Nick had also promised to come. Mrs Gresham was romantic; she wondered what was the good of mere snippets and snatches, the chances that any one might have, when large, still days *à deux* were open to you – chances of which half the sanctity was in what they excluded. However, there were more unsettled matters between Mrs Dallow and her queer kinsman than even Mrs Gresham's fine insight could embrace. She was not present on the Sunday before Easter at the dinner in Great Stanhope Street; but if she had been Julia's singular indifference to observation would have stopped short of encouraging her to remain in the drawing-room with Nick after the others had gone. I may add that Mrs Gresham's extreme curiosity would have emboldened her as little to do so. She would have taken for granted that the pair wished to be alone together, though she would have regarded this only as a snippet.

The guests stayed late and it was nearly twelve o'clock when Nick, standing before the fire in the room they had quitted, broke out to his companion:

'See here, Julia, how long do you really expect me to endure this kind of thing?' Mrs Dallow made him no answer; she only leaned back in her chair with her eyes upon his. He met her gaze for a moment; then he turned round to the fire and for another moment looked into it. After this he faced Mrs Dallow again with the exclamation: 'It's so foolish – it's so damnably foolish!'

She still said nothing, but at the end of a minute she spoke without answering him. 'I shall expect you on Tuesday, and I hope you'll come by a decent train.'

'What do you mean by a decent train?'

'I mean I hope you'll not leave it till the last thing before dinner, so that we can have a little walk or something.'

'What's a little walk or something? Why, if you make such a point of my coming to Griffin, do you want me to come at all?'

Mrs Dallow hesitated an instant; then she exclaimed: 'I knew you hated it!'

'You provoke me so,' said Nick. 'You try to, I think.'

'And Severals still worse. You'll get out of that if you can,' Mrs Dallow went on.

'If I can? What's to prevent me?'

'You promised Lady Whiteroy. But of course that's nothing.'

'I don't care a straw for Lady Whiteroy.'

'And you promised me. But that's less still.'

'It *is* foolish – it's quite idiotic,' said Nick, with his hands in his pockets and his eyes on the ceiling.

There was another silence, at the end of which Mrs Dallow remarked: 'You might have answered Mr Macgeorge when he spoke to you.'

'Mr Macgeorge – what has he to do with it?'

'He has to do with your getting on a little. If you think that's the way!'

Nick broke into a laugh. 'I like lessons in getting on – in other words I suppose you mean in urbanity – from you, Julia!'

'Why not from me?'

'Because you can do nothing base. You're incapable of putting on a flattering manner, to get something by it: therefore why should you expect me to? You're unflattering – that is you're austere – in proportion as there may be something to be got.'

Mrs Dallow sprang up from her chair, coming towards him.

253

'There is only one thing I want in the world – you know very well.'

'Yes, you want it so much that you won't even take it when it's pressed upon you. How long do you seriously expect me to bear it?' Nick repeated.

'I never asked you to do anything base,' she said, standing in front of him. 'If I'm not clever about throwing myself into things, it's all the more reason you should be.'

'If you're not clever, my dear Julia?' Nick, standing close to her, placed his hands on her shoulders and shook her a little with a mixture of tenderness and passion. 'You're clever enough to make me furious, sometimes!'

She opened and closed her fan, looking down at it while she submitted to this attenuated violence. 'All I want is that when a man like Mr Macgeorge talks to you, you shouldn't appear to be bored to death. You used to be so charming in that sort of way. And now you appear to take no interest in anything. At dinner tonight you scarcely opened your lips; you treated them all as if you only wished they'd go.'

'I did wish they'd go. Haven't I told you a hundred times what I think of your salon?'

'How then do you want me to live?' Mrs Dallow asked. 'Am I not to have a creature in the house?'

'As many creatures as you like. Your freedom is complete, and as far as I am concerned always will be. Only when you challenge me and overhaul me – not justly I think – I must confess the simple truth, that there are many of your friends I don't delight in.'

'Oh, *your* idea of pleasant people!' Julia exclaimed. 'I should like once for all to know what it really is.'

'I can tell you what it really isn't: it isn't Mr Macgeorge. He's a being almost grotesquely limited.'

'He'll be where you'll never be – unless you change.'

'To be where Mr Macgeorge is not would be very much my desire. Therefore why should I change?' Nick demanded. 'However, I hadn't the least intention of being rude to him, and I don't think I was,' he went on. 'To the best of my ability I assume a virtue if I have it not; but apparently I'm not enough of a comedian.'

'If you have it not? It's when you say things like that that you're so dreadfully tiresome. As if there were anything that you haven't or mightn't have!'

Nick turned away from his hostess; he took a few impatient steps in the room, looking at the carpet, with his hands in his pockets again. Then he came back to the fire with the observation: 'It's rather hard to be found wanting when one has tried to play one's part so beautifully.' He paused, with his eyes on Mrs Dallow's; then continued, with a vibration in his voice: 'I've imperilled my immortal soul, or at least I've bemuddled my intelligence, by all the things I don't care for that I've tried to do, and all the things I detest that I've tried to be, and all the things I never can be that I've tried to look as if I were – all the appearances and imitations, the pretences and hypocrisies in which I've steeped myself to the eyes; and at the end of it (it serves me right!) my reward is simply to learn that I'm still not half humbug enough!'

Mrs Dallow looked away from him as soon as he had spoken these words; she attracted her eyes to the clock which stood behind him and observed irrelevantly:

'I'm very sorry, but I think you had better go. I don't like you to stay after midnight.'

'Ah, what you like and what you don't like, and where one begins and the other ends – all that's an impenetrable mystery!' the young man declared. But he took no further notice of her allusion to his departure, adding in a different tone: ' "A man like Mr Macgeorge!" When you say a thing of that sort, in a certain particular way, I should rather like to suffer you to perish.'

Mrs Dallow stared; it might have seemed for an instant that she was trying to look stupid. 'How can I help it if a few years hence he is certain to be at the head of any Liberal government?'

'We can't help it, of course, but we can help talking about it,' Nick smiled. 'If we don't mention it, it may not be noticed.'

'You're trying to make me angry. You're in one of your vicious moods,' observed Mrs Dallow, blowing out, on the chimney-piece, a guttering candle.

'That I'm exasperated I have already had the honour very

positively to inform you. All the same I maintain that I was irreproachable at dinner. I don't want you to think I shall always be so good as that.'

'You looked so out of it; you were as gloomy as if every earthly hope had left you, and you didn't make a single contribution to any discussion that took place. Don't you think I observe you?' Mrs Dallow asked, with an irony tempered by a tenderness that was unsuccessfully concealed.

'Ah, my darling, what you observe!' Nick exclaimed, laughing and stopping. But he added the next moment, more seriously, as if his tone had been disrespectful: 'You probe me to the bottom, no doubt.'

'You needn't come either to Griffin or to Severals if you don't want to.'

'Give them up yourself; stay here with me!'

She coloured quickly, as he said this, and broke out: 'Lord! how you hate political houses!'

'How can you say that, when from February to August I spend every blessed night in one?'

'Yes, and hate that worst of all.'

'So do half the people who are in it. You must have so many things, so many people, so much *mise-en-scène* and such a perpetual spectacle to live,' Nick went on. 'Perpetual motion, perpetual visits, perpetual crowds! If you go into the country you'll see forty people every day and be mixed up with them all day. The idea of a quiet fortnight in town, when by a happy if idiotic superstition everybody goes out of it, disconcerts and frightens me. It's the very time, it's the very place, to do a little work and possess one's soul.'

This vehement allocution found Mrs Dallow evidently somewhat unprepared; but she was sagacious enough, instead of attempting for the moment a general rejoinder, to seize on a single phrase and say: 'Work? What work can you do in London at such a moment as this?'

Nick hesitated a little. 'I might tell you that I wanted to get up a lot of subjects, to sit at home and read blue-books; but that wouldn't be quite what I mean.'

'Do you mean you want to paint?'

'Yes, that's it, since you drag it out of me.'

'Why do you make such a mystery about it? You're at perfect liberty,' said Mrs Dallow.

She extended her hand, to rest it on the mantel-shelf, but her companion took it on the way and held it in both his own. 'You're delightful, Julia, when you speak in that tone – then I know why it is I love you; but I can't do anything if I go to Griffin, if I go to Severals.'

'I see – I see,' said Julia, reflectively and kindly.

'I've scarcely been inside of my studio for months and I feel quite homesick for it. The idea of putting in a few quiet days there has taken hold of me: I rather cling to it.'

'It seems so odd, your having a studio!' Julia dropped, speaking so quickly that the words were almost incomprehensible.

'Doesn't it sound absurd, for all the good it does me, or I do in it? Of course one can produce nothing but rubbish on such terms – without continuity or persistence, with just a few days here and there. I ought to be ashamed of myself, no doubt; but even my rubbish interests me. "*Guenille si l'on veut, ma guenille m'est chère.*" But I'll go down to Harsh with you in a moment, Julia,' Nick pursued: 'that would do as well, if we could be quiet there, without people, without a creature; and I should really be perfectly content. You'd sit for me; it would be the occasion we've so often wanted and never found.'

Mrs Dallow shook her head slowly, with a smile that had a meaning for Nick. 'Thank you, my dear; nothing would induce me to go to Harsh with you.'

The young man looked at her. 'What's the matter, whenever it's a question of anything of that sort? Are you afraid of me?' She pulled her hand quickly out of his, turning away from him; but he went on: 'Stay with me here then, when everything is so right for it. We shall do beautifully – have the whole place, have the whole day to ourselves. Hang your engagements! Telegraph you won't come. We'll live at the studio – you'll sit to me every day. Now or never is our chance – when shall we have so good a one? Think how charming it will be! I'll make you wish awfully that I shall do something.'

'I can't get out of Griffin – it's impossible,' returned Mrs Dallow, moving further away, with her back presented to him.

'Then you *are* afraid of me – simply?'

She turned quickly round, very pale. 'Of course I am; you are welcome to know it.'

He went towards her and for a moment she seemed to make another slight movement of retreat. This however was scarcely perceptible, and there was nothing to alarm in the tone of reasonable entreaty in which Nick said to her as he went towards her: 'Put an end, Julia, to our absurd situation – it really can't go on: you have no right to expect a man to be happy or comfortable in so false a position. We're talked of odiously – of that we may be sure; and yet what good have we of it?'

'Talked of? Do I care for that?'

'Do you mean you're indifferent because there are no grounds? That's just why I hate it.'

'I don't know what you're talking about,' exclaimed Mrs Dallow, with quick disdain.

'Be my wife tomorrow – be my wife next week. Let us have done with this fantastic probation and be happy.'

'Leave me now – come back tomorrow. I'll write to you.' She had the air of pleading with him at present as he pleaded with her.

'You can't resign yourself to the idea of one's looking "out of it"!' laughed Nick.

'Come tomorrow, before lunch,' Mrs Dallow continued.

'To be told I must wait six months more and then be sent about my business? Ah, Julia, Julia!' murmured the young man.

Something in this simple exclamation – it sounded natural and perfectly unstudied – evidently on the instant made a great impression on his companion. 'You shall wait no longer,' she said after a short silence.

'What do you mean by no longer?'

'Give me about five weeks – say till the Whitsuntide recess.'

'Five weeks are a great deal,' smiled Nick.

'There are things to be done – you ought to understand.'

'I only understand how I love you.'

'Dearest Nick!' said Mrs Dallow; upon which he caught her in his arms.

'I have your promise then for five weeks hence, to a day?' he demanded, as she released herself.

'We'll settle that – the exact day: there are things to consider and to arrange. Come to luncheon tomorrow.'

'I'll come early – I'll come at one,' Nick said; and for a moment they stood smiling at each other.

'Do you think I *want* to wait, any more than you?' Mrs Dallow asked.

'I don't feel so much out of it now!' he exclaimed by way of answer. 'You'll stay, of course, now – you'll give up your visits?'

She had hold of the lappet of his coat; she had kept it in her hand even while she detached herself from his embrace. There was a white flower in his buttonhole which she looked at and played with a moment before she said: 'I have a better idea – you needn't come to Griffin. Stay in your studio – do as you like – paint dozens of pictures.'

'Dozens? Barbarian!' Nick ejaculated.

The epithet apparently had an endearing suggestion to Mrs Dallow; at any rate it led her to allow him to kiss her on her forehead – led her to say: 'What on earth do I want but that you should do absolutely as you please and be as happy as you can?'

Nick kissed her again, in another place, at this; but he inquired: 'What dreadful proposition is coming now?'

'I'll go off and do up my visits and come back.'

'And leave me alone?'

'Don't be affected!' said Mrs Dallow. 'You know you'll work much better without me. You'll live in your studio – I shall be well out of the way.'

'That's not what one wants of a sitter. How can I paint you?'

'You can paint me all the rest of your life. I shall be a perpetual sitter.'

'I believe I could paint you without looking at you,' said Nick, smiling down at her. 'You do excuse me, then, from those dreary places?'

'How can I insist, after what you said about the pleasure of keeping these days?' Mrs Dallow asked sweetly.

'You're the best woman on earth; though it does seem odd you should rush away as soon as our little business is settled.'

'We shall make it up. I know what I'm about. And now go!' Mrs Dallow terminated, almost pushing her visitor out of the room.

23

IT was certainly singular under the circumstances that on sitting down in his studio after Julia had left town Nick Dormer should not, as regards the effort to reproduce some beautiful form, have felt more chilled by the absence of a friend who was such an embodiment of beauty. She was away and he longed for her, and yet without her the place was more filled with what he wanted to find in it. He turned into it with confused feelings, the most definite of which was a sense of release and recreation. It looked blighted and lonely and dusty, and his old studies, as he rummaged them out, struck him as even clumsier than the last time he had ventured to drop his eyes on them. But amid this neglected litter, in the colourless and obstructed light of a high north window which needed washing, he tasted more sharply the possibility of positive happiness: it appeared to him that, as he had said to Julia, he was more in possession of his soul. It was frivolity and folly, it was puerility to spend valuable hours pottering over the vain implements of an art he had relinquished; and a certain shame that he had felt in presenting his plea to Julia Dallow that Sunday night arose from the sense not of what he clung to, but of what he had given up. He had turned his back upon serious work, so that pottering was now all he could aspire to. It couldn't be fruitful, it couldn't be anything but ridiculous, almost ignoble; but it soothed his nerves, it was in the nature of a secret dissipation. He had never suspected that he should ever have on his own part nerves to count with; but this possibility had been revealed to him on the day it became clear that he was letting something precious go. He was glad he had not to justify himself to the critical, for this might have been a delicate business. The critical were mostly absent; and besides, shut up all day in his studio, how should he ever meet them? It was the place in the world where he felt furthest away from his constituents. This was a part of the pleasure – the consciousness that for the hour the coast was clear and his mind was free. His mother and his sister had gone to Broadwood: Lady Agnes (the phrase sounds brutal, but it represents his state of mind) was well out of the way. He had written to her as soon as Julia left town – he had apprised her of the fact that his wedding-day was fixed: a relief, for poor Lady Agnes, to a period of intolerable mystification, of taciturn wondering and watching. She had said her say the day of the poll at Harsh; she was too proud to ask and too discreet to 'nag': so she could only wait for something that didn't arrive. The unconditioned loan of Broadwood had of course been something of a bribe to patience: she had at first felt that on the day she should take possession of that capital house Julia would indeed seem to have come into the family. But the gift had confirmed expectations just enough to make disappointment more bitter; and the discomfort was greater in proportion as Lady Agnes failed to discover what was the matter. Her daughter Grace was much occupied with this question and brought it up in conversation in a manner irritating to her ladyship, who had a high theory of being silent about it, but who however, in the long run, was more unhappy when, in consequence of a reprimand, the girl suggested no reasons at all than when she suggested stupid ones. It eased Lady Agnes a little to discuss the mystery when she could have the air of not having begun.

The letter Nick received from her the first day of Passion Week in reply to his important communication was the only one he read at that moment; not counting of course several notes that Mrs Dallow addressed to him from Griffin. There were letters piled up, as he knew, in Calcutta Gardens, which his servant had strict orders not to bring to the studio. Nick slept now in the bedroom attached to this retreat; got things as he wanted them from Calcutta Gardens; and dined at his club, where a stray surviving friend or two, seeing him prowl about the library in the evening, was free to suppose that such eccentricity had a crafty political basis. When he thought of his neglected letters he remembered Mr Carteret's convictions on the subject of not 'getting behind'; they made him laugh, in the slightly sonorous painting-room, as he bent over one of the old canvases that he had ventured to turn to the light. He was fully determined however to master his correspondence before going down, the last thing before Parliament should re-assemble, to spend another day at Beauclere. Mastering his correspondence meant in Nick's mind breaking open envelopes; writing answers was scarcely involved in the idea. But Mr Carteret would never guess that. Nick was not moved even to write to him that the affair with Mrs Dallow was on the point of taking the form he had been so good as to desire: he reserved the pleasure of this announcement for a personal interview.

The day before Good Friday, in the morning, his stillness was broken by a rat-tat-tat on the outer door of his studio, administered apparently by the knob of a walking-stick. His servant was out and he went to the door, wondering who his visitor could be at such a time, especially of the familiar class. The class was indicated by the visitor's failure to look for the bell; for there was a bell, though it required a little research. In a moment the mystery was solved: the gentleman who stood smiling at him from the threshold could only be Gabriel Nash. Dormer had not seen this whimsical personage for several months and had had no news of him beyond the general intimation that he was abroad. His old friend had sufficiently prepared him at the time of their reunion in Paris for the idea of the fitful in intercourse: and he had not been ignorant on his return from Paris that he should have had an opportunity to miss him if he had not been too busy to take advantage of it. In London, after the episode at Harsh, Gabriel had not reappeared: he had redeemed none of the pledges given the night they walked together to Notre Dame and conversed on important matters. He was to have interposed in Nick's destiny, but he had not interposed; he was to have dragged him in the opposite sense from Mrs Dallow, but there had been no dragging; he was to have saved him, as he called it, and yet Nick was lost. This circumstance indeed constituted his excuse: the member for Harsh had rushed so to perdition. Nick had for the hour seriously wished to keep hold of him: he valued him as a salutary influence. Yet when he came to his senses after his election our young man had recognized that Nash might very well have reflected on the thankfulness of such a slippery subject – might have considered that he was released from his vows. Of course it had been particularly in the event of a Liberal triumph that he had threatened to make himself felt; the effect of a brand plucked from the burning would be so much greater if the flames were already high. Yet Nick had not held him to the letter of this pledge, and had so fully admitted the right of a properly-constituted aesthete to lose patience with him that he was now far from greeting his visitor with a reproach. He felt much more thrown on his defence.

Gabriel did not attack him however. He brought in only blandness and benevolence and a great content at having obeyed the mystic voice – it was really a remarkable case of second sight – which had whispered to him that the recreant comrade of his prime was in town. He had just come back from Sicily, after a southern winter, according to a custom frequent with him, and had been moved by a miraculous prescience, unfavourable as the moment might seem, to go and ask for Nick in Calcutta Gardens, where he had extracted from his friend's servant an address not known to all the world. He showed Nick what a mistake it had been to fear a reproach from Gabriel Nash, and how he habitually ignored all lapses and kept up the standard only by taking a hundred fine things for granted. He also abounded more than ever in his own sense, reminding his friend how no recollection of him, no evocation of him in absence could do him justice. You couldn't recall him without seeming to exaggerate him, and then recognized when you saw him that your exaggeration had fallen short. He emerged out of vagueness (his Sicily might have been the Sicily of *A Winter's Tale*), and would evidently be reabsorbed in it; but his presence was positive and pervasive enough. He was very lively while he lasted. His connections were with beauty, urbanity and conversation, as usual, but it was a circle you couldn't find in the *Court Guide*. Nick had a sense that he knew 'a lot of aesthetic people', but he dealt in ideas much more than in names and addresses. He was genial and jocose, sunburnt and romantically allusive. Nick gathered that he had been living for many days in a Saracenic tower, where his principal occupation was to watch for the flushing of the west. He had retained all the serenity of his opinions, and made light, with a candour of which the only defect was apparently that it was not quite enough a conscious virtue, of many of the objects of common esteem. When Nick asked him what he had been doing he replied: 'Oh, living, you know'; and the tone of the words seemed to offer them as a record of magnificent success. He made a long visit, staying to luncheon and after luncheon, so that the little studio heard all at once more conversation, and of a wider scope, than in the several previous years of its history. With much of our story left to tell, it is a pity that so little of this rich colloquy may be transcribed here; because, as affairs took their course, it marked really (if it be a question of noting the exact point) a turn of the tide in Nick Dormer's personal situation. He was destined to remember the accent with which Nash exclaimed, on his drawing forth sundry specimens of amateurish earnestness: 'I say – I say – I say!'

Nick glanced round with a heightened colour. 'They're pretty bad, eh?'

'Oh, you're a deep one!' Nash went on.

'What's the matter?'

'Do you call your conduct that of a man of honour?'

'Scarcely, perhaps. But when no one has seen them!'

'That's your villainy. *C'est de l'exquis, du pur exquis*. Come, my dear fellow, this is very serious – it's a bad business,' said Gabriel Nash. Then he added, almost with austerity: 'You'll be so good as to place before me every patch of paint, every sketch and scrap that this room contains.'

Nick complied, in great good-humour. He turned out his boxes and drawers, shovelled forth the contents of bulging portfolios, mounted on chairs to unhook old canvases that had been severely 'skied'. He was modest and docile and patient and amused, and above all quite thrilled – thrilled with the idea of eliciting a note of appreciation so late in the day. It was the oddest thing how at present in fact he found himself attributing value to Gabriel Nash – attributing to him, among attributions more confused, the dignity of judgement, the authority of intelligence. Nash was an ambiguous being, but he was an excellent touchstone. The two said very little for a while, and they had almost half an hour's silence, during which, after Nick had hastily improvised a little exhibition, there was only a puffing of cigarettes. The visitor walked about, looking at this and that, taking up rough studies and laying them down, asking a question of fact, fishing with his umbrella, on the floor, amid a pile of unarranged sketches. Nick accepted jocosely the attitude of suspense, but there was even more of it in his heart than in his face. So few people had seen his young work – almost no one who really counted. He had been ashamed of it, never showing it to bring on a conclusion, inasmuch as it was precisely of a conclusion that he was afraid. He whistled now while he let his companion take time. He rubbed old panels with his sleeve and dabbed wet sponges on surfaces that had sunk. It was a long time since he had felt so gay, strange as such an assertion sounds in regard to a young man whose bridal day had at his urgent solicitation lately been fixed. He had stayed in town to be alone with his imagination, and suddenly, paradoxically, the sense of that result had arrived with Gabriel Nash.

'Nicholas Dormer,' this personage remarked at last, 'for grossness of immorality I think I have never seen your equal.'

'That sounds so well that I hesitate to risk spoiling it by wishing it explained.'

'Don't you recognize in *any* degree the elevated idea of duty?'

'If I don't grasp it with a certain firmness I'm a great failure, for I was quite brought up in it,' Nick said.

'Then you are the wretchedest failure I know. Life *is* ugly, after all.'

'Do I gather that you yourself recognize obligations of the order you allude to?' asked Nick.

'Do you "gather"?' Nash stared. 'Why, aren't they the very flame of my faith, the burden of my song?'

'My dear fellow, duty is doing, and I inferred that you think rather poorly of doing – that it spoils one's style.'

'Doing wrong, assuredly.'

'But what do you call right? What's your canon of certainty there?'

'The conscience that's in us – that charming, conversible, infinite thing, the intensest thing we know. But you must treat the oracle civilly if you wish to make it speak. You mustn't stride into the temple in muddy jack-boots, with your hat on your head, as the Puritan troopers tramped into the dear old abbeys. One must do one's best to find out the right, and your criminality appears to be that you have not taken common trouble.'

'I hadn't you to ask,' smiled Nick. 'But duty strikes me as doing *something*. If you are too afraid it may be the wrong thing, you may let everything go.'

'Being is doing, and if doing is duty, being is duty. Do you follow?'

'At a great distance.'

'To be what one *may* be, really and efficaciously,' Nash went on, 'to feel it and understand it, to accept it, adopt it, embrace it – that's conduct, that's life.'

'And suppose one's a brute or an ass, where's the efficacy?'

'In one's very want of intelligence. In such cases one is out of it – the question doesn't exist; one simply becomes a part of the duty of others. The brute, the ass, neither feels, nor understands, nor accepts, nor adopts. Those fine processes in themselves classify us. They educate, they exalt, they preserve; so that, to profit by them, we must be as perceptive as we can. We must recognize our particular form, the instrument that each of us – each of us who carries anything – carries in his being. Mastering the instrument, learning to play it in perfection – that's what I call duty, what I call conduct, what I call success.'

Nick listened with friendly attention and the air of general assent was in his face as he said: 'Everyone has it then, this individual pipe?'

'"Everyone," my dear fellow, is too much to say, for the world is full of the crudest *remplissage*. The book of life is padded, ah but padded – a deplorable want of editing. I speak of every one that is any one. Of course there are pipes and pipes – little quavering flutes for the concerted movements and big *cornets-à-piston* for the great solos.'

'I see, I see. And what might your instrument be?'

Nash hesitated not a moment; his answer was radiantly ready. 'To speak to people just as I am speaking to you. To prevent for instance a great wrong being done.'

'A great wrong?'

'Yes – to the human race. I talk – I talk; I say the things that other people don't, that they can't, that they won't,' Gabriel continued, with his inimitable candour.

'If it's a question of mastery and perfection, you certainly have them,' his companion replied.

'And you haven't, alas; that's the pity of it, that's the scandal. That's the wrong I want to set right, before it becomes too public a shame. I called you just now grossly immoral, on account of the spectacle you present – a spectacle to be hidden from the eye of ingenuous youth: that of a man neglecting his own fiddle to blunder away on that of one of his fellows. We can't afford such mistakes, we can't tolerate such licence.'

'You think then I *have* a fiddle?' asked Nick.

'A regular Stradivarius! All these things you have shown me are singularly interesting. You have a talent of a wonderfully pure strain.'

'I say – I say – I say!' Nick exclaimed, standing in front of his visitor with his hands in his pockets and a blush on his smiling face, and repeating with a change of accent Nash's exclamation of half an hour before.

'I like it, your talent; I measure it, I appreciate it, I insist upon it,' Nash went on, between the whiffs of his cigarette. 'I have to be accomplished to do so, but fortunately I am. In such a case that's my duty. I shall make you my business for a while. Therefore,' Nash added, piously, 'don't say I'm unconscious of the moral law.'

'A Stradivarius?' said Nick, interrogatively, with his eyes wide open and the thought in his mind of how different this was from having gone to Griffin.

24

GABRIEL NASH had plenty of further opportunity to elucidate this and other figurative remarks, for he not only spent several of the middle hours of the day with his friend, but came back with him in the evening (they dined together at a little foreign pot-house in Soho, revealed to Nick on this occasion) and discussed the great question far into the night. The great question was whether, on the showing of those examples of his ability with which the room in which they sat was now densely bestrewn, Nick Dormer would be justified in 'really going in' for the practice of pictorial art. This may strike many of my readers as a limited and even trivial inquiry, with little of the heroic or the romantic in it; but it was none the less carried to a very fine point by our clever young men. Nick suspected Nash of exaggerating his encouragement in order to play a malign trick on the political world, at whose expense it was his fancy to divert himself (without making that organization bankrupt assuredly), and reminded him that his present accusation of immorality was strangely inconsistent with the wanton hope expressed by him in Paris – the hope that the Liberal candidate at Harsh would be returned. Nash replied first: 'Oh, I hadn't been in this place then!' but he defended himself more effectually in saying that it was not of Nick's having got elected that he complained: it was of his visible hesitancy to throw up his seat. Nick requested that he wouldn't speak of this, and his gallantry failed to render him incapable of saying: 'The fact is I haven't the nerve for it.' They talked then for a while of what he could do, not of what he couldn't; of the mysteries and miracles of reproduction and representation; of the strong, sane joys of the artistic life. Nick made afresh, with more fullness, his great confession, that his private ideal of happiness was the life of a great painter of portraits. He uttered his thought about this so copiously and lucidly that Nash's own abundance was stilled, and he listened almost as if he had been listening to something new, difficult as it was to suppose that there could be a point of view in relation to such a matter with which he was unacquainted.

'There it is,' said Nick at last – 'there's the naked, preposterous truth: that if I were to do exactly as I liked I should spend my years copying the more or less vacuous countenances of my fellow-mortals. I should find peace and pleasure and wisdom and worth, I should find fascination and a measure of success in it: out of the din and the dust and the scramble, the world of party labels, party cries, party bargains and party treacheries – of humbuggery, hypocrisy and cant. The cleanness and quietness of it, the independent effort to do something, to leave something which shall give joy to man long after the howling has died away to the last ghost of an echo – such a vision solicits me at certain hours with an almost irresistible force.'

As he dropped these remarks Nick lolled on a big divan, with one of his long legs folded up; and his visitor stopped in front of him, after moving about the room vaguely and softly, almost on tiptoe, not to interrupt him. 'You speak with the eloquence that rises to a man's lips on a very particular occasion; when he has practically, whatever his theory may be, renounced the right and dropped hideously into the wrong. Then his regret for the right, a certain exquisite appreciation of it, takes on an accent which I know well how to recognize.'

Nick looked up at him a moment. 'You've hit it, if you mean by that that I haven't resigned my seat and that I don't intend to.'

'I thought you took it only to give it up. Don't you remember our talk in Paris?'

'I like to be a part of the spectacle that amuses you, but I scarcely have taken so much trouble as that for it.'

'But isn't it an absurd comedy, the life you lead?'

'Comedy or tragedy – I don't know which; whatever it is I appear to be capable of it to please two or three people.'

'Then you *can* take trouble,' said Nash.

'Yes, for the woman I'm to marry.'

'Ah, you're to marry?'

'That's what has come on since we met in Paris, and it makes just the difference.'

'Ah, my poor friend,' smiled Gabriel, standing there, 'no wonder you have an eloquence, an accent!'

'It's a pity I have them in the wrong place. I'm expected to have them in the House of Commons.'

'You will when you make your farewell speech there – to announce that you chuck it up. And may I venture to ask who's to be your wife?' Gabriel went on.

'Mrs Dallow has kindly consented. I think you saw her in Paris.'

'Ah, yes: you spoke of her to me. I remember asking you if you were in love with her.'

'I wasn't then.'

Nash hesitated a moment. 'And are you now?'

'Oh dear, yes,' said Nick.

'That would be better if it wasn't worse.'

'Nothing could be better; it's the best thing that can happen to me.'

'Well,' said Nash, 'you must let me very respectfully approach her. You must let me bring her round.'

'Bring her round?'

'Talk her over.'

'Over to what?' Nick repeated his companion's words, a little as if it were to gain time, remembering the effect Gabriel Nash had produced upon Julia – an effect which scantily ministered to the idea of another meeting. Julia had had no occasion to allude again to Nick's imperturbable friend; he had passed out of her life at once and forever; but there flickered up a vivid recollection of the contempt he had led her to express, together with a sense of how odd she would think it that her intended should have thrown over two pleasant visits to cultivate such company.

'Over to a proper pride in what you may do – what you may do above all if she will help you.'

'I scarcely see how she can help me,' said Nick, with an air of thinking.

'She's extremely handsome, as I remember her: you could do great things with her.'

'Ah, there's the rub,' Nick went on. 'I wanted her to sit for me this week, but she wouldn't.'

Elle a bien tort. You should do some fine strong type. Is Mrs Dallow in London?' Nash inquired.

'For what do you take her? She's paying visits.'

'Then I have a model for you.'

'Then you have –?' Nick stared. 'What has that to do with Mrs Dallow's being away?'

'Doesn't it give you more time?'

'Oh, the time flies!' sighed Nick, in a manner that caused his companion to break into a laugh – a laugh in which for a moment he himself joined, blushing a little.

'Does she like you to paint?' Nash continued, with one of his candid intonations.

'So she says.'

'Well, do something fine to show her.'

'I'd rather show it to you,' Nick confessed.

'My dear fellow, I see it from here, if you do your duty. Do you remember the Tragic Muse?' Nash pursued, explicatively.

'The Tragic Muse?'

'That girl in Paris, whom we heard at the old actress's and whom we afterwards met at the charming entertainment given by your cousin (isn't he?) the secretary of embassy.'

'Oh, Peter's girl: of course I remember her.'

'Don't call her Peter's; call her rather mine,' Nash said with good-humoured dissuasiveness. 'I invented her, I introduced her, I revealed her.'

'I thought on the contrary you ridiculed and repudiated her.'

'As an individual, surely not; I seem to myself to have been all the while rendering her services. I said I disliked tea-party ranters, and so I do; but if my estimate of her powers was below the mark she has more than punished me.'

'What has she done?' asked Nick.

'She has become interesting, as I suppose you know.'

'How should I know?'

'You must see her, you must paint her,' said Nash. 'She tells me that something was said about it that day at Madame Carré's.'

'Oh, I remember – said by Peter.'

'Then it will please Mr Sherringham – you'll be glad to do that. I suppose you know all he has done for Miriam?'

'Not a bit. I know nothing about Peter's affairs, unless it be in general that he goes in for mountebanks and mimes and that it occurs to me I have heard one of my sisters mention – the rumour had come to her – that he has been backing Miss Rooth.'

'Miss Rooth delights to talk of his kindness: she's charming when she speaks of it. It's to his good offices that she owes her appearing here.'

'Here! Is she in London?' Nick inquired.

'*D'où tombez-vous?* I thought you people read the papers.'

'What should I read, when I sit (sometimes!) through the stuff they put into them?'

'Of course I see that – that your engagement at your own theatre keeps you from going to others. Learn then,' said Gabriel Nash, 'that you have a great competitor and that you are distinctly not, much as you may suppose it, *the* rising comedian. The Tragic Muse is the great modern personage. Haven't you heard people speak of her, haven't you been taken to see her?'

'I dare say I've heard of her; but with a good many other things on my mind I had forgotten it.'

'Certainly I can imagine what has been on your mind. She

remembers you at any rate; she repays neglect with sympathy. She wants to come and see you.'

'To see me?'

'To be seen by you – it comes to the same thing. She's worth seeing: you must let me bring her; you'll find her very suggestive. That idea that you should paint her – she appears to consider it a sort of bargain.'

'A bargain? What will she give me?' Nick asked.

'A splendid model. She *is* splendid.'

'Oh, then bring her,' said Nick.

25

NASH brought her, the great modern personage as he had described her, the very next day, and it took Nick Dormer but a short time to appreciate his declaration that Miriam Rooth was splendid. She had made an impression upon him ten months before, but it had haunted him only for a day, immediately overlaid with other images. Yet after Nash had spoken of her a few moments he evoked her again; some of her attitudes, some of her tones began to hover before him. He was pleased in advance with the idea of painting her. When she stood there in fact however it seemed to him that he had remembered her wrong: the brilliant young lady who instantly filled his studio with a presence that it had never known was exempt from the curious clumsiness which had interfused his former admiration of her with a certain pity. Miriam Rooth was light and bright and straight today – straight without being stiff and bright without being garish. To Nick's perhaps inadequately sophisticated mind the model, the actress were figures with a vulgar setting; but it would have been impossible to show that taint less than his present extremely natural yet extremely distinguished visitor. She was more natural even than Gabriel Nash ('nature' was still Nick's formula for his old friend), and beside her he appeared almost commonplace.

Nash recognized her superiority with a frankness that was honourable to both of them, testifying in this manner to his sense that they were all three serious beings, worthy to deal with realities. She attracted crowds to her theatre, but to his appreciation of such a fact as that, important doubtless in its way, there were limits which he had already expressed. What he now felt bound in all integrity to express was his perception that she had, in general and quite apart from the question of the box-office, a remarkable, a very remarkable artistic nature. He confessed that she had surprised him there; knowing of her in other days mainly that she was hungry to adopt an overrated profession, he had not imputed to her the normal measure of intelligence. Now he saw – he had had some talks with her – that she *was* intelligent; so much so that he was sorry for the embarrassment it would be to her. Nick could imagine the discomfort of having that sort of commodity to dispose of in such conditions. 'She's a distinguished woman – really a distinguished woman,' Nash explained, kindly and lucidly, almost paternally; 'and the head you can see for yourself.'

Miriam, smiling, as she sat on an old Venetian chair, held aloft, with the noblest effect, that portion of her person to which this patronage was extended, and remarked to Nick that, strange as it might appear, she had got quite to like poor Mr Nash: she could make him go about with her; it was a relief to her mother.

'When I take him she has perfect peace,' the girl said; 'then she can stay at home and see the interviewers. She delights in that and I hate it, so our friend here is a great comfort. Of course a *femme de théâtre* is supposed to be able to go out alone, but there's a kind of appearance, an added *chic*, in having some one. People think he's my companion; I'm sure they fancy I pay him. I would pay him rather than give him up, for it doesn't matter that he's not a lady. He *is* one in tact and sympathy, as you see. And base as he thinks the sort of thing I do, he can't keep away from the theatre. When you're celebrated, people will look at you who before could never find out for themselves *why* they should.'

'When you're celebrated you become handsomer; at least that's what has happened to you, though you were pretty too of old,' Gabriel argued. 'I go to the theatre to look at your head; it gives me the greatest pleasure. I take up anything of that sort as soon as I find it; one never knows how long it may last.'

'Are you speaking of my appearance?' Miriam asked.

'Dear no, of my own pleasure, the first freshness,' Nash went on. 'Dormer at least, let me tell you in justice to him, hasn't waited till you were celebrated to want to see you again (he stands there open-eyed); for the simple reason that he hadn't the least idea of your renown. I had to announce it to him.'

'Haven't you seen me act?' Miriam asked, without reproach, of her host.

'I'll go tonight,' said Nick.

'You have your Parliament, haven't you? What do they call it – the demands of public life?' Miriam continued: to which Gabriel Nash rejoined that he had the demands of private as well, inasmuch as he was in love – he was on the point of being married. Miriam listened to this with participation; then she said: 'Ah, then, do bring your – what do they call her in English? I'm always afraid of saying something improper – your *future*. I'll send you a box, under the circumstances; you'd like that better.' She added that if he were to paint her he would have to see her often on the stage, wouldn't he? to profit by the *optique de la scène* (what did they call *that* in English?) studying her and fixing his impression. Before he had time to respond to this proposition she asked him if it disgusted him to hear her speak like that, as if she were always posing and thinking about herself, living only to be looked at, thrusting forward her person. She often got sick of doing so, already; but *à la guerre comme à la guerre*.

'That's the fine artistic nature, you see – a sort of divine disgust breaking out in her,' Nash expounded.

'If you want to paint me at all, of course. I'm struck with the way I'm taking that for granted,' Miriam continued. 'When Mr Nash spoke of it to me I jumped at the idea. I remembered our meeting in Paris and the kind things you said to me. But no doubt one oughtn't to jump at ideas when they represent serious sacrifices on the part of others.'

'Doesn't she speak well!' Nash exclaimed to Nick. 'Oh, she'll go far!'

'It's a great privilege to me to paint you; what title in the world have I to pretend to such a model?' Nick replied to Miriam. 'The sacrifice is yours – a sacrifice of time and good-nature and credulity. You come in your beauty and your genius to this shabby place where I've nothing to show, not a guarantee to offer you; and I wonder what I've done to deserve such a gift of the gods.'

'Doesn't *he* speak well?' Nash demanded, smiling, of Miriam.

She took no notice of him, but she repeated to Nick that she hadn't forgotten his friendly attitude in Paris; and when he answered that he surely had done very little she broke out, first resting her eyes on him a moment with a deep, reasonable smile and then springing up quickly: 'Ah, well, if I must justify myself, I liked you!'

'Fancy my appearing to challenge you!' laughed Nick. 'To see you again is to want tremendously to try something; but you must have an infinite patience, because I'm an awful duffer.'

Miriam looked round the walls. 'I see what you have done – *bien des choses*.'

'She understands – she understands,' Gabriel dropped. And he added to Miriam: 'Imagine, when he might do something, his choosing a life of shams! At bottom he's like you – a wonderful artistic nature.'

'I'll have patience,' said the girl, smiling at Nick.

'Then, my children, I leave you – the peace of the Lord be with you.' With these words Nash took his departure.

The others chose a position for Miriam's sitting, after she had placed herself in many different attitudes and different lights; but an hour had elapsed before Nick got to work – began, on a large canvas, to knock her in, as he called it. He was hindered a little even by a certain nervousness, the emotion of finding himself, out of a clear sky, confronted with such a sitter and launched in such a task. The situation was incongruous, just after he had formally renounced all manner of 'art' – the renunciation taking effect not a bit the less from the whim that he had consciously treated himself to *as* a whim (the last he should ever indulge), the freak of relapsing for a fortnight into a fingering of old sketches, for the purpose, as he might have said, of burning them up, of clearing out his studio and terminating his lease. There were both embarrassment and inspiration in the strange chance of snatching back for an hour a relinquished joy: the jump with which he found he could still rise to such an occasion took away his breath a little, at the same time that the idea – the idea of what one might make of such material – touched him with an irresistible wand. On the spot, to his inner vision, Miriam became a magnificent result, drawing a hundred formative instincts out of their troubled sleep, defying him where he privately felt strongest and imposing herself triumphantly in her own strength. He had the good fortune to see her, as a subject, without striking matches, in a vivid light, and his quick attempt was as exciting as a sudden gallop – it was almost the sense of riding a runaway horse.

She was in her way so fine that he could only think how to 'do' her: that hard calculation soon flattened out the consciousness, lively in him at first, that she was a beautiful woman who had sought him out in his retirement. At the end of their first sitting her having sought him out appeared the most natural thing in the world: he had a perfect right to entertain her there – explanations and complications were engulfed in the productive mood. The business of 'knocking her in' held up a lamp to her beauty, showed him how much there was of it and that she was infinitely interesting. He didn't want to fall in love with her (*that* would be a sell! as he said to himself), and she promptly became much too interesting for that. Nick might have reflected, for simplification's sake, as his cousin Peter had done, but with more validity, that he was engaged with Miss Rooth in an undertaking that didn't in the least refer to themselves, that they were working together seriously and that work was a suspension of sensibility. But after her first sitting (she came, poor girl, but twice), the need of such exorcisms passed from his spirit: he had so thoroughly, practically taken her up. As to whether Miriam had the same bright, still sense of co-operation to a definite end, the sense of the distinctively technical nature of the answer to every question to which the occasion might give birth, that mystery would be cleared up only if it were open to us to regard this young lady through some other medium than the mind of her friends. We have chosen, as it happens, for some of the advantages it carries with it, the indirect vision; and it fails as yet to tell us (what Nick of course wondered about before he ceased to care, as indeed he intimated to his visitor) why a young person crowned with success should have taken it into her head that there was something for her in so blighted a spot. She should have gone to one of the regular people, the great people: they would have welcomed her with open arms. When Nick asked her if some of the R.A.s hadn't expressed a desire to have a crack at her she said: 'Oh, dear, no, only the tiresome photographers; and fancy *them*, in the future! If mamma could only do *that* for me!' And she added, with the charming fellowship for which she was conspicuous on this occasion: 'You know I don't think any one yet has been quite so much struck with me as you.'

'Not even Peter Sherringham?' asked Nick, laughing and stepping back to judge of the effect of a line.

'Oh, Mr Sherringham's different. You're an artist.'

'For heaven's sake, don't say that!' cried Nick. 'And as regards your art I thought Peter knew more than any one.'

'Ah, you're severe,' said Miriam.

'Severe?'

'Because that's what he thinks. But he does know a lot – he has been a providence to me.'

'And why hasn't he come here to see you act?'

Miriam hesitated a moment. 'How do you know he hasn't come?'

'Because I take for granted he would have called on me if he had.'

'Does he like you very much?' asked Miriam.

'I don't know. I like him.'

'He's a gentleman – *pour cela*,' said Miriam.

'Oh, yes, for that!' Nick went on absently, sketching hard.

'But he's afraid of me – afraid to see me.'

'Doesn't he think you're good enough?'

'On the contrary – he believes I shall carry him away and he's in a terror of my doing it.'

'He ought to like that,' said Nick.

'That's what I mean when I say he's not an artist. However, he declares he does like it, only it appears it is not the right thing for him. Oh, the right thing – he's bent upon getting that. But it's not for me to blame him, for I am too. He's coming some night, however: he shall have a dose!'

'Poor Peter!' Nick exclaimed, with a compassion none the less real because it was mirthful: the girl's tone was so expressive of good-humoured, unscrupulous power.

'He's such a curious mixture,' Miriam went on; 'sometimes I lose patience with him. It isn't exactly trying to serve both God and Mammon, but it's muddling up the stage and the world. The world be hanged; the stage, or anything of that sort (I mean one's faith), comes first.'

'Brava, brava, you do me good,' Nick murmured, still hilarious and at his work. 'But it's very kind of you, when I was in this absurd state of ignorance, to attribute to me the honour of having been more struck with you than any one else,' he continued, after a moment.

'Yes, I confess I don't quite see – when the shops were full of my photographs.'

'Oh, I'm so poor – I don't go into shops,' returned Nick.

'Are you very poor?'

'I live on alms.'

'And don't they pay you – the government, the ministry?'

'Dear young lady, for what? – for shutting myself up with beautiful women?'

'Ah, you have others, then?' asked Miriam.

'They are not so kind as you, I confess.'

'I'll buy it from you – what you're doing; I'll pay you well when it's done,' said the girl. 'I've got money now; I make it, you know – a good lot of it. It's too delightful, after scraping and starving. Try it and you'll see. Give up the base, bad world.'

'But isn't it supposed to be the base, bad world that pays?'

'Precisely; make it pay, without mercy – squeeze it dry. That's what it's meant for – to pay for art. Ah, if it wasn't for that! I'll bring you a quantity of photographs, tomorrow – you must let me come back tomorrow: it's so amusing to have them, by the hundred, all for nothing, to give away. That's what takes mamma most: she can't get over it. That's luxury and glory; even at Castle Nugent they didn't do that. People used to sketch me, but not so much as mamma *veut bien le dire*; and in all my life I never had but one poor little *carte-de-visite*, when I was sixteen, in a plaid frock, with the banks of a river, at three francs the dozen.'

26

IT was success, Nick felt, that had made Miriam finer – the full possession of her talent and the sense of the recognition of it. There was an intimation in her presence (if he had given his mind to it) that for him too the same cause would produce the same effect – that is would show him that there is nothing like being launched in the practice of an art to learn what it may do for one. Nick felt clumsy beside a person who manifestly now had such an extraordinary familiarity with the point of view. He remembered too the inferiority that had been in his visitor – something clumsy and shabby, of quite another quality from her actual smartness, as London people would call it, her well-appointedness and her evident command of more than one manner. Handsome as she had been the year before, she had suggested provincial lodgings, bread-and-butter, heavy tragedy and tears; and if then she was an ill-dressed girl with thick hair who wanted to be an actress, she was already in a few weeks an actress who could act even at not acting. She showed what a light hand she could have, forbore to startle and looked as well for unprofessional life as Julia: which was only the perfection of her professional character.

This function came out much in her talk, for there were many little bursts of confidence as well as many familiar pauses as she sat there; and she was ready to tell Nick the whole history of her début – the chance that had suddenly turned up and that she had caught with a jump as it passed. He missed some of the details, in his attention to his own task, and some of them he failed to understand, attached as they were to the name of Mr Basil Dashwood, which he heard for the first time. It was through Mr Dashwood's extraordinary exertions that a hearing – a morning performance at a London theatre – had been obtained for her. That had been the great step, for it had led to the putting on at night of the play at the same theatre, in place of a wretched thing they were trying (it was no use) to keep on its feet, and to her engagement for the principal part. She had made a hit in it (she couldn't pretend not to know that); but she was already tired of it, there were so many other things she wanted to do; and when she thought it would probably run a month or two more she was in the humour to curse the odious conditions of artistic production in such an age. The play was a simplified version of a new French piece, a thing that had taken in Paris, at a third-rate theatre, and had now, in London, proved itself good enough for houses mainly made up of ten-shilling stalls. It was Dashwood who had said it would go, if they could get the rights and a fellow to make some changes: he had discovered it at a nasty little theatre she had never been to, over the Seine. They had got the rights and the fellow who had made the changes was practically Dashwood himself; there was another man, in London, Mr Gushmore – Miriam didn't know whether Nick would ever have heard of him (Nick hadn't), who had done some of it. It had been awfully chopped down, to a mere bone, with the meat all gone; but that was what people in London seemed to like. They were very innocent, like little dogs amusing themselves with a bone. At any rate, she had made something, she had made a figure of the woman (a dreadful idiot, really, especially in what Dashwood had muddled her into); and Miriam added, in the complacency of her young expansion: 'Oh, give me fifty words, any time, and the ghost of a situation, and I'll set you up a figure. Besides, I mustn't abuse poor Yolande – she has saved us,' she said.

'Yolande?'

'Our ridiculous play. That's the name of the impossible woman. She has put bread into our mouths and she's a loaf on the shelf for the future. The rights are mine.'

'You're lucky to have them,' said Nick, a little vaguely, troubled about his sitter's nose, which was somehow Jewish without the convex arch.

'Indeed I am. He gave them to me. Wasn't it charming?'

'He gave them – Mr Dashwood?'

'Dear me, no; where would poor Dashwood have got them? He hasn't a penny in the world. Besides, if he had got them he would have kept them. I mean your blessed cousin.'

'I see – they're a present from Peter.'

'Like many other things. Isn't he a dear? If it hadn't been for him the shelf would have remained bare. He bought the play for this country and America for four hundred pounds, and on the chance: fancy! There was no rush for it, and how could he tell? And then he gracefully handed it to me. So I have my little capital. Isn't he a duck? You have nice cousins.'

Nick assented to the proposition, only putting in an amendment to the effect that surely Peter had nice cousins also, and making, as he went on with his work, a tacit preoccupied reflection or two; such as that it must be pleasant to render little services like that to youth, beauty and genius (he rather wondered how Peter could afford them), and that, 'duck' as he was, Miss Rooth's benefactor was rather taken for granted. *Sic vos non vobis* faintly murmured itself in Nick's brain. This community of interests, or at least of relations, quickened the flight of time, so that he was still fresh when the sitting came to an end. It was settled that Miriam should come back on the morrow, to enable her portrayer to make the most of the few days of the parliamentary recess; and just before she left she asked –

'Then you *will* come tonight?'

'Without fail. I hate to lose an hour of you.'

'Then I'll place you. It will be my affair.'

'You're very kind,' he responded. 'Isn't it a simple matter for me to take a stall? This week I suppose they're to be had.'

'I'll send you a box,' said Miriam. 'You shall do it well. There are plenty now.'

'Why should I be lost, all alone, in the grandeur of a box?'

'Can't you bring your friend?'

'My friend?'

'The lady you are engaged to.'

'Unfortunately she's out of town.'

Miriam looked at him with a grand profundity. 'Does she leave you alone like that?'

'She thought I should like it – I should be more free to paint. You see I am.'

'Yes, perhaps it's good for *me*. Have you got her portrait?' Miriam asked.

'She doesn't like me to paint her.'

'Really? Perhaps then she won't like you to paint me.'

'That's why I want to be quick,' laughed Nick.

'Before she knows it?'

'She'll know it tomorrow. I shall write to her.'

Miriam gave him another of her special looks; then she said: 'I see; you're afraid of her.' And she added, 'Mention my name: they'll give you the box at the theatre.'

Whether or no Nick was afraid of Mrs Dallow, he still protested against receiving this bounty from the hands of Miss Rooth – repeated that he would rather take a stall according to his wont and pay for it. This led her to declare with a sudden flicker of passion that if he didn't do as she wished she would never sit to him again.

'Ah then, you have me,' returned Nick. 'Only I *don't* see why you should give me so many things.'

'What in the world have I given you?'

'Why, an idea.' And Nick looked at his picture a little ruefully. 'I don't mean to say I haven't let it fall and smashed it.'

'Ah, an idea – that *is* a great thing for people in our line. But you'll see me much better from the box, and I'll send you Gabriel Nash,' Miriam added, getting into the hansom which her host's servant had fetched for her. As Nick turned back into his studio after watching her drive away he laughed at the conception that they were in the same 'line'.

Nick shared his box at the theatre with Gabriel Nash, who talked during the entr'actes not in the least about the performance or the performer, but about the possible greatness of the art of the portraitist – its reach, its range, its fascination, the magnificent examples it had left us in the past: windows open into history, into psychology, things that were among the most precious possessions of the human race. He insisted, above all, on the interest, the richness arising from this great peculiarity of it: that, unlike most other forms, it was a revelation of two realities, the man whom it was the artist's conscious effort to reveal and the man (the interpreter) expressed in the very quality and temper of that effort. It offered a double vision, the strongest dose of life that art could give, the strongest dose of art that life could give. Nick Dormer had already become aware that he had two states of mind in listening to Gabriel Nash: one of them in which he laughed, doubted, sometimes even reprobated, and at any rate failed to follow or to accept; the other in which this contemplative genius seemed to take the words out of his mouth, to utter for him, better and more completely, the very things he was on the point of saying. Nash's saying them at such moments appeared to make them true, to set them up in the world, and tonight he said a good many, especially as to the happiness of cultivating one's own garden; growing there in stillness and freedom certain strong, pure flowers that would bloom forever, long after the rank weeds of the hour were withered and blown away.

It was to keep Miriam Rooth in his eye for his object that Nick had come to the play; and she dwelt there all the evening, being constantly on the stage. He was so occupied in watching her face (for he now saw pretty clearly what he should attempt to make of it) that he was conscious only in a secondary degree of the story she illustrated, and in regard to her acting in particular had mainly a surprised sense that she was extraordinarily quiet. He remembered her loudness, her violence in Paris, at Peter Sherringham's, her wild wails, the first time, at Madame Carré's; compared with which her present manner was eminently temperate and modern. Nick Dormer was not critical at the theatre; he believed what he saw and had a pleasant sense of the inevitable; therefore he would not have guessed what Gabriel Nash had to tell him – that for Miriam, with her tragic cast and her peculiar attributes, her present performance, full of actuality, of light, fine indications and, in parts, of pointed touches of comedy, was a rare *tour de force*. It went on altogether in a register that he had not supposed her to possess; in which, as he said, she didn't touch her capital, doing it wholly with her little savings. It gave him the idea that she was capable of almost anything.

In one of the intervals they went round to see her; but for Nick this purpose was partly defeated by the wonderful amiability with which he was challenged by Mrs Rooth, whom they found sitting with her daughter and who attacked him with a hundred questions about his dear mother and his charming sisters. She maintained that that day in Paris they had shown her a kindness she should never forget. She abounded also in gracious expressions in regard to the portrait he had so cleverly begun, declaring that she was so eager to see it, however little he might as yet have accomplished, that she should do herself the honour to wait upon him in the morning, when Miriam came to sit.

'I'm acting for *you* tonight,' the girl said to Nick, before he returned to his place.

'No, that's exactly what you are not doing,' Nash interposed, with one of his intellectual superiorities. 'You have stopped acting, you have reduced it to the least that will do, you simply *are* – you are just the visible image, the picture on the wall. It keeps you wonderfully in focus. I have never seen you so beautiful.'

Miriam stared at this; then it could be seen that she coloured. 'What a luxury in life to have everything explained! He's the great explainer,' she said, turning to Nick.

He shook hands with her for goodnight. 'Well then, we must give him lots to do.'

She came to his studio in the morning, but unaccompanied by her mother; in allusion to whom she simply said: 'Mamma wished to come, but I wouldn't let her.' They proceeded promptly to business. The girl divested herself of her hat and coat, taking the position already established for her. After they had worked for more than an hour, with much less talk than the day before, Nick being extremely absorbed and Miriam wearing in silence the kindest, most religious air of consideration for the sharp tension she imposed upon him – at the end of this period of patience, pervaded by a holy calm, our young lady suddenly got up and exclaimed, 'I say, I *must* see it!' with which, quickly, she stepped down from her place and came round to the canvas. She had, at Nick's request, not looked at his work the day before. He fell back, glad to rest, and put down his palette and brushes.

'*Ah bien, c'est tapé!*' Miriam cried, as she stood before the easel. Nick was pleased with her ejaculation, he was even pleased with what he had done; he had had a long, happy spurt and felt excited and sanctioned. Miriam, retreating also a little, sank into a high-backed, old-fashioned chair that stood two or three yards from the picture, and reclined in it, with her head on one side, looking at the rough resemblance. She made a remark or two about it, to which Nick replied standing behind her and after a moment leaning on the top of the chair. He was away from his work and his eyes searched it with a kind of fondness of hope. They rose, however, as he presently became conscious that the door of the large room opposite to him had opened without making a sound and that some one stood upon the threshold. The person on the threshold was Julia Dallow.

As soon as he perceived her Nick wished he had posted a letter to her the night before. He had written only that morning. Nevertheless there was genuine joy in the words with which he bounded towards her – 'Ah, my dear Julia, what a jolly surprise!' – for her unannounced descent spoke to him above all of an irresistible desire to see him again sooner than they had arranged. She had taken a step forward, but she had done no more, stopping short at the sight of the strange woman, so divested of visiting-gear that she looked half undressed, who lounged familiarly in the middle of the room and over whom Nick had been still more familiarly hanging. Julia's eyes rested on this embodied unexpectedness, and as they did so she grew pale – so pale that Nick, observing it, instinctively looked back to see what Miriam had done to produce such an effect. She had done nothing at all, which was precisely what was embarrassing; only staring at the intruder, motionless and superb. She seemed, somehow, in indolent possession of the place, and even in that instant Nick noted how handsome she looked; so that he exclaimed somewhere, inaudibly, in a region beneath his other emotions: 'How I should like to paint her *that* way!' Mrs Dallow transferred her eyes for a single moment to Nick's; then they turned away – away from Miriam, ranging over the room.

'I've got a sitter, but you mustn't mind that; we're taking a rest. I'm delighted to see you,' said Nick. He closed the door of the studio behind her; his servant was still at the outer door, which was open and through which he saw Julia's carriage drawn up. This made her advance a little further, but still she said nothing; she dropped no answer even when Nick went on, with a sense of awkwardness: 'When did you come back? I hope nothing has gone wrong. You come at a very interesting moment,' he continued, thinking as soon as he had spoken that they were such words as might have made her laugh. She was far from laughing; she only managed to look neither at him nor at Miriam and to say, after a little, when he had repeated his question about her return:

'I came back this morning – I came straight here.'

'And nothing's wrong, I hope?'

'Oh, no – everything's all right,' she replied very quickly and without expression. She vouchsafed no explanation of her premature return and took no notice of the seat Nick offered her; neither did she appear to hear him when he begged her not to look yet at the work on the easel – it was in such a dreadful state. He was conscious, as he phrased it, that his request gave to Miriam's position directly in front of his canvas an air of privilege which her neglect to recognize in any way Mrs Dallow's entrance or her importance did nothing to correct. But that mattered less if the appeal failed to reach Julia's intelligence, as he judged, seeing presently how deeply she was agitated. Nothing mattered in face of the sense of danger which took possession of him after she had been in the room a few moments. He wanted to say: 'What's the difficulty? Has anything happened?' but he felt that she would not like him to utter words so intimate in presence of the person she had been rudely startled to find between them. He pronounced Miriam's name to Mrs Dallow and Mrs Dallow's to Miriam, but Julia's recognition of the ceremony was so slight as to be scarcely perceptible. Miriam had the air of waiting for something more before she herself made a sign; and as nothing more came she continued to be silent and not to budge. Nick added a remark to the effect that Mrs Dallow would remember to have had the pleasure of meeting Miss Rooth the year before – in Paris, that day, at her brother Peter's; to which Mrs Dallow rejoined, 'Ah, yes,' without any qualification, while she looked down at some rather rusty studies, on panels, which were ranged along the floor, resting against the base of the wall. Her agitation was evidently a pain to herself; she had had a shock of extreme violence, and Nick saw that as Miriam showed no symptom of offering to give up her sitting her stay would be of the briefest. He wished Miriam would do something – say she would go, get up, move about; as it was she had the appearance of watching from her point of vantage Mrs Dallow's discomfiture. He made a series of inquiries about Julia's doings in the country, to two or three of which she gave answers monosyllabic and scarcely comprehensible, while she turned her eyes round and round the room as if she were looking for something she couldn't find – for an escape, for something that was not Miriam. At last she said – it was at the end of a very few minutes:

'I didn't come to stay – when you're so busy. I only looked in to see if you were here. Good-bye.'

'It's charming of you to have come. I'm so glad you've seen for yourself how well I'm occupied,' Nick replied, not unaware that he was very red. This made Mrs Dallow look at him, while Miriam considered them both. Julia's eyes had something in them that he had never seen before – a flash of fright by which he was himself frightened. 'Of course I'll see you later,' he added, laughing awkwardly, while she reached the door, while she opened it herself and got off without a good-day to Miriam. 'I wrote to you this morning – you've missed my letter,' he repeated behind her, having already given her this information. The door of the studio was very near that of the house, but before Mrs Dallow had reached the street the visitors' bell was set ringing. The passage was narrow and she kept in advance of Nick, anticipating his motion to open the street-door. The bell was tinkling still when, by the action of her own hand, a gentleman on the step stood revealed.

'Ah my dear, don't go!' Nick heard pronounced in quick, soft dissuasion and in the now familiar accents of Gabriel Nash. The rectification followed more quickly still, if that were a rectification which scarcely improved the matter: 'I beg a thousand pardons. I thought you were Miriam.'

Gabriel gave way, and Mrs Dallow dashed out of the house. Her carriage, a victoria with a pair of horses who had got hot, had taken a turn up the street, but the coachman had already perceived his mistress and was rapidly coming back. He drew near; not so fast however but that Gabriel Nash had time to accompany Mrs Dallow to the edge of the pavement with an apology for the freedom into which he had blundered. Nick was at her other hand, waiting to put her into her carriage and freshly disconcerted by the encounter with Nash, who somehow, as he stood making Julia an explanation that she didn't listen to, looked less eminent than usual, though not more conscious of difficulties. Nick coloured deeper and watched the footman spring down as the victoria drove up; he heard Nash say something about the honour of having met Mrs Dallow in Paris. Nick wanted him to go into the house; he damned inwardly his want of delicacy. He desired a word with Julia alone – as much alone as the two inconvenient servants would allow. But Nash was not too much discouraged to say: 'You came for a glimpse of the great model? *Doesn't* she sit? That's what I wanted too, this morning – just a look, for a blessing on the day. Ah, but *you*, madam –'

Julia had sprung into the carriage while he was still speaking and had flashed out to the coachman a 'Home!' which of itself set the vehicle in motion. The carriage went a few yards, but while Gabriel, with a magnificent bow, turned away, Nick Dormer, with his hand on the edge of the hood, moved with it.

'You don't like it, but I'll explain,' he said, laughing and in a low tone.

'Explain what?' Mrs Dallow asked, still very pale and grave, but showing nothing in her voice. She was thinking of the servants. She could think of them even then.

'Oh, it's all right. I'll come in at five,' Nick returned, gallantly jocular, while the carriage rolled away.

Gabriel had gone into the studio and Nick found him standing in admiration before Miriam, who had resumed the position in which she was sitting.

'Lord, she's good today! Isn't she good today?' Nash broke out, seizing Nick by the arm to give him a certain view. Miriam looked indeed still handsomer than before, and she had taken up her attitude again and with a splendid sphinx-like air of being capable of keeping it forever. Nick said nothing, but he went back to work with a tingle of confusion, which proved in fact, when he resumed his palette, to be a sharp and, after a moment, a delightful stimulus. Miriam spoke never a word, but she was doubly grand, and for more than an hour, till Nick, exhausted, declared he must stop, the industrious silence was broken only by the desultory discourse of Gabriel Nash.

27

NICK DORMER went to Great Stanhope Street at five o'clock and learned, rather to his surprise, that Mrs Dallow was not at home – to his surprise because he had told her he would come at that hour, and he attributed to her, with a certain simplicity, an eager state of mind in regard to his explanation. Apparently she was not eager; the eagerness was his own – he was eager to explain. He recognized, not without a certain consciousness of magnanimity in doing so, that there had been some reason for her quick withdrawal from his studio, or at any rate for her extreme discomposure there. He had a few days before put in a plea for a snatch of worship in that sanctuary, and she had accepted and approved it; but the worship, when the curtain happened to blow back, proved to be that of a magnificent young woman, an actress with disordered hair, who wore in a singular degree the aspect of a person settled to spend the day. The explanation was easy: it resided in the circumstance that when one was painting, even very badly and only for a moment, one had to have models. Nick was impatient to give it, with frank, affectionate lips and a full, jocose admission that it was natural Julia should have been startled; and he was the more impatient that, though he would not in the least have expected her to like finding a strange woman domesticated for the hour under his roof, she had disliked it even more than would have seemed probable. That was because, not having heard from him about the matter, the impression was for the moment irresistible with her that a trick had been played her. But three minutes with him alone would make the difference.

They would indeed have a considerable difference to make, Nick reflected, as minutes much more numerous elapsed without bringing Mrs Dallow home. For he had said to the butler that he would come in and wait (though it was odd she should not have left a message for him): she would doubtless return from one moment to the other. Nick had of course full licence to wait, anywhere he preferred; and he was ushered into Julia's particular sitting-room and supplied with tea and the evening papers. After a quarter of an hour however he gave little attention to these beguilements, owing to the increase of his idea that it was odd that when she definitely knew he was coming she should not have taken more pains to be at home. He walked up and down and looked out of the window, took up her books and dropped them again, and then, as half an hour had elapsed, began to feel rather angry. What could she be about when, at a moment when London was utterly empty, she could not be paying visits? A footman came in to attend to the fire; whereupon Nick questioned him as to the manner in which Mrs Dallow was probably engaged. The man revealed the fact that his mistress had gone out only a quarter of an hour before Nick arrived, and, as if he appreciated the opportunity for a little decorous conversation, gave him still more information than he asked for. From this it appeared that, as Nick knew or could surmise, she had the evening previous, in the country, telegraphed for the victoria to meet her in the morning at Paddington and had gone straight from the station to the studio, while her maid, with her luggage, proceeded in a cab to Great Stanhope Street. On leaving the studio however she had not come directly home; she had chosen this unusual season for an hour's drive in the Park. She had finally re-entered her house, but had remained up-stairs all day, seeing no one and not coming down to luncheon. At four o'clock she had ordered the brougham for four forty-five, and had got into it punctually, saying 'To the Park!' as she did so.

Nick, after the footman had left him, felt himself much mystified by Julia's sudden passion for the banks of the Serpentine, forsaken and foggy now, inasmuch as the afternoon had come on gray and the light was waning. She usually hated the Park and she hated a closed carriage. He had a discomfortable vision of her, shrunken into a corner of her brougham and veiled as if she had been crying, revolving round the solitude of the Drive. She had of course been deeply disconcerted, and she was nervous and upset: the motion of the carriage soothed her and made her fidget less. Nick remembered that in the morning, at his door, she had appeared to be going home; so she had turned into the Park on second thoughts, as she passed. He lingered another half hour, walked up and down the room many times and thought of many things. Had she misunderstood him when he said he would come at five? Couldn't she be sure, even if she had, that he would come early rather than late, and might she not have left a message for him on the chance? Going out that way a few minutes before he was to come had even a little the air of a thing done on purpose to offend him; as if she had been so displeased that she had taken the nearest occasion of giving him a sign that she meant to break. But were these the things that Julia did and was that the way she did them – his fine, proud, delicate, generous Julia?

When six o'clock came poor Nick felt distinctly resentful; but he stayed ten minutes longer, on the possibility that Mrs Dallow would in the morning have understood him to mention that hour. The April dusk began to gather and the unsociability of her behaviour, especially if she were still rumbling about the Park, became absurd. Anecdotes came back to Nick, vaguely remembered, heard he couldn't have said when or where, of poor artists for whom life had been rendered difficult by wives who wouldn't allow them the use of the living female model and who made scenes if, on the staircase, they encountered such sources of inspiration. These ladies struck him as vulgar and odious persons, with whom it seemed grotesque that Julia should have anything in common. Of course she was not his wife yet, and of course if she were he should have washed his hands of every form of activity requiring the services of the sitter; but even these qualifications left him with a capacity to shudder at the way Julia just escaped ranking herself with the heavy-handed.

At a quarter-past six he rang a bell and told the servant who answered it that he was going and that Mrs Dallow was to be informed as soon as she came in that he had expected to find her and had waited an hour and a quarter for her. But he had just reached the doorstep, on his departure, when her brougham, emerging from the evening mist, stopped in front of the house. Nick stood at the door, hanging back till she got out, allowing the servants to help her. She saw him – she was not veiled, like his mental image of her; but this did not prevent her from pausing to give an order to the coachman, a matter apparently requiring some discussion. When she came to the door Nick remarked to her that he had been waiting an eternity for her; to which she replied that he must not make a grievance to her of that – she was too unwell to do justice to it. He immediately professed regret and sympathy, adding, however, that in that case she had much better not have gone out. She made no answer to this – there were three servants in the hall who looked as if they might understand at least what was *not* said to them: only when he followed her in she asked if his idea had been to stay longer.

'Certainly, if you're not too ill to see me.'

'Come in, then,' Julia said, turning back after having gone to the foot of the stairs.

This struck him immediately as a further restriction of his visit: she would not readmit him to the drawing-room or to her boudoir; she would receive him in an impersonal apartment down-stairs, in which she saw people on business. What did she want to do to him? He was prepared by this time for a scene of jealousy; for he was sure he had learned to read her character justly in feeling that if she had the appearance of a cold woman she had also on certain occasions a liability to extreme emotion. She was very still, but every now and then she would fire off a pistol. As soon as Nick had closed the door she said, without sitting down:

'I dare say you saw I didn't like that at all.'

'My having a sitter, that way? I was very much annoyed at it myself,' Nick answered.

'Why were you annoyed? She's very handsome,' said Mrs Dallow, perversely.

'I didn't know you looked at her!' Nick laughed.

Julia hesitated a moment. 'Was I very rude?'

'Oh, it was all right. It was only awkward for me, because you didn't know,' Nick replied.

'I did know; that's why I came.'

'How do you mean? My letter couldn't have reached you.'

'I don't know anything about your letter,' said Mrs Dallow, casting about her for a chair and then seating herself on the edge of a sofa, with her eyes on the floor.

'She sat to me yesterday; she was there all the morning; but I didn't write to tell you. I went at her with great energy and, absurd as it may seem to you, found myself very tired afterwards. Besides, in the evening I went to see her act.'

'Does she act?' asked Mrs Dallow.

'She's an actress; it's her profession. Don't you remember her that day at Peter's, in Paris? She's already a celebrity; she has great talent; she's engaged at a theatre here and is making a sensation. As I tell you, I saw her last night.'

'You needn't tell me,' Mrs Dallow replied, looking up at him with a face of which the intense, the tragic sadness startled him.

He had been standing before her, but at this he instantly sat down beside her, taking her passive hand. 'I want to, please; otherwise it must seem so odd to you. I knew she was coming when I wrote to you the day before yesterday. But I didn't tell you then, because I didn't know how it would turn out and I didn't want to exult in advance over a poor little attempt that might come to nothing. Moreover it was no use speaking of the matter at all unless I told you exactly how it came about,' Nick went on, explaining kindly, copiously. 'It was the result of a visit unexpectedly paid me by Gabriel Nash.'

'That man – the man who spoke to me?' Julia asked, startled into a shuddering memory.

'He did what he thought would please you, but I dare say it didn't. You met him in Paris and didn't like him; so I thought it best to hold my tongue about him.'

'Do *you* like him?'

'Very much.'

'Great heaven!' Julia ejaculated, almost under her breath.

'The reason I was annoyed was because, somehow, when you came in, I suddenly had the air of having got out of those visits and shut myself up in town to do something that I had kept from you. And I have been very unhappy till I could explain.'

'You don't explain – you can't explain,' Mrs Dallow declared, turning on her companion eyes which, in spite of her studied stillness, expressed deep excitement. 'I knew it – I knew everything; that's why I came.'

'It was a sort of second-sight – what they call a brainwave,' Nick smiled.

'I felt uneasy, I felt a kind of call; it came suddenly, yesterday. It was irresistible; nothing could have kept me this morning.'

'That's very serious, but it's still more delightful. You mustn't go away again,' said Nick. 'We must stick together – forever and ever.'

He put his arm round her, but she detached herself as soon as she felt its pressure. She rose quickly, moving away, while, mystified, he sat looking up at her as she had looked a few moments before at him. 'I've thought it all over; I've been thinking of it all day,' she began. 'That's why I didn't come in.'

'Don't think of it too much; it isn't worth it.'

'You like it more than anything else. You do – you can't deny it,' she went on.

'My dear child, what are you talking about?' Nick asked, gently.

'That's what you like – doing what you were this morning; with women lolling, with their things off, to be painted, and people like that man.'

Nick slowly got up, hesitating. 'My dear Julia, apart from the surprise, this morning, do you object to the living model?'

'Not a bit, for you.'

'What's the inconvenience, then, since in my studio they are only for me?'

'You love it, you revel in it; that's what you want, and that's the only thing you want!' Julia broke out.

'To have models, lolling women, do you mean?'

'That's what I felt, what I knew, what came over me and haunted me yesterday, so that I couldn't throw it off. It seemed to me that if I could see it with my eyes and have the perfect proof I should feel better, I should be quiet. And now I *am* – after a struggle of some hours, I confess. I *have* seen; the whole thing's clear and I'm satisfied.'

'I'm not, and to me the whole thing isn't clear. What exactly are you talking about?' Nick demanded.

'About what you were doing this morning. That's your innermost preference, that's your secret passion.'

'A little go at something serious? Yes, it was almost serious,' said Nick. 'But it was an accident, this morning and yesterday: I got on better than I intended.'

'I'm sure you have immense talent,' Mrs Dallow remarked, with a joylessness that was almost droll.

'No, no, I might have had. I've plucked it up: it's too late for it to flower. My dear Julia, I'm perfectly incompetent and perfectly resigned.'

'Yes, you looked so this morning, when you hung over her. Oh, she'll bring back your talent!'

'She's an obliging and even an intelligent creature, and I've no doubt she would if she could. But I've received from *you* all the help that any woman is destined to give me. No one can do for me again what you have done.'

'I shouldn't try it again; I acted in ignorance. Oh, I've thought it all out!' Julia declared. Then, with a strange face of anguish resting on his, she said: 'Before it's too late – before it's too late!'

'Too late for what?'

'For you to be free – for you to be free. And for me – for me to be free too. You hate everything I like!' she exclaimed, with a trembling voice. 'Don't pretend, don't pretend!' she went on, as a sound of protest broke from him.

'I thought you wanted me to paint,' protested Nick, flushed and staring.

'I do – I do. That's why you must be free, why we must part.'

'Why we must part?'

'Oh, I've turned it over. I've faced the truth. It wouldn't do at all,' said Mrs Dallow.

'I like the way you talk of it, as if it were a trimming for your dress!' Nick rejoined, with bitterness. 'Won't it do for you to be loved and cherished as well as any woman in England?'

Mrs Dallow turned away from him, closing her eyes as if not to see something that would be dangerous to her. 'You mustn't give anything up for me. I should feel it all the while and I should hate it. I'm not afraid of the truth, but you are.'

'The truth, dear Julia? I only want to know it,' said Nick. 'It seems to me I've got hold of it. When two persons are united by the tenderest affection and are sane and generous and just, no difficulties that occur in the union their life makes for them are insurmountable, no problems are insoluble.'

Mrs Dallow appeared for a moment to reflect upon this; it was spoken in a tone that might have touched her. At any rate at the end of the moment, lifting her eyes, she announced: 'I hate art, as you call it. I thought I did, I knew I did; but till this morning I didn't know how much.'

'Bless your soul, *that* wasn't art,' pleaded Nick. 'The real thing will be a thousand miles away from us; it will never come into the house, *soyez tranquille*. Why then should you worry?'

'Because I want to understand, I want to know what I'm doing. You're an artist: you are, you are!' Mrs Dallow cried, accusing him passionately.

'My poor Julia, it isn't so easy as that, nor a character one can take on from one day to the other. There are all sorts of things; one must be caught young and put through the mill and see things as they are. There would be sacrifices I never can make.'

'Well then, there are sacrifices for both of us, and I can't make them either. I dare say it's all right for you, but for me it would be a terrible mistake. When I think I'm doing something I mustn't do just the opposite,' Julia went on, as if she wished to explain and be clear. 'There are things I've thought of, the things I like best; and they are not what you mean. It would be a great deception, and it's not the way I see my life, and it would be misery if we don't understand.'

Nick looked at her in hard perplexity, for she did not succeed in explaining as well as she wished. 'If we don't understand what?'

'That we are awfully different – that you are doing it all for me.'

'And is that an objection to me – what I do for you?' asked Nick.

'You do too much. You're awfully good, you're generous, you're a dear fellow; but I don't believe in it. I didn't, at bottom, from the first – that's why I made you wait, why I gave you your freedom. Oh, I've suspected you! I had my ideas. It's all right for you, but it won't do for me: I'm different altogether. Why should it always be put upon me, when I hate it? What have I done? I was drenched with it, before.' These last words, as they broke forth, were accompanied, even as the speaker uttered them, with a quick blush; so that Nick could as quickly discern in them the uncalculated betrayal of an old irritation, an old shame almost – her late husband's flat, inglorious taste for pretty things, his indifference to every chance to play a public part.

This had been the mortification of her youth, and it was indeed a perversity of fate that a new alliance should contain for her even an oblique demand for the same spirit of accommodation, impose on her the secret bitterness of the same concessions. As Nick stood there before her, struggling sincerely with the force that he now felt to be strong in her, the intense resolution to break with him, a force matured in a few hours, he read a riddle that hitherto had baffled him, saw a great mystery become simple. A personal passion for him had all but thrown her into his arms (the sort of thing that even a vain man – and Nick was not especially vain – might hesitate to recognize the strength of); held in check, with a tension of the cord at moments of which he could still feel the vibration, by her deep, her rare ambition, and arrested at the last only just in time to save her calculations. His present glimpse of the immense extent of these calculations did not make him think her cold or poor; there was in fact a positive strange heat in them and they struck him rather as grand and high. The fact that she could drop him even while she longed for him – drop him because it was now fixed in her mind that he would not after all serve her determination to be associated, so far as a woman could, with great affairs; that she could postpone, and postpone to an uncertainty, the satisfaction of a gnawing tenderness and judge for the long run – this exhibition of will and courage, of the large plan that possessed her, commanded his admiration on the spot. He paid the heavy penalty of being a man of imagination; he was capable of far excursions of the spirit, disloyalties to habit and even to faith, and open to wondrous communications. He ached for the moment to convince her that he would achieve what he wouldn't, for the vision of his future that she had tried to entertain shone before him as a bribe and a challenge. It seemed to him there was nothing he couldn't fancy enough, to be so fancied by her. Presently he said:

'You want to be sure the man you marry will be prime minister of England. But how can you be sure, with any one?'

'I can be sure some men won't,' Mrs Dallow replied.

'The only safe thing, perhaps, would be to marry Mr Macgeorge,' Nick suggested.

'Possibly not even him.'

'You're a prime minister yourself,' Nick answered. 'To hold fast to you as I hold, to be determined to be of your party – isn't that political enough, since you are the incarnation of politics?'

'Ah, how you hate them!' Julia moaned. 'I saw that when I saw you this morning. The whole place reeked of it.'

'My dear child, the greatest statesmen have had their distractions. What do you make of my hereditary talent? That's a tremendous force.'

'It wouldn't carry you far.' Then Mrs Dallow added: 'You must be a great artist.' Nick gave a laugh at the involuntary contempt of this, but she went on: 'It's beautiful of you to want to give up anything, and I like you for it. I shall always like you. We shall be friends, and I shall always take an interest –'

He stopped her at this, made a movement which interrupted her phrase, and she suffered him to hold her hand as if she were not afraid of him now. 'It isn't only for you,' he argued gently; 'you're a great deal, but you're not everything. Innumerable vows and pledges repose upon my head. I'm inextricably committed and dedicated. I was brought up in the temple; my father was a high priest and I'm a child of the Lord. And then the life itself – when *you* speak of it I feel stirred to my depths: it's like a herald's trumpet. Fight *with* me, Julia – not against me! Be on my side, and we shall do everything. It *is* fascinating, to be a great man before the people – to be loved by them, to be followed by them. An artist isn't – never, never. Why *should* he be? Don't forget how clever I am.'

'Oh, if it wasn't for that!' she rejoined, flushed with the effort to resist his tone. She asked abruptly: 'Do you pretend that if I were to die tomorrow you would stay in the House?'

'If you were to die? God knows! But you do singularly little justice to my incentives,' Nick continued. 'My political career is everything to my mother.'

Julia hesitated a moment; then she inquired: 'Are you afraid of your mother?'

'Yes, particularly; for she represents infinite possibilities of disappointment and distress. She represents all my father's as well as all her own; and in them my father tragically lives again. On the other hand I see him in bliss, as I see my mother, over our marriage and our life of common aspirations; though of course that's not a consideration that I can expect to have power with you.'

Mrs Dallow shook her head slowly, even smiling a little with an air of recovered calmness and lucidity. 'You'll never hold high office.'

'But why not take me as I am?'

'Because I'm abominably keen about that sort of thing; I must recognize it. I must face the ugly truth. I've been through the worst; it's all settled.'

'The worst, I suppose, was when you found me this morning.'

'Oh, that was all right – for you.'

'You're magnanimous, Julia; but evidently what's good enough for me isn't good enough for you.' Nick spoke with bitterness.

'I don't like you enough – that's the obstacle,' said Mrs Dallow bravely.

'You did a year ago; you confessed to it.'

'Well, a year ago was a year ago. Things are changed today.'

'You're very fortunate – to be able to throw away a devotion,' Nick replied.

Julia had her pocket-handkerchief in her hand, and at this she quickly pressed it to her lips, as if to check an exclamation. Then for an instant she appeared to be listening as if for a sound from outside. Nick interpreted her movement as an honourable impulse to repress the words: 'Do you mean the devotion that I was witness of this morning?' But immediately afterwards she said something very different: 'I thought I heard a ring. I've telegraphed for Mrs Gresham.'

'Why did you do that?' asked Nick.

'Oh, I want her.'

He walked to the window, where the curtains had not been drawn, and saw in the dusk a cab at the door. When he turned back he said: 'Why won't you trust me to make you like me, as you call it, better? If I make you like me as well as I like you, it will be about enough, I think.'

'Oh, I like you enough for *your* happiness. And I don't throw away a devotion,' Mrs Dallow continued. 'I shall be constantly kind to you. I shall be beautiful to you.'

'You'll make me lose a fortune,' declared Nick.

Julia stared, then she coloured. 'Ah, you may have all the money you want.'

'I don't mean yours,' he answered, flushing in his turn. He had determined on the instant, since it might serve, to tell her what he had never spoken of to her before. 'Mr Carteret last year promised me a pot of money on the day I should stand up with you. He has his heart set on our marriage.'

'I'm sorry to disappoint Mr Carteret,' said Julia. 'I'll go and see him. I'll make it all right,' she went on. 'Besides, you'll make a fortune by your portraits. The great men get a thousand, just for a head.'

'I'm only joking,' Nick returned, with sombre eyes that contradicted this profession. 'But what things you deserve I should do!'

'Do you mean striking likenesses?'

'You do hate it! Pushed to that point, it's curious,' the young man audibly mused.

'Do you mean you're joking about Mr Carteret's promise?'

'No, the promise is real; but I don't seriously offer it as a reason.'

'I shall go to Beauclere,' said Mrs Dallow. 'You're an hour late,' she added in a different tone; for at that moment the door of the room was thrown open and Mrs Gresham, the butler pronouncing her name, was ushered in.

'Ah, don't impugn my punctuality; it's my character!' the useful lady exclaimed, putting a sixpence from the cabman into her purse. Nick went off, at this, with a simplified farewell – went off foreseeing exactly what he found the next day, that Mrs Gresham would have received orders not to budge from her hostess's side. He called on the morrow, late in the afternoon, and Julia saw him liberally, in pursuance of her assertion that she would be 'beautiful' to him, that she had not thrown away his devotion; but Mrs Gresham remained immutably a spectator of her liberality. Julia looked at him kindly, but her companion was more benignant still; so that what Nick did with his own eyes was not to appeal to Mrs Dallow to see him for a moment alone, but to solicit, in the name of this luxury, the second occupant of the drawing-room. Mrs Gresham seemed to say, while Julia said very little: 'I understand, my poor friend, I know everything (she has told me only *her* side, but I'm so competent that I know yours too), and I enter into the whole thing deeply. But it would be as much as my place is worth to accommodate you.' Still, she did not go so far as to give him an inkling of what he learned on the third day and what he had not gone so far as to suspect – that the two ladies had made rapid arrangements for a scheme of foreign travel. These arrangements had already been carried out when, at the door of the house in Great Stanhope Street, the fact was imparted to Nick that Mrs Dallow and her friend had started that morning for Paris.

28

ON their way to Florence, Julia Dallow and Mrs Gresham spent three days in Paris, where Peter Sherringham had as much conversation with his sister as it often befell one member of that family to have with another. That is, on two different occasions he enjoyed half an hour's gossip with her in her sitting-room at the hotel. On one of these occasions he took the liberty of asking her whether or no, definitely, she meant to marry Nick Dormer. Julia expressed to him that she was much obliged for his interest, but that Nick and she were nothing more than relations and good friends. 'He wants to marry you, tremendously,' Peter remarked; to which Mrs Dallow simply made answer: 'Well, then, he may want!'

After this they sat silent for some moments, as if the subject had been quite threshed out between them. Peter felt no impulse to penetrate further, for it was not a habit of the Sherringhams to talk with each other of their love-affairs; and he was conscious of the particular deterrent that he and Julia had in general so different a way of feeling that they could never go far together in discussion. He liked her and was sorry for her, thought her life lonely and wondered she didn't make a 'great' marriage. Moreover he pitied her for being without the interests and consolations that he had found substantial: those of the intellectual, the studious order he considered these to be, not knowing how much she supposed that she reflected and studied or what an education she had found in her political aspirations, regarded by him as scarcely more a personal part of her than the livery of her servants or the jewels George Dallow's money had bought. Her relations with Nick were unfathomable to him; but they were not his affair. No affair of Julia's was sufficiently his to justify him in an attempt to understand it. That there should have been any question of her marrying Nick was the anomaly to him, rather than that the question should have been dropped. He liked his clever cousin very well as he was – enough to have a vague sense that he might be spoiled by being altered into a brother-in-law. Moreover, though he was not perhaps distinctly conscious of this, Peter pressed lightly on Julia's doings from a tacit understanding that in this case she would let him off as easily. He could not have said exactly what it was that he judged it pertinent to be let off from: perhaps from irritating inquiry as to whether he had given any more tea-parties for young ladies connected with the theatre.

Peter's forbearance however did not bring him all the security he prefigured. After an interval he indeed went so far as to ask Julia if Nick had been wanting in respect to her; but this was a question intended for sympathy, not for control. She answered: 'Dear, no – though he's very provoking.' Thus Peter guessed that they had had a quarrel in which it didn't concern him to interpose: he added the epithet and her flight from England together and they made up, to his perception, one of the little magnified embroilments which do duty for the real in superficial lives. It was worse to provoke Julia than not, and Peter thought Nick's doing so not particularly characteristic of his versatility for good. He might wonder why she didn't marry the member for Harsh if the subject had come up; but he wondered still more why Nick didn't marry her. Julia said nothing, again, as if to give him a chance to make some inquiry which would save her from gushing; but as his idea appeared to be to change the subject, and as he changed it only by silence, she was reduced to resuming presently:

'I should have thought you would have come over to see your friend the actress.'

'Which of my friends? I know so many actresses,' Peter rejoined.

'The woman you inflicted on us in this place a year ago – the one who is in London now.'

'Oh, Miriam Rooth! I should have liked to come over, but I've been tied fast. Have you seen her?'

'Yes, I've seen her.'

'Do you like her?'

'Not at all.'

'She has a lovely voice,' Peter hazarded, after a moment.

'I don't know anything about her voice – I haven't heard it.'

'But she doesn't act in pantomime, does she?'

'I don't know anything about her acting. I saw her in private – in Nick Dormer's studio.'

'In Nick Dormer's studio? What was she doing there?'

'She was sprawling over the room and staring at me.'

If Mrs Dallow had wished to 'draw' her brother it is probable that at this point she suspected she had succeeded, in spite of the care he took to divest his tone of everything like emotion in uttering the words: 'Why, does he know her so well? I didn't know.'

'She's sitting to him for her portrait; at least she was then.'

'Oh, yes, I remember: I put him up to that. I'm greatly interested. Is the portrait good?'

'I haven't the least idea – I didn't look at it. I dare say it's clever,' Julia added.

'How in the world does Nick find time to paint?'

'I don't know. That horrid man brought her.'

'What horrid man?' Peter demanded.

'The one Nick thinks so clever – the vulgar little man who was at your place that day and tried to talk to me. I remember he abused theatrical people to me – as if I cared anything about them. But he has apparently something to do with this girl.'

'Oh, I recollect him – I had a discussion with him,' Peter said.

'How could you? I must go and dress,' Julia went on.

'He *was* clever, remarkably. Miss Rooth and her mother were old friends of his, and he was the first person to speak of them to me.'

'What a distinction! I thought him disgusting!' exclaimed

Mrs Dallow, who was pressed for time and who had now got up.

'Oh, you're severe,' said Peter; but as they separated she had given him something to think of.

That Nick was painting a beautiful actress was no doubt in part at least the reason why he was provoking and why his most intimate female friend had come abroad. The fact did not render him provoking to Peter Sherringham: on the contrary Peter had been quite sincere when he qualified it as interesting. It became indeed on reflection so interesting that it had perhaps almost as much to do with Sherringham's rush over to London as it had to do with Julia's coming away. Reflection taught Peter further that the matter was altogether a delicate one, and suggested that it was odd he should be mixed up with it in fact, when, as Julia's business, he had wished only to keep out of it. It was his own business a little too: there was somehow a still more pointed implication of that in his sister's saying to him the next day that she wished immensely he would take a fancy to Biddy Dormer. She said more: she said there had been a time when she believed he *had* done so – believed too that the poor child herself had believed the same. Biddy was far away the nicest girl she knew – the dearest, sweetest, cleverest, *best*, and one of the prettiest creatures in England, which never spoiled anything. She would make as charming a wife as ever a man had, suited to any position, however high, and (Julia didn't mind mentioning it, since Peter would believe it whether she mentioned it or no) was so predisposed in his favour that he would have no trouble at all. In short she herself would see him through – she would answer for it that he would only have to speak. Biddy's life at home was horrid; she was very sorry for her – the child was worthy of a better fate. Peter wondered what constituted the horridness of Biddy's life, and perceived that it mainly arose from the fact that Julia disliked Lady Agnes and Grace; profiting comfortably by the freedom to do so conferred upon her by her having given them a house of which she had perhaps not felt the want till they were in possession of it. He knew she had always liked Biddy, but he asked himself (this was the rest of his wonder) why she had taken to liking her so extraordinarily just now. He liked her himself – he even liked to be talked to about her and he could believe everything Julia said: the only thing that mystified him was her motive for suddenly saying it. He assured her that he was infinitely indebted to her for her expenditure of imagination on his behalf, but that he was sorry if he had put it into any one's head (most of all into the girl's own) that he had looked at Biddy with a covetous eye. He knew not whether she would make a good wife, but he liked her quite too much to wish to put such a ticklish matter to the test. She was surely not intended for cruel experiments. As it happened he was not thinking of marrying any one – he had ever so many reasons against it. Of course one was never safe against accidents, but one could at least take precautions, and he didn't mind telling her that there were several he *had* taken.

'I don't know what you mean, but it seems to me quite the best precaution would be to care for a charming, steady girl like Biddy,' Mrs Dallow replied. 'Then you would be quite in shelter, you would know the worst that can happen to you, and it wouldn't be bad.' The objection Peter had made to this argument is not important, especially as it was not remarkably candid; it need only be mentioned that before he and Julia parted she said to him, still in reference to Bridget Dormer: 'Do go and see her and be nice to her: she'll save you disappointments.'

These last words reverberated in Sherringham's mind; there was a shade of the portentous in them and they seemed to proceed from a larger knowledge of the subject than he himself as yet possessed. They were not absent from his memory when, in the beginning of May, availing himself, to save time, of the night-service, he crossed from Paris to London. He arrived before the breakfast-hour and went to his sister's house in Great Stanhope Street, where he always found quarters whether she were in town or not. If she were at home she welcomed him, and if she were not the relaxed servants hailed him for the chance he gave them to recover their 'form'. In either case his allowance of space was large and his independence complete. He had obtained permission this year to take in fractions instead of as a single draught the leave of absence to which he was entitled; and there was moreover a question of his being transferred to another Embassy, in which event he believed that he might count upon a month or two in England before proceeding to his new post.

He waited after breakfast but a very few minutes before jumping into a hansom and rattling away to the north. A part of his waiting indeed consisted of a fidgety walk up Bond Street, during which he looked at his watch three or four times while he paused at shop-windows for fear of being a little early. In the cab, as he rolled along, after having given an address – Balaklava Place, St John's Wood – the fear that he should be too early took curiously at moments the form of a fear that he should be too late: a symbol of the inconsistencies of which his spirit at present was full. Peter Sherringham was nervous, too nervous for a diplomatist, and haunted with inclinations, and indeed with purposes, which contradicted each other. He wanted to be out of it and yet he dreaded not to be in it, and on this particular occasion the sense of exclusion made him sore. At the same time he was not unconscious of the impulse to stop his cab and make it turn round and drive due south. He saw himself launched in the breezy fact while, morally speaking, he was hauled up on the hot sand of the principle, and he had the intelligence to perceive how little these two faces of the same idea had in common. However, as the sense of movement encouraged him to reflect, a principle was a poor affair if it remained mere inaction. Yet from the moment it turned to action it manifestly could only be the particular action in which he was engaged; so that he was in the absurd position of thinking his behaviour more consummate for the reason that it was directly opposed to his intentions.

He had kept away from London ever since Miriam Rooth came over; resisting curiosity, sympathy, importunate haunting passion and considering that his resistance, founded, to be salutary, on a general scheme of life, was the greatest success he had yet achieved. He was deeply occupied with plucking up the feeling that attached him to her, and he had already, by various little ingenuities, loosened some of its roots. He suffered her to make her first appearance on any stage without the comfort of his voice or the applause of his hand; saying to himself that the man who could do the more could do the less and that such an act of fortitude was a proof he should keep straight. It was not exactly keeping straight to run over to London three months later and, the hour he arrived, scramble off to Balaklava

Place; but after all he pretended only to be human and aimed in behaviour only at the heroic, not at the monstrous. The highest heroism was three parts tact. He had not written to Miriam that he was coming to England and would call upon her at eleven o'clock in the morning, because it was his secret pride that he had ceased to correspond with her. Sherringham took his prudence where he could find it, and in doing so was rather like a drunkard who should flatter himself that he had forsworn liquor because he didn't touch lemonade.

It is an example of how much he was drawn in different directions at once that when, on reaching Balaklava Place and alighting at the door of a small much-ivied house which resembled a gate-lodge bereft of its park, he learned that Miss Rooth had only a quarter of an hour before quitted the spot with her mother (they had gone to the theatre, to rehearsal, said the maid who answered the bell he had set tinkling behind a dingy plastered wall): when at the end of his pilgrimage he was greeted by a disappointment he suddenly found himself relieved and for the moment even saved. Providence was after all taking care of him and he submitted to Providence. He would still be watched over doubtless, even if he should follow the two ladies to the theatre, send in his card and obtain admission to the histrionic workshop. All his old technical interest in the girl's development flamed up again, and he wondered what she was rehearsing, what she was to do next. He got back into his hansom and drove down the Edgware Road. By the time he reached the Marble Arch he had changed his mind again – he had determined to let Miriam alone for that day. The day would be over at eight o'clock in the evening (he hardly played fair), and then he should consider himself free. Instead of going to the theatre he drove to a shop in Bond Street, to take a place for the play. On first coming out he had tried, at one of those establishments strangely denominated 'libraries', to get a stall, but the people to whom he applied were unable to accommodate him – they had not a single seat left. His second attempt, at another 'library', was more successful: he was unable to obtain a stall, but by a miracle he might have a box. There was a certain wantonness in paying for a box to see a play on which he had already expended four hundred pounds; but while he was mentally measuring this abyss an idea came into his head which flushed the extravagance with a slight rose-tint.

Peter came out of the shop with the voucher for the box in his pocket, turned into Piccadilly, noted that the day was growing warm and fine, felt glad that this time he had no business, unless it were business to leave a card or two on official people, and asked himself where he should go if he didn't go after Miriam. Then it was that it struck him as most acutely desirable, and even most important, that he should see Nick Dormer's portrait of her. He wondered which would be the natural place at that hour of the day to look for the artist. The House of Commons was perhaps the nearest one, but Nick, incongruous as his proceedings certainly were, probably didn't keep the picture there; and moreover it was not generally characteristic of him to be in the natural place. The end of Peter's debate was that he again entered a hansom and drove to Calcutta Gardens. The hour was early for calling, but cousins with whom one's intercourse was mainly a conversational scuffle would accept it as a practical illustration of that method. And if Julia wanted him to be nice to Biddy (which was exactly, though with a different view, what he wanted himself), what could be nicer than to pay his visit to Lady Agnes (he would have in decency to go to see her some time) at a friendly, fraternizing hour, when they would all be likely to be at home?

Unfortunately, as it turned out, they were not at home, so that Peter had to fall back on neutrality and the butler, who was however, more luckily, an old friend. Her ladyship and Miss Dormer were absent from town, paying a visit; and Mr Dormer was also away, or was on the point of going away for the day. Miss Bridget was in London, but was out: Peter's informant mentioned with earnest vagueness that he thought she had gone somewhere to take a lesson. On Peter's asking what sort of a lesson he meant, he replied, 'Oh, I think the a-sculpture, you know, sir.' Peter knew, but Biddy's lesson in a-sculpture (it sounded on the butler's lips like a fashionable new art) struck him a little as a mockery of the benevolent spirit in which he had come to look her up. The man had an air of participating respectfully in his disappointment and, to make up for it, added that he might perhaps find Mr Dormer at his other address. He had gone out early and had directed his servant to come to Rosedale Road in an hour or two with a portmanteau: he was going down to Beauclere in the course of the day, Mr Carteret being ill – perhaps Mr Sherringham didn't know it. Perhaps too Mr Sherringham would catch him in Rosedale Road before he took his train – he was to have been busy there for an hour. This was worth trying, and Peter immediately drove to Rosedale Road; where, in answer to his ring, the door was opened to him by Biddy Dormer.

29

WHEN Biddy saw him her cheek exhibited the prettiest pleased, surprised red that he had ever observed there, though he was not unacquainted with its fluctuations, and she stood still, smiling at him with the outer dazzle in her eyes, making no motion for him to enter. She only said: 'Oh, Peter!' And then: 'I'm all alone.'

'So much the better, dear Biddy. Is that any reason I shouldn't come in?'

'Dear, no – do come in. You've just missed Nick; he has gone to the country – half an hour ago.' She had on a large apron, and in her hand she carried a small stick, besmeared, as his quick eye saw, with modelling-clay. She dropped the door and fled back before him into the studio, where, when he followed her, she was in the act of flinging a cloth over a rough head, in clay, which, in the middle of the room, was supported on a high wooden stand. The effort to hide what she had been doing before he caught a glimpse of it made her redder still and led to her smiling more, to her laughing with a charming confusion of shyness and gladness. She rubbed her hands on her apron, she pulled it off, she looked delightfully awkward, not meeting Peter's eye, and she said: 'I'm just scraping here a little – you mustn't mind me. What I do is awful, you know. Peter, please don't look. I've been coming here lately to make my little mess, because mamma doesn't particularly like it at home. I've had a lesson from a lady who exhibits; but you wouldn't suppose it, to see what I do. Nick's so kind; he lets me come here; he uses the studio so little; I do what I please. What a pity he's gone – he would have been so glad. I'm really alone – I hope you don't mind. Peter, *please* don't look.'

Peter was not bent upon looking; his eyes had occupation enough in Biddy's own agreeable aspect, which was full of an unusual element of domestication and responsibility. Though she had taken possession, by exception, of her brother's quarters, she struck her visitor as more at home and more herself than he had ever seen her. It was the first time she had been to his vision so separate from her mother and sister. She seemed to know this herself and to be a little frightened by it – just enough to make him wish to be reassuring. At the same time Peter also on this occasion found himself touched with diffidence, especially after he had gone back and closed the door and settled down to a regular visit; for he became acutely conscious of what Julia had said to him in Paris and was unable to rid himself of the suspicion that it had been said with Biddy's knowledge. It was not that he supposed his sister had told the girl that she meant to do what she could to make him propose to her: that would have been cruel to her (if she liked him enough to consent), in Julia's uncertainty. But Biddy participated by imagination, by divination, by a clever girl's secret tremulous instincts, in her good friend's views about her, and this probability constituted for Sherringham a sort of embarrassing publicity. He had impressions, possibly gross and unjust, in regard to the way women move constantly together amid such considerations and subtly intercommunicate, when they do not still more subtly dissemble, the hopes or fears of which persons of the opposite sex form the subject. Therefore poor Biddy would know that if she failed to strike him in the right light it would not be for want of his attention's having been called to her claims. She would have been tacitly rejected, virtually condemned. Peter could not, without a slight sense of fatuity, endeavour to make up for this to her by kindness; he was aware that if any one knew it a man would be ridiculous who should take so much as that for granted. But no one would know it: oddly enough, in this calculation of security he left Biddy herself out. It did not occur to him that she might have a secret small irony to spare for his ingenious and magnanimous impulse to show her how much he liked her in order to make her forgive him for not liking her more. This magnanimity at any rate coloured the whole of Sherringham's visit to Rosedale Road, the whole of the pleasant, prolonged chat that kept him there for more than an hour. He begged the girl to go on with her work and not to let him interrupt it; and she obliged him at last, taking the cloth off the lump of clay and giving him a chance to be delightful by guessing that the shapeless mass was intended, or would be intended after a while, for Nick. He saw that she was more comfortable when she began to smooth it and scrape it with her little stick again, to manipulate it with an ineffectual air of knowing how; for this gave her something to do, relieved her nervousness and permitted her to turn away from him when she talked.

Peter walked about the room and sat down; got up and looked at Nick's things; watched her at moments in silence (which made her always say in a minute that he was not to look at her so or she could do nothing); observed how her position, before her high stand, her lifted arms, her turns of the head, considering her work this way and that, all helped her to be pretty. She repeated again and again that it was an immense pity about Nick, till he was obliged to say he didn't care a straw for Nick: he was perfectly content with the company he found. This was not the sort of thing he thought it right, under the circumstances, to say; but then even the circumstances did not require him to pretend he liked her less than he did. After all she was his cousin; she would cease to be so if she should become his wife; but one advantage of her not entering into that relation was precisely that she would remain his cousin. It was very pleasant to find a young, bright, slim, rose-coloured kinswoman all ready to recognize consanguinity when one came back from cousinless foreign lands. Peter talked about family matters; he didn't know, in his exile, where no one took an interest in them, what a fund of latent curiosity about them was in him. It was in him to gossip about them and to enjoy the sense that he and Biddy had indefeasible properties in common – ever so many things as to which they would understand each other *à demi-mot*. He smoked a cigarette, because she begged him to, said that people always smoked in studios – it made her feel so much more like an artist. She apologized for the badness of her work on the ground that Nick was so busy he could scarcely ever give her a sitting; so that she had to do the head from photographs and occasional glimpses. They had hoped to be able to put in an hour that morning, but news had suddenly come that Mr Carteret was worse, and Nick had hurried down to Beauclere. Mr Carteret was very ill, poor old dear, and Nick and he were immense friends. Nick had always been charming to him. Peter and Biddy took the concerns of the houses of Dormer and Sherringham in order, and the young man felt after a little as if they were as wise as a French *conseil de famille*, settling what was best for everyone. He heard all about Lady Agnes and manifested an interest in the detail of her existence that he had not supposed himself to possess, though indeed Biddy threw out intimations which excited his curiosity, presenting her mother in a light that might call upon his sympathy.

'I don't think she has been very happy or very pleased, of late,' the girl said. 'I think she has had some disappointments, poor dear mamma; and Grace has made her go out of town for three or four days, for a little change. They have gone down to see an old lady, Lady St Dunstans, who never comes to London now, and who, you know – she's tremendously old – was papa's godmother. It's not very lively for Grace, but Grace is such a dear she'll do anything for mamma. Mamma will go anywhere to see people she can talk with about papa.'

Biddy added, in reply to a further inquiry from Peter, that what her mother was disappointed about was – well, themselves, her children and all their affairs; and she explained that Lady Agnes wanted all kinds of things for them that didn't come to them, that they didn't get or seem likely to get, so that their life appeared altogether a failure. She wanted a great deal, Biddy admitted; she really wanted everything and she had thought in her happier days that everything was to be hers. She loved them all so much, and then she was proud: she couldn't get over the thought of their not being successful. Sherringham was unwilling to press, at this point, for he suspected one of the things that Lady Agnes wanted; but Biddy relieved him a little by saying that one of these things was that Grace should get married.

'That's too unselfish of her,' rejoined Peter, who didn't care for Grace. 'Cousin Agnes ought to keep her near her always, if Grace is so obliging and devoted.'

'Oh, mamma would give up anything of that sort for our good; she wouldn't sacrifice us that way!' Biddy exclaimed. 'Besides, I'm the one to stay with mamma; not that I can manage and look after her and do everything so well as Grace. But, you know, I want to,' said Biddy, with a liquid note in her voice, giving her lump of clay a little stab.

'But doesn't your mother want the rest of you to get married – Percival and Nick and you?' Peter asked.

'Oh, she has given up Percy. I don't suppose she thinks it would do. Dear Nick, of course – that's just what she does want.'

Sherringham hesitated. 'And you, Biddy?'

'Oh, I dare say; but that doesn't signify – I never shall.'

Peter got up, at this; the tone of it set him in motion and he took a turn round the room. He said something to her about her being too proud; to which she replied that that was the only thing for a girl to be, to get on.

'What do you mean by getting on?' Peter demanded, stopping, with his hands in his pockets, on the other side of the studio.

'I mean crying one's eyes out!' Biddy unexpectedly exclaimed; but she drowned the effect of this pathetic paradox in a foolish laugh and in the quick declaration: 'Of course it's about Nick that poor mother's really broken-hearted.'

'What's the matter with Nick?' Sherringham asked, diplomatically.

'Oh, Peter, what's the matter with Julia?' Biddy quavered softly, back to him, with eyes suddenly frank and mournful. 'I dare say you know what we all hoped – what we all supposed, from what they told us. And now they won't!' said Biddy.

'Yes, Biddy, I know. I had the brightest prospect of becoming your brother-in-law: wouldn't that have been it – or something like that? But it is indeed visibly clouded. What's the matter with them? May I have another cigarette?' Peter came back to the wide, cushioned bench where he had been lounging: this was the way they took up the subject he wanted most to look into. 'Don't they know how to love?' he went on, as he seated himself again.

'It seems a kind of fatality!' sighed Biddy.

Peter said nothing for some moments, at the end of which he

inquired whether his companion were to be quite alone during her mother's absence. She replied that her mother was very droll about that – she would never leave her alone: she thought something dreadful would happen to her. She had therefore arranged that Florence Tressilian should come and stay in Calcutta Gardens for the next few days, to look after her and see she did no wrong. Peter asked who Florence Tressilian might be: he greatly hoped, for the success of Lady Agnes's precautions, that she was not a flighty young genius like Biddy. She was described to him as tremendously nice and tremendously clever, but also tremendously old and tremendously safe; with the addition that Biddy was tremendously fond of her and that while she remained in Calcutta Gardens they expected to enjoy themselves tremendously. She was to come that afternoon, before dinner.

'And are you to dine at home?' said Peter.

'Certainly; where else?'

'And just you two, alone? Do you call that enjoying yourselves tremendously?'

'It will do for me. No doubt I oughtn't, in modesty, to speak for poor Florence.'

'It isn't fair to her; you ought to invite some one to meet her.'

'Do you mean you, Peter?' the girl asked, turning to him quickly, with a look that vanished the instant he caught it.

'Try me; I'll come like a shot.'

'That's kind,' said Biddy, dropping her hands and now resting her eyes on him gratefully. She remained in this position a moment, as if she were under a charm; then she jerked herself back to her work with the remark: 'Florence will like that immensely.'

'I'm delighted to please Florence, your description of her is so attractive!' Sherringham laughed. And when the girl asked him if he minded if there were not a great feast, because when her mother went away she allowed her a fixed amount for that sort of thing and, as he might imagine, it wasn't millions – when Biddy, with the frankness of their pleasant kinship, touched anxiously on this economical point (illustrating, as Peter saw, the lucidity with which Lady Agnes had had in her old age to learn to recognize the occasions when she could be conveniently frugal), he answered that the shortest dinners were the best, especially when one was going to the theatre. That was his case tonight, and did Biddy think he might look to Miss Tressilian to go with him? They would have to dine early; he wanted not to miss a moment.

'The theatre – Miss Tressilian?' Biddy stared, interrupted and in suspense again.

'Would it incommode you very much to dine say at 7.15 and accept a place in my box? The finger of Providence was in it when I took a box an hour ago. I particularly like your being free to go – if you are free.'

Biddy became fairly incoherent with pleasure. 'Dear Peter, how good you are! They'll have it at any hour. Florence will be so glad.'

'And has Florence seen Miss Rooth?'

'Miss Rooth?' the girl repeated, redder than before. He perceived in a moment that she had heard that he had devoted much time and attention to that young lady. It was as if she were conscious that he would be conscious in speaking of her, and there was a sweetness in her allowance for him on that score. But Biddy was more confused for him than he was for himself. He guessed in a moment how much she had thought over what she had heard; this was indicated by her saying vaguely: 'No, no, I've not seen her.' Then she became aware that she was answering a question he had not asked her, and she went on: 'We shall be too delighted. I saw her – perhaps you remember – in your rooms in Paris. I thought her so wonderful then! Everyone is talking of her here. But we don't go to the theatre much, you know: we don't have boxes offered us except when *you* come. Poor Nick is too much taken up in the evening. I've wanted awfully to see her. They say she's magnificent.'

'I don't know,' said Peter. 'I haven't seen her.'

'You haven't seen her?'

'Never, Biddy. I mean on the stage. In private, often – yes,' Sherringham added, conscientiously.

'Oh!' Biddy exclaimed, bending her face on Nick's bust again. She asked him no question about the new star, and he offered her no further information. There were things in his mind that pulled him different ways, so that for some minutes silence was the result of the conflict. At last he said, after an hesitation caused by the possibility that she was ignorant of the fact he had lately elicited from Julia, though it was more probable she might have learned it from the same source:

'Am I perhaps indiscreet in alluding to the circumstance that Nick has been painting Miss Rooth's portrait?'

'You are not indiscreet in alluding to it to me, because I know it.'

'Then there's no secret nor mystery about it?'

Biddy considered a moment. 'I don't think mamma knows it.'

'You mean you have been keeping it from her because she wouldn't like it?'

'We're afraid she may think papa wouldn't have liked it.'

This was said with an absence of humour which for an instant moved Sherringham to mirth; but he quickly recovered himself, repenting of any apparent failure of respect to the high memory of his late celebrated relative. He rejoined quickly, but rather vaguely: 'Ah, yes, I remember that great man's ideas'; and then he went on: 'May I ask if you know it, the fact that we are talking of, through Julia or through Nick?'

'I know it from both of them.'

'Then, if you're in their confidence, may I further ask whether this undertaking of Nick's is the reason why things seem to be at an end between them?'

'Oh, I don't think she likes it,' returned Biddy.

'Isn't it good?'

'Oh, I don't mean the picture – she hasn't seen it; but his having done it.'

'Does she dislike it so much that that's why she won't marry him?'

Biddy gave up her work, moving away from it to look at it. She came and sat down on the long bench on which Sherringham had placed himself. Then she broke out: 'Oh, Peter, it's a great trouble – it's a very great trouble; and I can't tell you, for I don't understand it.'

'If I ask you, it's not to pry into what doesn't concern me; but Julia is my sister, and I can't, after all, help taking some interest in her life. But she tells me very little. She doesn't think me worthy.'

'Ah, poor Julia!' Biddy murmured, defensively. Her tone recalled to him that Julia had thought him worthy to unite himself to Bridget Dormer, and inevitably betrayed that the girl was thinking of that also. While they both thought of it they sat looking into each other's eyes.

'Nick, I'm sure, doesn't treat *you* that way. I'm sure he confides in you; he talks to you about his occupations, his ambitions,' Peter continued. 'And you understand him, you enter into them, you are nice to him, you help him.'

'Oh, Nick's life – it's very dear to me,' said Biddy.

'That must be jolly for him.'

'It makes *me* very happy.'

Peter uttered a low, ambiguous groan; then he exclaimed, with irritation: 'What the deuce is the matter with them then? Why can't they hit it off and be quiet and rational and do what every one wants them to do?'

'Oh, Peter, it's awfully complicated,' said Biddy, with sagacity.

'Do you mean that Nick's in love with her?'

'In love with Julia?'

Biddy shook her head slowly; then with a smile which struck him as one of the sweetest things he had ever seen (it conveyed, at the expense of her own prospects, such a shy, generous little mercy of reassurance): 'He isn't, Peter,' she declared. 'Julia thinks it's trifling – all that sort of thing,' she added. 'She wants him to go in for different honours.'

'Julia's the oddest woman. I thought she loved him,' Sherringham remarked. 'And when you love a person –' He continued to reflect, leaving his sentence impatiently unfinished, while Biddy, with lowered eyes, sat waiting (it interested her) to learn what you did when you loved a person. 'I can't conceive her giving him up. He has great ability, besides being such a good fellow.'

'It's for his happiness, Peter – that's the way she reasons,' Biddy explained. 'She does it for an idea; she has told me a great deal about it, and I can see the way she feels.'

'You try to, Biddy, because you are such a dear good-natured girl, but I don't believe you do in the least. It's too little the way you yourself would feel. Julia's idea, as you call it, must be curious.'

'Well, it is, Peter,' Biddy mournfully admitted. 'She won't risk not coming out at the top.'

317

'At the top of what?'

'Oh, of everything.' Biddy's tone showed a trace of awe of such high views.

'Surely one's at the top of everything when one's in love.'

'I don't know,' said the girl.

'Do you doubt it?' Sherringham demanded.

'I've never been in love and I never shall be.'

'You're as perverse in your way as Julia. But I confess I don't understand Nick's attitude any better. He seems to me, if I may say so, neither fish nor flesh.'

'Oh, his attitude is very noble, Peter; his state of mind is wonderfully interesting,' Biddy pleaded. 'Surely *you* must be in favour of art,' she said.

Sherringham looked at her a moment. 'Dear Biddy, your little digs are as soft as zephyrs.'

She coloured, but she protested. 'My little digs? What do you mean? Are you not in favour of art?'

'The question is delightfully simple. I don't know what you're talking about. Everything has its place. A parliamentary life scarcely seems to me the situation for portrait-painting.'

'That's just what Nick says.'

'You talk of it together a great deal?'

'Yes, Nick's very good to me.'

'Clever Nick! And what do you advise him?'

'Oh, to *do* something.'

'That's valuable,' Peter laughed. 'Not to give up his sweetheart for the sake of a paint-pot, I hope?'

'Never, never, Peter! It's not a question of his giving up, for Julia has herself drawn back. I think she never really felt safe; she loved him, but she was afraid of him. Now she's only afraid – she has lost the confidence she tried to have. Nick has tried to hold her, but she has jerked herself away. Do you know what she said to me? She said: "My confidence has gone forever."'

'I didn't know she was such a prig!' Sherringham exclaimed. 'They're queer people, verily, with water in their veins instead of blood. You and I wouldn't be like that, should we? though you *have* taken up such a discouraging position about caring for a fellow.'

'I care for art,' poor Biddy returned.

'You do, to some purpose,' said Peter, glancing at the bust.

'To that of making you laugh at me.'

'Would you give a good man up for that?'

'A good man? What man?'

'Well, say me – if I wanted to marry you.'

Biddy hesitated a little. 'Of course I would, in a moment. At any rate, I'd give up the House of Commons. That's what Nick's going to do now – only you mustn't tell any one.'

Sherringham stared. 'He's going to chuck up his seat?'

'I think his mind is made up to it. He has talked me over – we have had some deep discussions. Yes, I'm on the side of art!' said Biddy, ardently.

'Do you mean in order to paint – to paint Miss Rooth?' Peter went on.

'To paint every one – that's what he wants. By keeping his seat he hasn't kept Julia, and she was the thing he cared most for, in public life. When he has got out of the whole thing his attitude, as he says, will be at least clear. He's tremendously interesting about it, Peter; he has talked to me wonderfully; he *has* won me over. Mamma's heart-broken; telling her will be the hardest part.'

'If she doesn't know, why is she heart-broken?'

'Oh, at the marriage not coming off – she knows that. That's what she wanted. She thought it perfection. She blames Nick fearfully. She thinks he held the whole thing in his hand and that he has thrown away a magnificent opportunity.'

'And what does Nick say to her?'

'He says, "Dear old mummy!"'

'That's good,' said Sherringham.

'I don't know what will become of her when this other blow arrives,' Biddy pursued. 'Poor Nick wants to please her – he does, he does. But, as he says, you can't please everyone, and you must, before you die, please yourself a little.'

Peter Sherringham sat looking at the floor; the colour had risen to his face while he listened to the girl. Then he sprang up and took another turn about the room. His companion's artless but vivid recital had set his blood in motion. He had taken Nick Dormer's political prospects very much for granted, thought of them as definite and brilliant and seductive. To learn there was something for which he was ready to renounce such honours, and to recognize the nature of that bribe, affected Sherringham powerfully and strangely. He felt as if he had heard the sudden blare of a trumpet, and he felt at the same time as if he had received a sudden slap in the face. Nick's bribe was 'art' – the strange temptress with whom he himself had been wrestling and over whom he had finally ventured to believe that wisdom and training had won a victory. There was something in the conduct of his old friend and playfellow that made all his reasonings small. Nick's unexpected choice acted on him as a reproach and a challenge. He felt ashamed at having placed himself so unromantically on his guard, rapidly saying to himself that if Nick could afford to allow so much for 'art' he might surely exhibit some of the same confidence. There had never been the least avowed competition between the cousins – their lines lay too far apart for that; but nevertheless they rode in sight of each other, and Sherringham had at present the sensation of suddenly seeing Nick Dormer give his horse the spur, bound forward and fly over a wall. He was put on his mettle and he had not to look long to spy an obstacle that he too might ride at. High rose his curiosity to see what warrant his kinsman might have for such risks – how he was mounted for such exploits. He really knew little about Nick's talent – so little as to feel no right to exclaim 'What an ass!' when Biddy gave him the news which only the existence of real talent could redeem from absurdity. All his eagerness to ascertain what Nick had been able to make of such a subject as Miriam Rooth came back to him; though it was what mainly had brought him to Rosedale Road he had forgotten it in the happy accident of his encounter with Biddy. He was conscious that if the surprise of a revelation of power were in store for him Nick would be justified more than he himself would feel reinstated in his self-respect. For the courage of renouncing the forum for the studio hovered before him as greater than the courage of marrying an actress whom one was in love with: the reward in the latter case was so much more immediate. Peter asked Biddy what Nick had done with his portrait of Miriam. He hadn't seen it anywhere in rummaging about the room.

'I think it's here somewhere, but I don't know,' Biddy replied, getting up and looking vaguely round her.

'Haven't you seen it? Hasn't he shown it to you?'

The girl rested her eyes on him strangely a moment; then she turned them away from him with a mechanical air of seeking for the picture. 'I think it's in the room, put away with its face to the wall.'

'One of those dozen canvases with their backs to us?'

'One of those perhaps.'

'Haven't you tried to see?'

'I haven't touched them,' said Biddy, colouring.

'Hasn't Nick had it out to show you?'

'He says it's in too bad a state – it isn't finished – it won't do.'

'And haven't you had the curiosity to turn it round for yourself?'

The embarrassed look in poor Biddy's face deepened, and it seemed to Sherringham that her eyes pleaded with him a moment, that there was a menace of tears in them, a gleam of anguish. 'I've had an idea he wouldn't like it.'

Her visitor's own desire however had become too lively for easy forbearance. He laid his hand on two or three canvases which proved, as he extricated them, to be either blank or covered with rudimentary forms. 'Dear Biddy, are you as docile, as obliging as that?' he asked, pulling out something else.

The inquiry was meant in familiar kindness, for Peter was struck, even to admiration, with the girl's having a sense of honour which all girls have not. She must in this particular case have longed for a sight of Nick's work – the work which had brought about such a crisis in his life. But she had passed hours in his studio alone, without permitting herself a stolen peep; she was capable of that if she believed it would please him. Sherringham liked a charming girl's being capable of that (he had known charming girls who would not have been), and his question was really an expression of respect. Biddy, however, apparently discovered some light mockery in it, and she broke out incongruously:

'I haven't wanted so much to see it. I don't care for her as much as that.'

'So much as that?'

'I don't care for his actress — for that vulgar creature. I don't like her!' said Biddy, unexpectedly.

Peter stared. 'I thought you hadn't seen her.'

'I saw her in Paris — twice. She was wonderfully clever, but she didn't charm me.'

Sherringham quickly considered, and then he said benevolently: 'I won't inflict the picture upon you then; we'll leave it alone for the present.' Biddy made no reply to this at first, but after a moment she went straight over to the row of stacked canvases and exposed several of them to the light. 'Why did you say you wished to go to the theatre tonight?' her companion continued.

Still the girl was silent; then she exclaimed, with her back turned to him and a little tremor in her voice, while she drew forth one of her brother's studies after the other: 'For the sake of your company, Peter! Here it is, I think,' she added, moving a large canvas with some effort. 'No, no, I'll hold it for you. Is that the light?'

She wouldn't let him take it; she bade him stand off and allow her to place it in the right position. In this position she carefully presented it, supporting it, at the proper angle, from behind and showing her head and shoulders above it. From the moment his eyes rested on the picture Sherringham accepted this service without protest. Unfinished, simplified and in some portions merely suggested, it was strong, brilliant and vivid and had already the look of life and the air of an original thing. Sherringham was startled, he was strangely affected — he had no idea Nick moved with that stride. Miriam was represented in three-quarters, seated, almost down to her feet. She leaned forward, with one of her legs crossed over the other, her arms extended and foreshortened, her hands locked together round her knee. Her beautiful head was bent a little, broodingly, and her splendid face seemed to look down at life. She had a grand appearance of being raised aloft, with a wide regard, from a height of intelligence, for the great field of the artist, all the figures and passions he may represent. Peter wondered where his kinsman had learned to paint like that. He almost gasped at the composition of the thing, at the drawing of the moulded arms. Biddy

Dormer abstained from looking round the corner of the canvas as she held it; she only watched, in Peter's eyes, for *his* impression of it. This she easily caught, and he could see that she had done so when after a few minutes he went to relieve her. She let him lift the thing out of her grasp; he moved it and rested it, so that they could still see it, against the high back of a chair.

'It's tremendously good,' he said.

'Dear, dear Nick,' Biddy murmured, looking at it now.

'Poor, poor Julia!' Sherringham was prompted to exclaim, in a different tone. His companion made no rejoinder to this, and they stood another minute or two side by side in silence, gazing at the portrait. Then Sherringham took up his hat – he had no more time, he must go. 'Will you come tonight all the same?' he asked, with a laugh that was somewhat awkward, putting out his hand to Biddy.

'All the same?'

'Why, you say she's a terrible creature,' Peter went on with his eyes on the painted face.

'Oh, anything for art!' said Biddy, smiling.

'Well, at seven o'clock then.' And Sherringham went away immediately, leaving the girl alone with the Tragic Muse and feeling again, with a quickened rush, a sense of the beauty of Miriam, as well as a new comprehension of the talent of Nick.

30

IT was not till noon, or rather later, the next day, that Sherringham saw Miriam Rooth. He wrote her a note that evening, to be delivered to her at the theatre, and during the performance she sent round to him a card with 'All right – come to luncheon tomorrow,' scrawled upon it in pencil.

When he presented himself in Balaklava Place he learned that the two ladies had not come in – they had gone again, early, to rehearsal; but they had left word that he was to be pleased to wait – they would come in from one moment to the other. It was further mentioned to him, as he was ushered into the drawing-room, that Mr Dashwood was on the ground. This circumstance however Sherringham barely noted: he had been soaring so high for the past twelve hours that he had almost lost consciousness of the minor differences of earthly things. He had taken Biddy Dormer and her friend Miss Tressilian home from the play, and after leaving them he had walked about the streets, he had roamed back to his sister's house, in a state of exaltation intensified by the fact that all the evening he had contained himself, thinking it more decorous and considerate, less invidious not to 'rave'. Sitting there in the shade of the box with his companions, he had watched Miriam in attentive but inexpressive silence, glowing and vibrating inwardly, but, for these fine, deep reasons, not committing himself to the spoken rapture. Delicacy, it appeared to him, should rule the hour; and indeed he had never had a pleasure more delicate than this little period of still observation and repressed ecstasy. Miriam's art lost nothing by it, and Biddy's mild nearness only gained. This young lady was silent also – wonderingly dauntedly, as if she too were conscious in relation to the actress of various other things beside her mastery of her art. To this mastery Biddy's attitude was a candid and liberal tribute: the poor girl sat quenched and pale, as if in the blinding light of a comparison by which it would be presumptuous even to be annihilated. Her subjection however was a gratified, a charmed subjection: there was a beneficence in such beauty – the beauty of the figure that moved before the footlights and spoke in music – even if it deprived one of hope. Peter didn't say to her, in vulgar elation and in reference to her whimsical profession of dislike at the studio: 'Well, do you find this performer so disagreeable now?' and she was grateful to him for his forbearance, for the tacit kindness of which the idea seemed to be: 'My poor child, I would prefer you if I could; but – judge for yourself – how can I? Expect of me only the possible. Expect that certainly, but only that.' In the same degree Peter liked Biddy's sweet, hushed air of judging for herself, of recognizing his discretion and letting him off, while she was lost in the illusion, in the convincing picture of the stage. Miss Tressilian did most of the criticism: she broke out cheerfully and sonorously from time to time, in reference to the actress: 'Most striking, certainly,' or, 'She *is* clever, isn't she?' It was a manner to which her companions found it impossible to respond. Miss Tressilian was disappointed in nothing but their enjoyment: they didn't seem to think the exhibition as amusing as she.

Walking away through the ordered void of Lady Agnes's quarter, with the four acts of the play glowing again before him in the smokeless London night, Sherringham found the liveliest thing in his impression the certitude that if he had never seen Miriam before and she had had for him none of the advantages of association, he would still have recognized in her performance the most interesting thing that the theatre had ever offered him. He floated in a sense of the felicity of it, in the general encouragement of a thing perfectly done, in the almost aggressive bravery of still larger claims for an art which could so triumphantly, so exquisitely render life. 'Render it?' Peter said to himself. 'Create it and reveal it, rather; give us something new and large and of the first order!' He had *seen* Miriam now; he had never seen her before; he had never seen her till he saw her in her conditions. Oh, her conditions – there were many things to be said about them; they were paltry enough as yet, inferior, inadequate, obstructive, as compared with the right, full, finished setting of such a talent; but the essence of them was now irremovably in Sherringham's eye, the vision of how the uplifted stage and the listening house transformed her. That idea of her having no character of her own came back to him with a force that made him laugh in the empty street: this was a disadvantage she was so exempt from that he appeared to himself not to have known her till tonight. Her character was simply to hold you by the particular spell; any other – the good-nature of home, the relation to her mother, her friends, her lovers, her debts, the practice of virtues or industries or vices – was not worth speaking of. These things were the fictions and shadows; the representation was the deep substance.

Sherringham had, as he went, an intense vision (he had often had it before) of the conditions which were still absent, the great and complete ones, those which would give the girl's talent a superior, glorious stage. More than ever he desired them, mentally invoked them, filled them out in imagination, cheated himself with the idea that they were possible. He saw them in a momentary illusion and confusion: a great academic, artistic theatre, subsidized and unburdened with money-getting, rich in its repertory, rich in the high quality and the wide array of its servants, and above all in the authority of an impossible administrator – a manager personally disinterested, not an actor with an eye to the main chance, pouring forth a continuity of tradition, striving for perfection, laying a splendid literature under contribution. He saw the heroine of a hundred 'situations', variously dramatic and vividly real; he saw the comedy and drama and passion and character and English life; he saw all humanity and history and poetry, and perpetually, in the midst of them, shining out in the high relief of some great moment, an image as fresh as an unveiled statue. He was not unconscious that he was taking all sorts of impossibilities and miracles for granted; but it really seemed to him for the time that the woman he had been watching three hours, the incarnation of the serious drama, would be a new and vivifying force. The world was just then so bright to him that Basil Dashwood struck him at first as an harmonious minister of that force.

It must be added that before Miriam arrived the breeze that filled Sherringham's sail began to sink a little. He passed out of the eminently 'let' drawing-room, where twenty large photographs of the young actress bloomed in the desert; he went into the garden by a glass-door that stood open, and found Mr Dashwood reclining on a bench and smoking cigarettes. This young man's conversation was a different music – it took him down, as he felt; showed him, very sensibly and intelligibly, it must be confessed, the actual theatre, the one they were all concerned with, the one they would have to make the miserable best of. It was fortunate for Sherringham that he kept his intoxication mainly to himself: the Englishman's habit of not being effusive still prevailed with him, even after his years of exposure to the foreign infection. Nothing could have been less exclamatory than the meeting of the two men, with its question or two, its remark or two about Sherringham's arrival in London; its offhand 'I noticed you last night – I was glad you turned up at last,' on one side, and its attenuated 'Oh, yes, it was the first time – I was very much interested,' on the other. Basil Dashwood played a part in *Yolande*, and Sherringham had had the satisfaction of taking the measure of his aptitude. He judged it to be of the small order, as indeed the part, which was neither that of the virtuous nor that of the villainous hero, restricted him to two or three inconspicuous effects and three or four changes of dress. He represented an ardent but respectful young lover whom the distracted heroine found time to pity a little and even to rail at; but it was impressed upon Sherringham that he scarcely represented young love. He looked very well, but Peter had heard him already in a hundred contemporary pieces; he never got out of rehearsal. He uttered sentiments and breathed vows with a nice voice, with a shy, boyish tremor in it, but as if he were afraid of being chaffed for it afterwards; giving the spectator in the stalls the feeling of holding the prompt-book and listening to a recitation. He made one think of country-houses and lawn-tennis and private theatricals; than which there could not be, to Sherringham's sense, an association more disconnected with the actor's art.

Dashwood knew all about the new thing, the piece in rehearsal; he knew all about everything – receipts and salaries and expenses and newspaper articles, and what old Baskerville said and what Mrs Ruffler thought: matters of superficial concern to Sherringham, who wondered, before Miriam appeared, whether she talked with her 'walking-gentleman' about them by the hour, deep in them and finding them not vulgar and boring, but the natural air of her life and the essence of her profession. Of course she did – she naturally would; it was all in the day's work and he might feel sure she wouldn't turn up her nose at the shop. He had to remind himself that he didn't care if she didn't – that he would think worse of her if she should. She certainly had much confabulation with her competent playfellow, talking shop by the hour: Sherringham could see that from the familiar, customary way Dashwood sat there with his cigarette, as if he were in possession and on his own ground. He divined a great intimacy between the two young artists, but asked himself at the same time what he, Peter Sherringham, had to say about it. He didn't pretend to control Miriam's intimacies, it was to be supposed; and if he had encouraged her to adopt a profession which abounded in opportunities for comradeship it was not for him to cry out because she had taken to it kindly. He had already descried a fund of utility in Mrs Lovick's light brother; but it irritated him all the same, after a while, to hear Basil

Dashwood represent himself as almost indispensable. He was practical – there was no doubt of that; and this idea added to Sherringham's paradoxical sense that as regards the matters actually in question he himself had not this virtue. Dashwood had got Mrs Rooth the house; it happened by a lucky chance that Laura Lumley, to whom it belonged (Sherringham would know Laura Lumley?) wanted to get rid, for a mere song, of the remainder of a lease. She was going to Australia with a troupe of her own. They just stepped into it; it was good air – the best sort of air to live in, to sleep in, in London, for people in their line. Sherringham wondered what Miriam's personal relations with this deucedly knowing gentleman might be, and was again able to assure himself that they might be anything in the world she liked, for any stake he, Peter, had in them. Dashwood told him of all the smart people who had tried to take up the new star – the way the London world had already held out its hand; and perhaps it was Sherringham's irritation, the crushed sentiment I just mentioned, that gave a little heave in the exclamation: 'Oh, that – that's all rubbish; the less of that the better!' At this Basil Dashwood stared; he evidently felt snubbed; he had expected his interlocutor to be pleased with the names of the eager ladies who had 'called' – which proved to Sherringham that he took a low view of his art. The secretary of embassy explained, it is to be hoped not pedantically, that this art was serious work and that society was humbug and imbecility; also that of old the great comedians wouldn't have known such people. Garrick had essentially his own circle.

'No, I suppose they didn't call, in the old narrow-minded time,' said Basil Dashwood.

'Your profession didn't call. They had better company – that of the romantic, gallant characters they represented. They lived with *them*, and it was better all round.' And Peter asked himself – for the young man looked as if that struck him as a dreary period – if *he* only, for Miriam, in her new life, or among the futilities of those who tried to find her accessible, expressed the artistic idea. This at least, Sherringham reflected, was a situation that could be improved.

He learned from Dashwood that the new play, the thing they were rehearsing, was an old play, a romantic drama of thirty years before, covered, from infinite queer handling, with a sort of dirty glaze. Dashwood had a part in it, but there was an act in which he didn't appear, and that was the act they were doing that morning. *Yolande* had done all *Yolande* could do: Sherringham was mistaken if he supposed *Yolande* was such a tremendous hit. It had done very well, it had run three months, but they were by no means coining money with it. It wouldn't take them to the end of the season; they had seen for a month past that they would have to put on something else. Miss Rooth moreover wanted a new part; she was impatient to show what a range she was capable of. She had grand ideas; she thought herself very good-natured to repeat the same thing for three months. Basil Dashwood lighted another cigarette and described to his companion some of Miss Rooth's ideas. He gave Sherringham a great deal of information about her – about her character, her temper, her peculiarities, her little ways, her manner of producing some of her effects. He spoke with familiarity and confidence, as if he knew more about her than any one else – as if he had invented or discovered her, were in a sense her proprietor or guarantor. It was the talk of the shop, with a perceptible shrewdness in it and a touching young candour; the expansiveness of the commercial spirit, when it relaxes and generalizes, is conscious it is safe with another member of the guild.

Sherringham could not help protesting against the lame old war-horse whom it was proposed to bring into action, who had been ridden to death and had saved a thousand desperate fields; and he exclaimed on the strange passion of the good British public for sitting again and again through expected situations, watching for speeches they had heard and surprises that struck the hour. Dashwood defended the taste of London, praised it as loyal, constant, faithful; to which Sherringham retorted with some vivacity that it was faithful to rubbish. He justified this sally by declaring that the play in rehearsal *was* rubbish, clumsy mediocrity which had outlived its convenience, and that the fault was the want of life in the critical sense of the public, which was ignobly docile, opening its mouth for its dose, like the pupils of Dotheboys Hall; not insisting on something different, on a fresh preparation. Dashwood asked him if he then wished

Miss Rooth to go on playing for ever a part she had repeated more than eighty nights on end: he thought the modern 'run' was just what he had heard him denounce, in Paris, as the disease the theatre was dying of. This imputation Sherringham gainsaid; he wanted to know if she couldn't change to something less stale than the piece in question. Dashwood opined that Miss Rooth must have a strong part and that there happened to be one for her in the before-mentioned venerable novelty. She had to take what she could get; she wasn't a girl to cry for the moon. This was a stop-gap – she would try other things later; she would have to look round her: you couldn't have a new piece left at your door every day with the milk. On one point Sherringham's mind might be at rest: Miss Rooth was a woman who would do every blessed thing there was to do. Give her time and she would walk straight through the repertory. She was a woman who would do this – she was a woman who would do that: Basil Dashwood employed this phrase so often that Sherringham, nervous, got up and threw an unsmoked cigarette away. Of course she was a woman: there was no need of Dashwood's saying it a hundred times!

As for the repertory, the young man went on, the most beautiful girl in the world could give but what she had. He explained, after Sherringham sat down again, that the noise made by Miss Rooth was not exactly what this admirer appeared to suppose. Sherringham had seen the house the night before; would recognize that, though good, it was very far from great. She had done very well, very well indeed, but she had never gone above a point which Dashwood expressed in pounds sterling, to the edification of his companion, who vaguely thought the figure high. Sherringham remembered that he had been unable to get a stall, but Dashwood insisted that the girl had not leaped into commanding fame: that was a thing that never happened in fact – it happened only in pretentious works of fiction. She had attracted notice, unusual notice for a woman whose name the day before had never been heard of: she was recognized as having, for a novice, extraordinary cleverness and confidence – in addition to her looks of course, which were the thing that had really fetched the crowd. But she hadn't been the talk of London; she had only been the talk of Gabriel Nash. He wasn't

London, more was the pity. He knew the aesthetic people – the worldly, semi-smart ones, not the frumpy, sickly lot who wore dirty drapery; and the aesthetic people had run after her. Basil Dashwood instructed Sherringham sketchily as to the different sects in the great religion of beauty, and was able to give him the particular 'note' of the critical clique to which Miriam had begun so quickly to owe it that she had a vogue. The information made the secretary of embassy feel very ignorant of the world, very uninitiated and buried in his little professional hole. Dashwood warned him that it would be a long time before the general public would wake up to Miss Rooth, even after she had waked up to herself: she would have to do some really big thing first. *They* knew it was in her, the big thing – Sherringham and he, and even poor Nash – because they had seen her as no one else had; but London never took any one on trust – it had to be cash down. It would take their young lady two or three years to pay out her cash and get her equivalent. But of course the equivalent would be simply a gold-mine. Within its limits however, her success was already quite a fairy-tale: there was magic in the way she had concealed, from the first, her want of experience. She absolutely made you think she had a lot of it, more than any one else. Mr Dashwood repeated several times that she was a cool hand – a deucedly cool hand; and that he watched her himself, saw ideas come to her, saw her try different dodges on different nights. She was always alive – she liked it herself. She gave *him* ideas, long as he had been on the stage. Naturally she had a great deal to learn – a tremendous lot to learn: a cosmopolite like Sherringham would understand that a girl of that age, who had never had a friend but her mother – her mother was greater fun than ever now – naturally *would* have. Sherringham winced at being called a 'cosmopolite' by his young companion, just as he had winced a moment before at hearing himself lumped, in esoteric knowledge, with Dashwood and Gabriel Nash; but the former of these gentlemen took no account of his sensibility while he enumerated a few of the things that the young actress had to learn. Dashwood was a mixture of acuteness and innocent fatuity; and Sherringham had to recognize that he had some of the elements of criticism in him when he said that the wonderful thing in the girl was that she learned so fast – learned something every night, learned from the same old piece a lot more than any one else would have learned from twenty. 'That's what it is to be a genius,' Sherringham remarked. 'Genius is only the art of getting your experience fast, of stealing it, as it were; and in this sense Miss Rooth's a regular brigand.' Dashwood assented good-humouredly; then he added, 'Oh, she'll do!' It was exactly in these simple words, in speaking to her, that Sherringham had phrased the same truth; yet he didn't enjoy hearing them on his neighbour's lips: they had a profane, patronizing sound, suggestive of displeasing equalities.

The two men sat in silence for some minutes, watching a fat robin hop about on the little seedy lawn; at the end of which they heard a vehicle stop on the other side of the garden wall and the voices of people descending from it. 'Here they come, the dear creatures,' said Basil Dashwood, without moving; and from where they sat Sherringham saw the small door in the wall pushed open. The dear creatures were three in number, for a gentleman had added himself to Mrs Rooth and her daughter. As soon as Miriam's eyes fell upon her Parisian friend she stopped short, in a large, droll theatrical attitude, and, seizing her mother's arm, exclaimed passionately: 'Look where he sits, the author of all my woes – cold, cynical, cruel!' She was evidently in the highest spirits; of which Mrs Rooth partook as she cried indulgently, giving her a slap: 'Oh, get along, you gipsy!'

'She's always up to something,' Basil Dashwood commented, as Miriam, radiant and with a conscious stage tread, glided toward Sherringham as if she were coming to the footlights. He rose slowly from his seat, looking at her and struck with her beauty; he had been impatient to see her, yet in the act his impatience had had a disconcerting check.

Sherringham had had time to perceive that the man who had come in with her was Gabriel Nash, and this recognition brought a low sigh to his lips as he held out his hand to her – a sigh expressive of the sudden sense that his interest in her now could only be a gross community. Of course that didn't matter, since he had set it, at the most, such rigid limits; but none the less he stood vividly reminded that it would be public and notorious, that inferior people would be inveterately mixed up with it, that she had crossed the line and sold herself to the vulgar, making him indeed only one of an equalized multitude. The way Gabriel Nash turned up there just when he didn't want to see him made Peter feel that it was a complicated thing to have a friendship with an actress so clearly destined to be famous. He quite forgot that Nash had known Miriam long before his own introduction to her and had been present at their first meeting, which he had in fact in a measure brought about. Had Sherringham not been so cut out to make trouble of this particular joy he might have found some adequate assurance that she distinguished him in the way in which, taking his hand in both of hers, she looked up at him and murmured, 'Dear old master!' Then, as if this were not acknowledgement enough, she raised her head still higher and, whimsically, gratefully, charmingly, almost nobly, she kissed him on the lips, before the other men, before the good mother whose 'Oh, you honest creature!' made everything regular.

31

IF Peter Sherringham was ruffled by some of Miriam's circumstances there was comfort and consolation to be drawn from others, beside the essential fascination (there was no doubt about that now) of the young lady's own society. He spent the afternoon, they all spent the afternoon, and the occasion reminded him of a scene in *Wilhelm Meister*. Mrs Rooth had little resemblance to Mignon, but Miriam was remarkably like Philina. Luncheon was delayed two or three hours; but the long wait was a positive source of gaiety, for they all smoked cigarettes in the garden and Miriam gave striking illustrations of the parts she was studying. Sherringham was in the state of a man whose toothache has suddenly stopped – he was exhilarated by the cessation of pain. The pain had been the effort to remain in Paris after Miriam came to London, and the balm of seeing her now was the measure of the previous soreness.

Gabriel Nash had, as usual, plenty to say, and he talked of Nick Dormer's picture so long that Sherringham wondered whether he did it on purpose to vex him. They went in and out of the house; they made excursions to see how lunch was coming on; and Sherringham got half an hour alone, or virtually alone, with the object of his unsanctioned passion – drawing her publicly away from the others and making her sit with him in the most sequestered part of the little gravelled grounds. There was summer enough in the trees to shut out the adjacent villas, and Basil Dashwood and Gabriel Nash lounged together at a convenient distance, while Nick's whimsical friend tried experiments upon the histrionic mind. Miriam confessed that, like all comedians, they ate at queer hours; she sent Dashwood in for biscuits and sherry – she proposed sending him round to the grocer's in the Circus Road for superior wine. Sherringham judged him to be the factotum of the little household: he knew where the biscuits were kept and the state of the grocer's account. When Peter congratulated the young actress on having so useful an associate she said genially, but as if the words disposed of him: 'Oh, he's awfully handy!' To this she added: 'You're not, you know'; resting the kindest, most pitying eyes on him. The sensation they gave him was as sweet as if she had stroked his cheek, and her manner was responsive even to tenderness. She called him 'Dear master' again, and sometimes *'Cher maître'*, and appeared to express gratitude and reverence by every intonation.

'You're doing the humble dependant now,' he said: 'you do it beautifully, as you do everything.' She replied that she didn't make it humble enough – she couldn't; she was too proud, too insolent in her triumph. She liked that, the triumph, too much, and she didn't mind telling him that she was perfectly happy. Of course as yet the triumph was very limited; but success was success, whatever its quantity; the dish was a small one, but it had the right taste. Her imagination had already bounded beyond the first phase, unexpectedly brilliant as this had been: her position struck her as modest compared with a future that was now vivid to her. Sherringham had never seen her so soft and sympathetic; she had insisted, in Paris, that her personal character was that of the good girl (she used the term in a fine loose way), and it was impossible to be a better girl than she showed herself this pleasant afternoon. She was full of gossip and anecdote and drollery; she had exactly the air that he would have liked her to have – that of thinking of no end of things to tell him. It was as if she had just returned from a long journey, had had strange adventures and made wonderful discoveries. She began to speak of this and that, and broke off to speak of something else; she talked of the theatre, of the newspapers and then of London, of the people she had met and the extraordinary things they said to her, of the parts she was going to take up, of lots of new ideas that had come to her about the art of comedy. She wanted to do comedy now – to do the comedy of London life. She was delighted to find that seeing more of the world suggested things to her; they came straight from the fact, from nature, if you could call it nature: so that she was convinced more than ever that the artist ought to *live*, to get on with his business, gather ideas, lights from experience – ought to welcome any experience that would give him lights. But work, of course, *was* experience, and everything in one's life that was good was work. That was the jolly thing in the actor's trade – it made up for other elements that were odious: if you only kept your eyes open nothing could happen to you that wouldn't be food for observation and grist to your mill, showing you how people looked and moved and spoke – cried and grimaced, or writhed and dissimulated, in given situations. She saw all round her things she wanted to 'do' – London was full of them, if you had eyes to see. Miriam demanded imperiously why people didn't take them up, put them into plays and parts, give one a chance with them: she expressed her sharp impatience of the general literary stupidity. She had never been chary of this particular displeasure, and there were moments (it was an old story and a subject of frank raillery to Sherringham) when to hear her you might have thought there was no cleverness anywhere but in her disdainful mind. She wanted tremendous things done, that she might use them, but she didn't pretend to say exactly what they were to be, nor even approximately how they were to be handled: her ground was rather that if *she* only had a pen – it was exasperating to have to explain! She mainly contented herself with declaring that nothing had really been touched: she felt that more and more as she saw more of people's goings-on.

Sherringham went to her theatre again that evening and he made no scruple of going every night for a week. Rather, perhaps I should say, he made a scruple; but it was a part of the pleasure of his life during these arbitrary days to overcome it. The only way to prove to himself that he could overcome it was to go; and he was satisfied, after he had been seven times, not only with the spectacle on the stage but with his own powers of demonstration. There was no satiety however with the spectacle on the stage, inasmuch as that only produced a further curiosity. Miriam's performance was a living thing, with a power to change, to grow, to develop, to beget new forms of the same life. Peter Sherringham contributed to it in his amateurish way, watching with solicitude the fate of his contributions. He talked it over in Balaklava Place, suggested modifications, variations worth trying. Miriam professed herself thankful for any refreshment that could be administered to her interest in *Yolande*, and, with an effectiveness that showed large resource, touched up her part and drew several new airs from it. Sherringham's suggestions bore upon her way of uttering certain speeches, the intonations that would have more beauty or make the words mean more. Miriam had her ideas, or rather she had her instincts, which she defended and illustrated, with a vividness superior to argument, by a happy pictorial phrase or a snatch of mimicry; but she was always for trying; she liked experiments and caught at them, and she was especially thankful when some one gave her a showy reason, a plausible formula, in a case where she only stood upon an intuition. She pretended to despise reasons and to like and dislike at her sovereign pleasure; but she always honoured the exotic gift, so that Sherringham was amused with the liberal way she produced it, as if she had been a naked islander rejoicing in a present of crimson cloth.

Day after day he spent most of his time in her society, and Miss Laura Lumley's recent habitation became the place in London to which his thoughts were most attached. He was highly conscious that he was not now carrying out that principle of abstention which he had brought to such maturity before leaving Paris; but he contented himself with a much cruder justification of this inconsequence than he would have thought adequate in advance. It consisted simply in the idea that to be identified with the first public steps of a young genius was a delightful experience. What was the harm of it, if the genius were real? Sherringham's main security was now that his relations with Miriam had been frankly placed under the protection of the idea of legitimate extravagance. In this department they made a very creditable figure and required much less watching and pruning than when it was his effort to fit them into a worldly plan. Sherringham had a sense of real wisdom when he said to himself that it surely should be enough that this momentary intellectual participation in the girl's dawning fame was a charming thing. Charming things, in a busy man's life, were not frequent enough to be kicked out of the way. Balaklava Place, looked at in this philosophic way, became almost idyllic: it gave Peter the pleasantest impression he had ever had of London.

The season happened to be remarkably fine; the temperature was high, but not so high as to keep people from the theatre. Miriam's 'business' visibly increased, so that the question of putting on the second play underwent some reconsideration. The girl insisted, showing in her insistence a temper of which Sherringham had already caught some splendid gleams. It was very evident that through her career it would be her expectation to carry things with a high hand. Her managers and agents would not find her an easy victim or a calculable force: but the public would adore her, surround her with the popularity that attaches to a humorous, good-natured princess, and her comrades would have a kindness for her because she wouldn't be selfish. They too would form in a manner a portion of her affectionate public. This was the way Sherringham read the signs, liking her whimsical tolerance of some of her vulgar playfellows almost well enough to forgive their presence in Balaklava Place, where they were a sore trial to her mother, who wanted her to multiply her points of contact only with the higher orders. There were hours when Sherringham thought he foresaw that her principal relation to the proper world would be to have, within two or three years, a grand battle with it, making it take her, if she let it have her at all, absolutely on her own terms: a picture which led our young man to ask himself, with a helplessness that was not exempt, as he perfectly knew, from absurdity, what part *he* should find himself playing in such a contest and if it would be reserved to him to be the more ridiculous as a peacemaker or as a heavy auxiliary.

'She might know any one she would, and the only person she

appears to take any pleasure in is that dreadful Miss Rover,' Mrs Rooth whimpered, more than once, to Sherringham, who recognized in the young lady so designated the principal complication of Balaklava Place.

Miss Rover was a little actress who played at Miriam's theatre, combining with an unusual aptitude for delicate comedy a less exceptional absence of rigour in private life. She was pretty and quick and clever, and had a fineness that Miriam professed herself already in a position to estimate as rare. She had no control of her inclinations; yet sometimes they were wholly laudable, like the devotion she had formed for her beautiful colleague, whom she admired not only as an ornament of the profession but as a being of a more fortunate essence. She had had an idea that real ladies were 'nasty'; but Miriam was not nasty, and who could gainsay that Miriam was a real lady? The girl justified herself to Sherringham, who had found no fault with her; she knew how much her mother feared that the proper world wouldn't come in if they knew that the improper, in the person of pretty Miss Rover, was on the ground. What did she care who came and who didn't, and what was to be gained by receiving half the snobs in London? People would have to take her exactly as they found her – that they would have to learn; and they would be much mistaken if they thought her capable of becoming a snob too for the sake of their sweet company. She didn't pretend to be anything but what she meant to be, the best general actress of her time; and what had that to do with her seeing or not seeing a poor ignorant girl who had lov— Well, she needn't say what Fanny had. She had met her in the way of business – she didn't say she would have run after her. She had liked her because she wasn't a stick, and when Fanny Rover had asked her quite wistfully if she mightn't come and see her, she hadn't bristled with scandalized virtue. Miss Rover was not a bit more stupid or more ill-natured than any one else: it would be time enough to shut the door when she should become so.

Sherringham commended even to extravagance the liberality of such comradeship; said that of course a woman didn't go into that profession to see how little she could swallow. She was right to live with the others so long as they were at all possible, and it was for her, and only for her, to judge how long that might be. This was rather heroic on Peter's part, for his assumed detachment from the girl's personal life still left him a margin for some forms of uneasiness. It would have made, in his spirit, a great difference for the worse that the woman he loved, and for whom he wished no baser lover than himself, should have embraced the prospect of consorting only with the cheaper kind. It was all very well, but Fanny Rover was simply a *cabotine*, and that sort of association was an odd training for a young woman who was to have been good enough (he couldn't forget that – he kept remembering it as if it might still have a future use) to be his wife. Certainly he ought to have thought of such things before he permitted himself to become so interested in a theatrical nature. His heroism did him service however for the hour: it helped him by the end of the week to feel tremendously broken in to Miriam's little circle. What helped him most indeed was to reflect that she would get tired of a good many of its members herself in time; for it was not that they were shocking (very few of them shone with that intense light), but that they could be trusted in the long run to bore you.

There was a lovely Sunday in particular that he spent almost wholly in Balaklava Place – he arrived so early – when, in the afternoon, all sorts of odd people dropped in. Miriam held a reception in the little garden and insisted on almost all the company's staying to supper. Her mother shed tears to Sherringham, in the desecrated house, because they had accepted, Miriam and she, an invitation – and in Cromwell Road too – for the evening. Miriam decreed that they shouldn't go: they would have much better fun with their good friends at home. She sent off a message – it was a terrible distance – by a cabman, and Sherringham had the privilege of paying the messenger. Basil Dashwood, in another vehicle, proceeded to an hotel that he knew, a mile away, for supplementary provisions, and came back with a cold ham and a dozen of champagne. It was all very Bohemian and journalistic and picturesque, very supposedly droll and enviable to outsiders; and Miriam told anecdotes and gave imitations of the people she would have met if she had gone out: so no one had a sense of loss – the two occasions were fantastically united. Mrs

Rooth drank champagne for consolation; though the consolation was imperfect when she remembered that she might have drunk it (not quite so much indeed) in Cromwell Road.

Taken in connection with the evening before, the day formed for Sherringham the most complete exhibition he had had of Miriam Rooth. He had been at the theatre, to which the Saturday night happened to have brought the fullest house she had yet played to, and he came early to Balaklava Place, to tell her once again (he had told her half a dozen times the evening before) that, with the excitement of her biggest audience, she had surpassed herself, acted with remarkable intensity. It pleased her to hear it, and the spirit with which she interpreted the signs of the future and, during an hour he spent alone with her, Mrs Rooth being upstairs and Basil Dashwood not arrived, treated him to specimens of fictive emotion of various kinds, was beyond any natural abundance that he had yet seen in a woman. The impression could scarcely have been other if she had been playing wild snatches to him at the piano: the bright, up-darting flame of her talk rose and fell like an improvisation on the keys. Later, all the rest of the day, he was fascinated by the good grace with which she fraternized with her visitors, finding the right words for each, the solvent of incongruities, the right ideas to keep vanity quiet and make humility gay. It was a wonderful expenditure of generous, nervous life. But what Sherringham read in it above all was the sense of success in youth, with the future large, and the action of that force upon all the faculties. Miriam's limited past had yet pinched her enough to make emancipation sweet, and the emancipation had come at last in an hour. She had stepped into her magic shoes, divined and appropriated everything they could give her, become in a day a really original contemporary. Sherringham was of course not less conscious of that than Nick Dormer had been when, in the cold light of his studio, he saw how she had altered.

But the great thing, to his mind and, these first days, the irresistible seduction of the theatre, was that she was a rare incarnation of beauty. Beauty was the principle of everything she did and of the way, unerringly, she did it – an exquisite harmony of line and motion and attitude and tone, what was most general and most characteristic in her performance. Accidents and instincts played together to this end and constituted something which was independent of her talent or of her merit, in a given case, and which in its influence, to Sherringham's imagination, was far superior to any merit and to any talent. It was a supreme infallible felicity, a source of importance, a stamp of absolute value. To see it in operation, to sit within its radius and feel it shift and revolve and change and never fail, was a corrective to the depression, the humiliation, the bewilderment of life. It transported Sherringham from the vulgar hour and the ugly fact; drew him to something which had no reason but its sweetness, no name nor place save as the pure, the distant, the antique. It was what most made him say to himself: 'Oh, hang it, what does it matter?' when he reflected that an *homme sérieux* (as they said in Paris) rather gave himself away (as they said in America) by going every night to the same theatre for all the world to stare. It was what kept him from doing anything but hover round Miriam – kept him from paying any other visits, from attending to any business, from going back to Calcutta Gardens. It was a spell which he shrank intensely from breaking, and the cause of a hundred postponements, confusions and incoherences. It made of the crooked little stucco villa in St John's Wood a place in the upper air, commanding the prospect; a nest of winged liberties and ironies, hanging far aloft above the huddled town. One should live at altitudes when one could – they braced and simplified; and for a happy interval Sherringham never touched the earth.

It was not that there were no influences tending at moments to drag him down – an abasement from which he escaped only because he was up so high. We have seen that Basil Dashwood could affect him at times like a piece of wood tied to his ankle, through the circumstance that he made Miriam's famous conditions – those of the public exhibition of her genius – seem small and prosaic; so that Sherringham had to remind himself that perhaps this smallness was involved in their being at all. She carried his imagination off into infinite spaces, whereas she carried Dashwood's only into the box-office and the revival of plays that were barbarously bad. The worst was that it was open to him to believe that a sharp young man who was in the business might know better than he. Another possessor of superior knowledge (he talked, that is, as if he knew better than any one) was Gabriel Nash, who appeared to have abundant leisure to haunt Balaklava Place, or in other words appeared to enjoy the same command of his time as Peter Sherringham. Our young diplomatist regarded him with mingled feelings, for he had not forgotten the contentious character of their first meeting or the degree to which he had been moved to urge upon Nick Dormer's consideration that his talkative friend was probably an ass. This personage turned up now as an admirer of the charming creature he had scoffed at, and there was something exasperating in the quietude of his inconsistency, of which he had not the least embarrassing consciousness. Indeed he had such arbitrary and desultory ways of looking at any question that it was difficult, in vulgar parlance, to have him; his sympathies hummed about like bees in a garden, with no visible plan, no economy in their flight. He thought meanly of the modern theatre and yet he had discovered a fund of satisfaction in the most promising of its exponents; so that Sherringham more than once said to him that he should really, to keep his opinions at all in hand, attach more value to the stage or less to the interesting actress. Miriam made infinitely merry at his expense and treated him as the most abject of her slaves: all of which was worth seeing as an exhibition, on Nash's part, of the imperturbable. When Sherringham mentally pronounced him impudent he felt guilty of an injustice – Nash had so little the air of a man with something to gain. Nevertheless he felt a certain itching in his boot-toe when his fellow-visitor exclaimed, explicatively (in general to Miriam herself), in answer to a charge of tergiversation: 'Oh, it's all right: it's the voice, you know – the enchanting voice!' He meant by this, as indeed he more fully set forth, that he came to the theatre, or to the villa in St John's Wood, simply to treat his ear to the sound (the richest then to be heard on earth, as he maintained) issuing from Miriam's lips. Its richness was quite independent of the words she might pronounce or the poor fable they might subserve, and if the pleasure of hearing her in public was the greater by reason of the larger volume of her utterance, it was still highly agreeable to see her at home, for it was there that the artistic nature that he freely conceded to her came out most. He spoke as if she had been formed by the bounty of nature to be his particular recreation, and as if, being an expert in innocent joys, he took his pleasure wherever he found it.

He was perpetually in the field, sociable, amiable, communicative, inveterately contradicted but never confounded, ready to talk to any one about anything and making disagreement (of which he left the responsibility wholly to others) a basis of intimacy. Everyone knew what he thought of the theatrical profession, and yet it could not be said that he did not regard its members as the exponents of comedy, inasmuch as he often elicited their foibles in a way that made even Sherringham laugh, notwithstanding his attitude of reserve where Nash was concerned. At any rate, though he had committed himself on the subject of the general fallacy of their attempt, he put up with their company, for the sake of Miriam's accents, with a practical philosophy that was all his own. Miriam pretended that he was her supreme, her incorrigible adorer, masquerading as a critic to save his vanity and tolerated for his secret constancy in spite of being a bore. To Sherringham he was not a bore, and the secretary of embassy felt a certain displeasure at not being able to regard him as one. He had seen too many strange countries and curious things, observed and explored too much to be void of illustration. Peter had a suspicion that if he himself was in the *grandes espaces* Gabriel Nash probably had a still wider range. If among Miriam's associates Basil Dashwood dragged him down, Gabriel challenged him rather to higher and more fantastic flights. If he saw the girl in larger relations than the young actor, who mainly saw her in ill-written parts, Nash went a step further and regarded her, irresponsibly and sublimely, as a priestess of harmony, with whom the vulgar ideas of success and failure had nothing to do. He laughed at her 'parts', holding that without them she would be great. Sherringham envied him his power to content himself with the pleasures he could get: he had a shrewd impression that contentment was not destined to be the sweetener of his own repast.

Above all Nash held his attention by a constant element of unstudied reference to Nick Dormer, who, as we know, had suddenly become much more interesting to his cousin. Sherringham found food for observation, and in some measure for perplexity, in the relations of all these clever people with each other. He knew why his sister, who had a personal impatience of unapplied ideas, had not been agreeably affected by Mr Nash and had not viewed with complacency a predilection for him in the man she was to marry. This was a side by which he had no desire to resemble Julia Dallow, for he needed no teaching to divine that Gabriel had not set her intelligence in motion. He, Peter, would have been sorry to have to confess that he could not understand him. He understood furthermore that Miriam, in Nick's studio, might very well have appeared to Julia a formidable power. She was younger, but she had quite as much her own form and she was beautiful enough to have made Nick compare her with Mrs Dallow even if he had been in love with that lady – a pretension as to which Peter had private ideas.

Sherringham for many days saw nothing of the member for Harsh, though it might have been said that, by implication, he participated in the life of Balaklava Place. Had Nick given Julia tangible grounds, and was his unexpectedly fine rendering of Miriam an act of virtual infidelity? In that case in what degree was Miriam to be regarded as an accomplice in his defection, and what was the real nature of this young lady's esteem for her new and (as he might be called) distinguished ally? These questions would have given Peter still more to think about if he had not flattered himself that he had made up his mind that they concerned Nick and Miriam infinitely more than they concerned him. Miriam was personally before him, so that he had no need to consult for his pleasure his fresh recollection of the portrait. But he thought of this striking production each time he thought of his enterprising kinsman. And that happened often, for in his hearing Miriam often discussed the happy artist and his possibilities with Gabriel Nash, and Gabriel broke out about them to Miriam. The girl's tone on the subject was frank and simple: she only said, with an iteration that was slightly irritating that Mr Dormer had been tremendously kind to her. She never mentioned Julia's irruption to Julia's brother; she only referred to the portrait, with inscrutable amenity, as a direct consequence of Peter's fortunate suggestion that first day at Madame Carré's. Gabriel Nash, however, showed such a disposition to expatiate sociably and luminously on the peculiarly interesting character of what he called Dormer's predicament and on the fine suspense which it was fitted to kindle in the breast of discerning friends, that Peter wondered, as I have already hinted, if this insistence were not a subtle perversity, a devilish little invention to torment a man whose jealousy was presumable. Yet on the whole Nash struck him as but scantily devilish and as still less occupied with the prefiguration of *his* emotions. Indeed, he threw a glamour of romance over Nick; tossed off such illuminating yet mystifying references to him that Sherringham found himself capable of a magnanimous curiosity, a desire to follow out the chain of events. He learned from Gabriel that Nick was still away, and he felt as if he could almost submit to instruction, to initiation. The rare charm of these unregulated days was troubled — it ceased to be idyllic — when, late on the evening of the second Sunday, he walked away with Gabriel southward from St John's Wood. For then something came out.

32

IT mattered not so much what the doctors thought (and Sir Matthew Hope, the greatest of them all, had been down twice in one week) as that Mr Chayter, the omniscient butler, declared with all the authority of his position and his experience that Mr Carteret was very bad indeed. Nick Dormer had a long talk with him (it lasted six minutes) the day he hurried to Beauclere in response to a telegram. It was Mr Chayter who had taken upon himself to telegraph, in spite of the presence in the house of Mr Carteret's nearest relation and only surviving sister, Mrs Lendon. This lady, a large mild, healthy woman, with a heavy tread, who liked early breakfasts, uncomfortable chairs and the advertisement-sheet of *The Times*, had arrived the week before and was awaiting the turn of events. She was a widow and lived in Cornwall, in a house nine miles from a station, which had, to make up for this inconvenience, as she had once told Nick, a delightful old herbaceous garden. She was extremely fond of an herbaceous garden; her principal interest was in that direction. Nick had often seen her — she came to Beauclere once or twice a year. Her sojourn there made no great difference; she was only an 'Urania dear,' for Mr Carteret to look across the table at

345

when, on the close of dinner, it was time for her to retire. She went out of the room always as if it were after some one else; and on the gentlemen 'joining' her later (the junction was not very close) she received them with an air of gratified surprise.

Chayter honoured Nick Dormer with a regard which approached, without improperly competing with it, the affection his master had placed on the same young head, and Chayter knew a good many things. Among them he knew his place; but it was wonderful how little that knowledge had rendered him inaccessible to other kinds. He took upon himself to send for Nick without speaking to Mrs Lendon, whose influence was now a good deal like that of a large occasional piece of furniture which had been introduced in case it should be required. She was one of the solid conveniences that a comfortable house would have; but you couldn't talk with a mahogany sofa or a folding-screen. Chayter knew how much she had 'had' from her brother, and how much her two daughters had each received on marriage; and he was of the opinion that it was quite enough, especially considering the society in which they (you could scarcely call it) moved. He knew beyond this that they would all have more, and that was why he hesitated little about communicating with Nick. If Mrs Lendon should be ruffled at the intrusion of a young man who neither was the child of a cousin nor had been formally adopted, Chayter was parliamentary enough to see that the forms of debate were observed. He had indeed a slightly compassionate sense that Mrs Lendon was not easily ruffled. She was always down an extraordinary time before breakfast (Chayter refused to take it as in the least admonitory), but she usually went straight into the garden (as if to see that none of the plants had been stolen in the night), and had in the end to be looked for by the footman in some out-of-the-way spot behind the shrubbery, where, plumped upon the ground, she was doing something 'rum' to a flower.

Mr Carteret himself had expressed no wishes. He slept most of the time (his failure at the last had been sudden, but he was rheumatic and seventy-seven), and the situation was in Chayter's hands. Sir Matthew Hope had opined, even on his second visit, that he would rally and go on, in rudimentary comfort, some time longer; but Chayter took a different and a still more intimate view. Nick was embarrassed: he scarcely knew what he was there for from the moment he could give his good old friend no conscious satisfaction. The doctors, the nurses, the servants, Mrs Lendon, and above all the settled equilibrium of the square, thick house, where an immutable order appeared to slant through the polished windows and tinkle in the quieter bells, all represented best the kind of supreme solace to which the master was most accessible.

For the first day it was judged better that Nick should not be introduced into the darkened chamber. This was the decision of the two decorous nurses, of whom the visitor had had a glimpse and who, with their black uniforms and fresh faces of business, suggested a combination of the barmaid and the nun. He was depressed, yet restless, felt himself in a false position and thought it lucky Mrs Lendon had powers of placid acceptance. They were old acquaintances: she treated him with a certain ceremony, but it was not the rigour of mistrust. It was much more an expression of remote Cornish respect for young abilities and distinguished connections, inasmuch as she asked him a great deal about Lady Agnes and about Lady Flora and Lady Elizabeth. He knew she was kind and ungrudging, and his principal chagrin was the sense of meagre information and of responding poorly in regard to his uninteresting aunts. He sat in the garden with newspapers and looked at the lowered blinds in Mr Carteret's windows; he wandered around the abbey with cigarettes and lightened his tread and felt grave, wishing that everything were over. He would have liked much to see Mr Carteret again, but he had no desire that Mr Carteret should see him. In the evening he dined with Mrs Lendon, and she talked to him, at his request and as much as she could, about her brother's early years, his beginnings of life. She was so much younger that they appeared to have been rather a tradition of her own youth; but her talk made Nick feel how tremendously different Mr Carteret had been at that period from what he, Nick, was today. He had published at the age of thirty a little volume (it was thought wonderfully clever) called *The Incidence of Rates*; but Nick had not yet collected the material for any such treatise. After dinner Mrs Lendon, who was in full dress, retired to the drawing-room, where at the end of ten minutes she was followed by Nick, who had remained behind only because he thought Chayter would expect it. Mrs Lendon almost shook hands with him again, and then Chayter brought in coffee. Almost in no time afterwards he brought in tea, and the occupants of the drawing-room sat for a slow half-hour, during which the lady looked round at the apartment with a sigh and said: 'Don't you think poor Charles had exquisite taste?'

Fortunately at this moment the 'local man' was ushered in. He had been upstairs and he entered, smiling, with the remark: 'It's quite wonderful – it's quite wonderful.' What was wonderful was a marked improvement in the breathing, a distinct indication of revival. The doctor had some tea and he chatted for a quarter of an hour in a way that showed what a 'good' manner and how large an experience a local man could have. When he went away Nick walked out with him. The doctor's house was near by and he had come on foot. He left Nick with the assurance that in all probability Mr Carteret, who was certainly picking up, would be able to see him on the morrow. Our young man turned his steps again to the abbey and took a stroll about it in the starlight. It never looked so huge as when it reared itself in the night, and Nick had never felt more fond of it than on this occasion, more comforted and confirmed by its beauty. When he came back he was readmitted by Chayter, who surveyed him in respectful deprecation of the frivolity which had led him to attempt to help himself through such an evening in such a way.

Nick went to bed early and slept badly, which was unusual with him; but it was a pleasure to him to be told almost as soon as he came out of his room that Mr Carteret had asked for him. He went in to see him and was struck with the change in his appearance. He had however spent a day with him just after the New Year, and another at the beginning of March, so that he had perceived the first symptoms of mortal alteration. A week after Julia Dallow's departure for the Continent Nick had devoted several hours to Beauclere and to the intention of telling his old friend how the happy event had been brought to naught – the advantage that he had been so good as to desire for him and to make the condition of a splendid gift. Before this, for a few days, Nick had been keeping back, to announce it personally, the good news that Julia had at last set their situation in order: he wanted to enjoy the old man's pleasure – so sore a trial had her arbitrary behaviour been for a year. Mrs Dallow had offered Mr Carteret a conciliatory visit before Christmas – had come down from London one day to lunch with him, but only with the effect of making him subsequently exhibit to poor Nick, as the victim of her whimsical hardness, a great deal of earnest commiseration in a jocose form. Upon his honour, as he said, she was as clever and 'specious' a woman (this was the odd expression he used) as he had ever seen in his life. The merit of her behaviour on this occasion, as Nick knew, was that she had not been specious at her lover's expense: she had breathed no doubt of his public purpose and had had the feminine courage to say that in truth she was older than he, so that it was only fair to give his affections time to mature. But when Nick saw their sympathizing host after the rupture that I lately narrated he found him in no state to encounter a disappointment: he was seriously ailing, it was the beginning of worse things and no time for trying on a sensation. After this excursion Nick went back to town saddened by Mr Carteret's now unmistakably settled decline, but rather relieved that he had not been forced to make his confession. It had even occurred to him that the need for making it might not come up if the ebb of his old friend's strength should continue unchecked. He might pass away in the persuasion that everything would happen as he wished it, though indeed without enriching Nick on his wedding-day to the tune that he had promised. Very likely he had made legal arrangements in virtue of which his bounty would take effect in the right conditions and in them alone. At present Nick had a larger confession to treat him to – the last three days had made the difference; but, oddly enough, though his responsibility had increased his reluctance to speak had vanished: he was positively eager to clear up a situation over which it was not consistent with his honour to leave a shade.

The doctor had been right when he came in after dinner; it was clear in the morning that they had not seen the last of Mr Carteret's power of picking up. Chayter, who had been in to see him, refused austerely to change his opinion with every change in his master's temperature; but the nurses took the cheering view that it would do their patient good for Mr Dormer to sit with him a little. One of them remained in the room, in the deep window-seat, and Nick spent twenty minutes by the bedside. It was not a case for much conversation, but Mr Carteret seemed to like to look at him. There was life in his kind old eyes, which would express itself yet in some further wise provision. He laid his liberal hand on Nick's with a confidence which showed it was not yet disabled. He said very little, and the nurse had recommended that the visitor himself should not overflow in speech; but from time to time he murmured with a faint smile: 'Tonight's division, you know – you mustn't miss it.' There was to be no division that night, as it happened, but even Mr Carteret's aberrations were parliamentary. Before Nick left him he had been able to assure him that he was rapidly getting better, that such valuable hours must not be wasted. 'Come back on Friday, if they come to the second reading.' These were the words with which Nick was dismissed, and at noon the doctor said the invalid was doing very well, but that Nick had better leave him alone for that day. Our young man accordingly determined to go up to town for the night and even, if he should receive no summons, for the next day. He arranged with Chayter that he should be telegraphed to if Mr Carteret were either better or worse.

'Oh, he can't very well be worse, sir,' Chayter replied, inexorably; but he relaxed so far as to remark that of course it wouldn't do for Nick to neglect the House.

'Oh, the House!' Nick sighed, ambiguously, avoiding the butler's eye. It would be easy enough to tell Mr Carteret, but nothing would have sustained him in the effort to make a clean breast to Chayter.

He might be ambiguous about the House, but he had the sense of things to be done awaiting him in London. He telegraphed to his servant and spent that night in Rosedale Road. The things to be done were apparently to be done in his studio: his servant met him there with a large bundle of letters. He failed that evening to stray within two miles of Westminster, and the legislature of his country reassembled without his support. The next morning he received a telegram from Chayter, to whom he had given Rosedale Road as an address. This missive simply informed him that Mr Carteret wished to see him, and it seemed to imply that he was better, though Chayter wouldn't say so. Nick again took his place in the train to Beauclere. He had been there very often, but it was present to him that now, after a little, he should go only once more, for a particular dismal occasion. All that was over – everything that belonged to it was over. He learned on his arrival – he saw Mrs Lendon immediately – that his old friend had continued to pick up. He had expressed a strong and a perfectly rational desire to talk with Nick, and the doctor had said that if it was about anything important it was much better not to oppose him. 'He says it's about something very important,' Mrs Lendon remarked, resting shy eyes on him while she added that *she* was looking after her brother for the hour. She had sent those wonderful young ladies out to see the abbey. Nick paused with her outside of Mr Carteret's door. He wanted to say something comfortable to her in return for her homely charity – give her a hint, which she was far from looking for, that practically he had now no interest in her brother's estate. This was impossible of course. Her absence of irony gave him no pretext, and such an allusion would be an insult to her simple discretion. She was either not thinking of his interest at all, or she was thinking of it with the tolerance of a mind trained to a hundred decent submissions. Nick looked for an instant into her mild, uninvestigating eyes, and it came over him supremely that the goodness of these people was singularly pure: they were a part of what was cleanest and sanest and dullest in humanity. There had been just a little mocking inflection in Mrs Lendon's pleasant voice; but it was dedicated to the young ladies in the black uniforms (she could perhaps be satirical about *them*), and not to the theory of the 'importance' of Nick's interview with her brother. Nick's arrested desire to let her know he was not dangerous translated itself into a vague friendliness and into the abrupt, rather bewildering words: 'I can't tell you half the good I think of you.' As he passed into Mr Carteret's room it occurred to him that she would perhaps interpret this speech as an acknowledgement of obligation – of her good-nature in not keeping him away from the rich old man.

33

MR CARTERET was propped up on pillows, and in this attitude, beneath the high, spare canopy of his bed, presented himself to Nick's picture-seeking vision as a figure in a clever composition or a novel. He had gathered strength, though this strength was not much in his voice; it was mainly in his brighter eye and his air of being pleased with himself. He put out his hand and said: 'I dare say you know why I sent for you'; upon which Nick sank into the seat he had occupied the day before, replying that he had been delighted to come, whatever the reason. Mr Carteret said nothing more about the division or the second reading; he only murmured that they were keeping the newspapers for him. 'I'm rather behind – I'm rather behind,' he went on; 'but two or three quiet mornings will make it all right. You can go back tonight, you know – you can easily go back.' This was the only thing not quite straight that Nick saw in him – his making light of his young friend's flying to and fro. Nick sat looking at him with a sense that was half compunction and half the idea of the rare beauty of his face, to which, strangely, the waste of illness now seemed to have restored some of its youth. Mr Carteret was evidently conscious that this morning he should not be able to go on long, so that he must be practical and concise. 'I dare say you know – you have only to remember,' he continued.

'You know what a pleasure it is to me to see you – there can be no better reason than that.'

'Hasn't the year come round – the year of that foolish arrangement?'

Nick thought a little, asking himself if it were really necessary to disturb his companion's earnest faith. Then the consciousness of the falsity of his own position surged over him again, and he replied: 'Do you mean the period for which Mrs Dallow insisted on keeping me dangling? Oh, that's over.'

'And are you married – has it come off?' the old man asked, eagerly. 'How long have I been ill?'

'We are uncomfortable, unreasonable people, not deserving of your interest. We are not married,' Nick said.

'Then I haven't been ill so long,' Mr Carteret sighed, with vague relief.

'Not very long – but things *are* different,' Nick continued.

The old man's eyes rested on his, and Nick noted how much larger they appeared. 'You mean the arrangements are made – the day is at hand?'

'There are no arrangements,' Nick smiled: 'but why should it trouble you?'

'What then will you do – without arrangements?' Mr Carteret's inquiry was plaintive and childlike.

'We shall do nothing – there is nothing to be done. We are not to be married – it's all off,' said Nick. Then he added: 'Mrs Dallow has gone abroad.'

The old man, motionless among his pillows, gave a long groan. 'Ah, I don't like that.'

'No more do I, sir.'

'What's the matter? It was so good – so good.'

'It wasn't good enough for her,' Nick Dormer declared.

'For her? Is she so great as that? She told me she had the greatest regard for you. You're good enough for the best, my dear boy,' Mr Carteret went on.

'You don't know me; I *am* disappointing. Mrs Dallow had, I believe, a great regard for me; but I have forfeited her regard.'

The old man stared at this cynical announcement: he searched his companion's face for some attenuation of the words. But Nick apparently struck him as unashamed; and a faint colour coming into his withered cheek indicated his mystification and alarm. 'Have you been unfaithful to her?' he demanded, considerately.

'She thinks so – it comes to the same thing. As I told you a year ago, she doesn't believe in me.'

'You ought to have made her – you ought to have made her,' said Mr Carteret. Nick was about to utter some rejoinder when he continued: 'Do you remember what I told you I would give you if you did? Do you remember what I told you I would give you on your wedding-day?'

'You expressed the most generous intentions; and I remember them as much as a man may do who has no wish to remind you of them.'

'The money is there – I have put it aside.'

'I haven't earned it – I haven't earned a penny of it. Give it to those who deserve it more.'

'I don't understand – I don't understand,' Mr Carteret murmured, with the tears of weakness coming into his eyes. His face flushed and he added: 'I'm not good for much discussion; I'm very much disappointed.'

'I think I may say it's not my fault – I have done what I can,' returned Nick.

'But when people are in love they do more than that.'

'Oh, it's all over!' Nick exclaimed; not caring much now, for the moment, how disconcerted his companion might be, so long as he disabused him of the idea that they were partners to a bargain. 'We've tormented each other and we've tormented you; and that is all that has come of it.'

'Don't you care for what I would have done for you – shouldn't you have liked it?'

'Of course one likes kindness – one likes money. But it's all over,' Nick repeated. Then he added: 'I fatigue you, I knock you up, with telling you these uncomfortable things. I only do so because it seems to me right you should know. But don't be worried – everything will be all right.'

He patted his companion's hand reassuringly, he leaned over him affectionately; but Mr Carteret was not easily soothed. He had practised lucidity all his life, he had expected it of others and he had never given his assent to an indistinct proposition. He was weak, but he was not too weak to perceive that he had formed a calculation which was now vitiated by a wrong factor – put his name to a contract of which the other side had not been carried out. More than fifty years of conscious success pressed him to try to understand; he had never muddled his affairs and he couldn't muddle them now. At the same time he was aware of the necessity of economizing his effort, and he evidently gathered himself, within, patiently and almost cunningly, for the right question and the right induction. He was still able to make his agitation reflective, and it could still consort with his high hopes of Nick that he should find himself regarding the declaration that everything would be all right as an inadequate guarantee. So, after he had looked a moment into his companion's eyes, he inquired:

'Have you done anything bad?'

'Nothing worse than usual,' laughed Nick.

'Everything should have been better than usual.'

'Ah, well, it hasn't been that – that I must say.'

'Do you sometimes think of your father?' Mr Carteret continued.

Nick hesitated a moment. '*You* make me think of him – you have always that pleasant effect.'

'His name would have lived – it mustn't be lost.'

'Yes, but the competition today is terrible,' Nick replied.

Mr Carteret considered this a moment, as if he found a serious flaw in it; after which he began again: 'I never supposed you were a trifler.'

'I'm determined not to be.'

'I thought her charming. Don't you love her?' Mr Carteret asked.

'Don't ask me that today, for I feel sore and resentful. I don't think she has treated me well.'

'You should have held her – you shouldn't have let her go,' the old man returned, with unexpected fire.

His companion flushed at this, so strange it seemed to him to receive a lesson in energy from a dying octogenarian. Yet after an instant Nick answered modestly enough: 'I haven't been clever enough, no doubt.'

'Don't say that – don't say that,' Mr Carteret murmured, looking almost frightened. 'Don't think I can allow you any mitigation of that sort. I know how well you've done. You're taking your place. Several gentlemen have told me. Hasn't she felt a scruple, knowing my settlement on you was contingent?' he pursued.

'Oh, she hasn't known – hasn't known anything about it.'

'I don't understand; though I think you explained somewhat, a year ago,' Mr Carteret said, with discouragement. 'I think she wanted to speak to me – of any intentions I might have in regard to you – the day she was here. Very nicely, very properly she would have done it, I'm sure. I think her idea was that I ought to make any settlement quite independent of your marrying her or not marrying her. But I tried to convey to her – I don't know whether she understood me – that I liked her too much for that, I wanted too much to make sure of her.'

'To make sure of me, you mean,' said Nick. 'And now, after all, you see you haven't.'

'Well, perhaps it was that,' sighed the old man, confusedly.

'All this is very bad for you – we'll talk again,' Nick rejoined.

'No, no – let us finish it now. I like to know what I'm doing. I shall rest better when I do know. There are great things to be done; the future will be full – the future will be fine,' Mr Carteret wandered.

'Let me say this for Julia: that if we hadn't been sundered her generosity to me would have been complete, she would have put her great fortune absolutely at my disposal,' Nick said, after a moment. 'Her consciousness of all that naturally carries her over any particular distress in regard to what won't come to me now from another source.'

'Ah, don't lose it,' pleaded the old man, painfully.

'It's in your hands, sir,' reasoned Nick.

'I mean Mrs Dallow's fortune. It will be of the highest utility. That was what your father missed.'

'I shall miss more than my father did,' said Nick.

'She'll come back to you – I can't look at you and doubt that.'

Nick shook his head slowly, smiling. 'Never, never, never! You look at me, my grand old friend, but you don't see me. I'm not what you think.'

'What is it – what is it? *Have* you been bad?' Mr Carteret panted.

'No, no; I'm not bad. But I'm different.'

'Different?'

'Different from my father – different from Mrs Dallow – different from you.'

'Ah, why do you perplex me?' moaned the old man. 'You've done something.'

'I don't want to perplex you, but I *have* done something,' said Nick, getting up.

He had heard the door open softly behind him and Mrs Lendon come forward with precautions. 'What has he done – what has he done?' quavered Mr Carteret to his sister. She however, after a glance at the patient, motioned Nick away and, bending over the bed, replied in a voice expressive at that moment of a sharply contrasted plenitude of vital comfort:

'He has only excited you, I'm afraid, a little more than is good for you. Isn't your dear old head a little too high?' Nick regarded himself as justly banished and he quitted the room with a ready acquiescence in any power to carry on the scene of which Mrs Lendon might find herself possessed. He felt distinctly brutal as he heard his host emit a soft, troubled exhalation of assent to some change of position. But he would have reproached himself more if he had wished less to guard against the acceptance of an equivalent for duties unperformed. Mr Carteret had had in his mind, characteristically, the idea of an enlightened agreement, and there was something more to be said about that.

Nick went out of the house and stayed away for two or three hours, quite ready to consider that the place was quieter and safer without him. He haunted the abbey, as usual, and sat a long time in its simplifying stillness, turning over many things. He came into the house again at the luncheon-hour, through the garden, and heard, somewhat to his surprise and greatly to his relief, that Mr Carteret had composed himself promptly enough after their agitating interview. Mrs Lendon talked at luncheon much as if she expected her brother to be, as she said, really quite fit again. She asked Nick no embarrassing questions; which was uncommonly good of her, he thought, considering that she might have said: 'What in the world were you trying to get out of him?' She only told our young man that the invalid had very little doubt he should be able to see him again, about half-past seven, for a *very* short time: this timid emphasis was Mrs Lendon's single tribute to the critical spirit. Nick divined that Mr Carteret's desire for further explanations was really strong and had been capable of sustaining him through a bad morning – capable even of helping him (it would be a secret and wonderful momentary victory over his weakness) to pass it off for a good one. He wished he might make a sketch of him from the life, as he had seen him after breakfast; he had a conviction he could make a strong one, and it would be a precious memento. But he shrank from proposing this – Mr Carteret might think it un-parliamentary. The doctor had called while Nick was out, and he came again at five o'clock, without our young man's seeing him. Nick was busy in his room at that hour: he wrote a short letter which took him a long time. But apparently there had been no veto on a resumption of talk, for at half-past seven the old man sent for him. The nurse, at the door, said: 'Only a moment, I hope, sir?' but she took him in and then withdrew.

The prolonged daylight was in the room, and Mr Carteret was again established on his pile of pillows, but with his head a little lower. Nick sat down by him and began to express the hope that he had not upset him in the morning; but the old man, with fixed, expanded eyes, took up their conversation exactly where they had left it.

'What have you done – what have you done? Have you associated yourself with some other woman?'

'No, no; I don't think she can accuse me of that.'

'Well, then, she'll come back to you, if you take the right way with her.'

It might have been droll to hear Mr Carteret, in his situation, giving his views on the right way with women; but Nick was not moved to enjoy that diversion. 'I've taken the wrong way. I've done something which will spoil my prospects in that direction for ever. I've written a letter,' Nick went on; but his companion had already interrupted him.

'You've written a letter?'

'To my constituents, informing them of my determination to resign my seat.'

'To resign your seat?'

'I've made up my mind, after no end of reflection, dear Mr Carteret, to work in a different line. I have a project of becoming a painter. So I've given up the idea of a political life.'

'A painter?' Mr Carteret seemed to turn whiter.

'I'm going in for the portrait in oils: it sounds absurd, I know, and I only mention it to show you that I don't in the least expect you to count upon me.' Mr Carteret had continued to stare, at first; then his eyes slowly closed and he lay motionless and blank. 'Don't let it trouble you now; it's a long story and rather a poor one; when you get better I'll tell you all about it. We'll talk it over amicably, and I'll bring you to my side,' Nick went on hypocritically. He had laid his hand on Mr Carteret's again: it felt cold, and as the old man remained silent he had a moment of exaggerated fear.

'This is dreadful news,' said Mr Carteret, opening his eyes.

'Certainly it must seem so to you, for I've always kept from you (I was ashamed, and my present confusion is a just chastisement) the great interest I have always taken in the' – Nick hesitated, and then added, with an intention of humour and a sense of foolishness – 'in the pencil and the brush.' He spoke of his present confusion; but it must be confessed that his manner showed it but little. He was surprised at his own serenity, and had to recognize that at the point things had come to now he was profoundly obstinate and quiet.

'The pencil – the brush? They're not the weapons of a gentleman,' said Mr Carteret.

'I was sure that would be your view. I repeat that I mention them only because you once said you intended to do something for me, as the phrase is, and I thought you oughtn't to do it in ignorance.'

'My ignorance was better. Such knowledge isn't good for me.'

'Forgive me, my dear old friend. When you're better you'll see it differently.'

'I shall never be better now.'

'Ah, no,' pleaded Nick, 'it will do you good after a little. Think it over quietly and you'll be glad I've stopped being a humbug.'

'I loved you – I loved you as my son,' moaned the old man.

Nick sank on his knee beside the bed and leaned over him tenderly. 'Get better, get better, and I'll be your son for the rest of your life.'

'Poor Dormer – poor Dormer!' Mr Carteret softly wailed.

'I admit that if he had lived I probably shouldn't have done it,' said Nick. 'I dare say I should have deferred to his prejudices, even if I thought them narrow.'

'Do you turn against your father?' Mr Carteret asked, making, to disengage his arm from the young man's touch, an effort in which Nick recognized the irritation of conscious weakness.

Nick got up, at this, and stood a moment looking down at him, while Mr Carteret went on: 'Do you give up your name, do you give up your country?'

'If I do something good my country may like it,' Nick contended.

'Do you regard them as equal, the two glories?'

'Here comes your nurse, to blow me up and turn me out,' said Nick.

The nurse had come in, but Mr Carteret managed to direct to her an audible, dry, courteous 'Be so good as to wait till I send for you,' which arrested her, in the large room, at some distance from the bed, and then had the effect of making her turn on her heel with a professional laugh. She appeared to think that an old gentleman with the fine manner of his prime might still be trusted to take care of himself. When she had gone Mr Carteret went on, addressing Nick, with the inquiry for which his deep displeasure lent him strength: 'Do you pretend there is a nobler life than a high political career?'

'I think the noble life is doing one's work well. One can do it very ill and be very base and mean in what you call a high political career. I haven't been in the House so many months without finding that out. It contains some very small souls.'

'You should stand against them – you should expose them!' stammered Mr Carteret.

'Stand against them – against one's own party?'

The old man looked bewildered a moment at this; then he broke out: 'God forgive you, are you a Tory – are you a Tory?'

'How little you understand me!' laughed Nick, with a ring of bitterness.

'Little enough – little enough, my boy. Have you sent your electors your dreadful letter?'

'Not yet; but it's ready, and I shan't change my mind.'

'You will – you will; you'll think better of it, you'll see your duty,' said the old man, almost coaxingly.

'That seems very improbable, for my determination, crudely and abruptly as, to my regret, it comes to you here, is the fruit of a long and painful struggle. The difficulty is that I see my duty just in this other effort.'

'An effort? Do you call it an effort to fall away, to sink far

360

down, to give up *every* effort? What does your mother say, heaven help her?' Mr Carteret pursued, before Nick could answer the other question.

'I haven't told her yet.'

'You're ashamed, you're ashamed!' Nick only looked out of the western window at this; he felt himself growing red. 'Tell her it would have been sixty thousand; I had the money all ready.'

'I shan't tell her that,' said Nick, redder still.

'Poor woman – poor dear woman!' Mr Carteret whimpered.

'Yes, indeed; she won't like it.'

'Think it all over again; don't throw away a splendid future!' These words were uttered with a recovering flicker of passion. Nick Dormer had never heard such an accent on his old friend's lips. But the next instant Mr Carteret began to murmur: 'I'm tired – I'm very tired,' and sank back with a groan and with closed lips.

Nick assured him tenderly that he had only too much cause to be exhausted, but that the worst was over now. He smoothed his pillows for him and said he must leave him, he would send in the nurse.

'Come back – come back,' Mr Carteret pleaded before he quitted him; 'come back and tell me it's a horrible dream.'

Nick did go back, very late that evening; Mr Carteret had sent a message to his room. But one of the nurses was on the ground this time and she remained there with her watch in her hand. The invalid's chamber was shrouded and darkened; the shaded candle left the bed in gloom. Nick's interview with his venerable host was the affair of but a moment; the nurse interposed, impatient and not understanding. She heard Nick tell Mr Carteret that he had posted his letter now, and Mr Carteret flashed out, with an acerbity which savoured still of the sordid associations of a world he had not done with: 'Then of course my settlement doesn't take effect!'

'Oh, that's all right,' Nick answered, kindly; and he went off the next morning by the early train – his injured host still sleeping. Mrs Lendon's habits made it easy for her to be present in matutinal bloom at the young man's hasty breakfast, and she sent a particular remembrance to Lady Agnes and (when Nick should see them) to the Ladies Flora and Elizabeth. Nick had a prevision of the spirit in which his mother at least would now receive hollow compliments from Beauclere.

The night before, as soon as he had quitted Mr Carteret, the old man said to the nurse that he wished her to tell Mr Chayter that, the first thing in the morning, he must go and fetch Mr Mitton. Mr Mitton was the first solicitor at Beauclere.

34

THE really formidable thing for Nick was to tell his mother: a truth of which he was so conscious that he had the matter out with her the very morning he returned from Beauclere. She and Grace had come back the afternoon before from Lady St Dunstans', and knowing this (she had written him her intention from the country), he drove straight from the station to Calcutta Gardens. There was a little room there, on the right of the house-door, which was known as his own room, but in which of a morning, when he was not at home, Lady Agnes sometimes wrote her letters. These were always numerous, and when she heard our young man's cab she happened to be engaged with them at the big brass-mounted bureau which had belonged to his father, where, behind an embankment of works of political reference, she seemed to herself to make public affairs feel the point of her elbow.

She came into the hall to meet her son and to hear about Mr Carteret, and Nick went straight back into the room with her and closed the door. It would be in the evening paper and she would see it, and he had no right to allow her to wait for that. It proved indeed a terrible hour; and when, ten minutes later, Grace, who learned upstairs that her brother had come back, went down for further news of him she heard from the hall a sound of voices which made her first pause and then retrace her steps on tiptoe. She mounted to the drawing-room and crept about there, palpitating, looking at moments into the dull street and wondering what on earth was going on. She had no one to express her wonder to, for Florence Tressilian had departed and Biddy, after breakfast, had betaken herself, in accordance with a custom now inveterate, to Rosedale Road. Her mother was

362

crying passionately – a circumstance tremendous in its significance, for Lady Agnes had not often been brought so low. Nick had seen her cry, but this almost awful spectacle had seldom been given to Grace; and it forced her to believe at present that some dreadful thing had happened.

That was of course in order, after Nick's mysterious quarrel with Julia, which had made his mother so ill and which now apparently had been followed up with new horrors. The row, as Grace mentally phrased it, had had something to do with this incident, some deeper depth of disappointment had opened up. Grace asked herself if they were talking about Broadwood; if Nick had demanded that, in the conditions so unpleasantly altered, Lady Agnes should restore that pretty property to its owner. This was very possible, but why should he so suddenly have broken out about it? And moreover their mother, though sore to bleeding about the whole business – for Broadwood, in its fresh comfort, was too delightful – would not have met this pretension with tears, inasmuch as she had already declared that they couldn't decently continue to make use of the place. Julia had said of course they must go on, but Lady Agnes was prepared with an effective rejoinder to this. It didn't consist of words – it was to be austerely practical, was to consist of letting Julia see, at the moment she should least expect it, that they quite wouldn't go on. Lady Agnes was ostensibly waiting for that moment – the moment when her renunciation would be most impressive.

Grace was conscious of how, for many days, her mother and she had been moving in darkness, deeply stricken by Nick's culpable (oh, he was culpable!) loss of his prize, but feeling there was an element in the matter they didn't grasp, an undiscovered explanation which would perhaps make it still worse, though it might make *them* a little better. Nick had explained nothing; he had simply said: 'Dear mother, we don't hit it off, after all; it's an awful bore, but we don't,' as if that were, under the circumstances, an adequate balm for two aching hearts. From Julia, naturally, satisfying attenuations were not to be looked for; and though Julia very often did the wrong thing you wouldn't suppose she was not unexpectedly apologetic in this case. Grace recognized that in such a position it would savour of apology for her to impart to Lady Agnes her grounds for letting Nick off; and she would not have liked to be the person to suggest to Julia that any one looked for anything from her. Neither of the disunited pair blamed the other or cast an aspersion, and it was all very magnanimous and superior and impenetrable and exasperating. With all this Grace had a suspicion that Biddy knew something more, that for Biddy the tormenting curtain had been lifted.

Biddy came and went in these days with a perceptible air of detachment from the tribulations of home. It made her fortunately very pretty – still prettier than usual; it sometimes happened that at moments when Grace was most angry she had a faint, sweet smile which might have been drawn from a source of private consolation. It was perhaps in some degree connected with Peter Sherringham's visit, as to which the girl was not silent. When Grace asked her if she had secret information and if it pointed to the idea that everything would be all right in the end, she pretended to know nothing (What should she know? she asked, with the loveliest candour), and begged her sister not to let Lady Agnes believe that she was any better off than they. She contributed nothing to their gropings towards the light save a better patience than theirs, but she went with noticeable regularity, on the pretext of her foolish modelling, to Rosedale Road. She was frankly on Nick's side; not going so far as to say he had been right, but saying distinctly that she was sure that, whatever had happened, he couldn't help it. This was striking, because, as Grace knew, the younger of the sisters had been much favoured by Julia and would not have sacrificed her easily. It associated itself in the irritated mind of the elder with Biddy's frequent visits to the studio and made Miss Dormer ask herself whether the crisis in Nick's and Julia's business had not somehow been linked to that unnatural spot.

She had gone there two or three times while Biddy was working, to pick up any clue to the mystery that might peep out. But she had put her hand upon nothing, save once on the personality of Gabriel Nash. She found this strange creature, to her surprise, paying a visit to her sister – he had come for Nick, who was absent: she remembered how they had met him in Paris and how he had frightened her. When she asked Biddy afterwards how she could receive him that way Biddy replied that even she, Grace, would have some charity for him if she could hear how fond he was of poor Nick. He talked to her only of Nick – of nothing else. Grace observed how she spoke of Nick as injured, and noted the implication that some one else had ceased to be fond of him and was thereby condemned in Biddy's eyes. It seemed to Grace that some one else had at least a right not to like some of his friends. The studio struck her as mean and horrid; and so far from suggesting to her that it could have played a part in making Nick and Julia fall out, she only felt how little its dusty want of consequence could count one way or the other for Julia. Grace, who had opinions on art, saw no merit whatever in those 'impressions' on canvas, from Nick's hand, with which the place was bestrewn. She didn't wish her brother to have talent in that direction; yet it was secretly humiliating to her that he had not more.

Nick felt a pang of almost horrified penitence, in the little room on the right of the hall, the moment after he had made his mother really understand that he had thrown up his seat and it would probably be in the evening papers. That she would take it badly was an idea that had pressed upon him hard enough; but she took it even worse than he had feared. He measured, in the look that she gave him when the full truth loomed upon her, the mortal cruelty of her discomfiture: her face was like that of a passenger on a ship who sees the huge bows of another vessel towering close out of the fog. There are visions of dismay before which the best conscience recoils; and though Nick had made his choice on all the grounds there were a few minutes in which he would gladly have admitted that his wisdom was a dark mistake. His heart was in his throat, he had gone too far; he had been ready to disappoint his mother – he had not been ready to destroy her.

Lady Agnes, I hasten to add, was not destroyed; she made, after her first drowning gasp, a tremendous scene of opposition, in the face of which Nick speedily fell back upon his intrenchments. She must know the worst, he had thought; so he told her everything, including the little story of the forfeiture of his 'expectations' from Mr Carteret. He showed her this time not only the face of the matter but what lay below it; narrated briefly the incident in his studio which had led to Julia Dallow's deciding that she couldn't after all put up with him. This was wholly new to Lady Agnes; she had had no clue to it, and he could instantly see how it made the case worse for her, adding a hideous positive to an abominable negative. He perceived now that, distressed and distracted as she had been by his rupture with Julia, she had still held to the faith that their engagement would come on again; believing evidently that he had a personal empire over the mistress of Harsh which would bring her back. Lady Agnes was forced to recognize that empire as precarious, to forswear the hope of a blessed renewal, from the moment it was a question of base infatuations on his own part. Nick confessed to an infatuation but did his best to show her it was not base; that it was not (since Julia had had faith in his loyalty) for the person of the young lady who had been discovered posturing to him and whom he had seen but half a dozen times in his life. He endeavoured to give his mother a notion of who this young lady was and to remind her of the occasion, in Paris, when they all had seen her together. But Lady Agnes's mind and memory were a blank on the subject of Miss Miriam Rooth, and she wanted to know nothing whatever about her: it was enough that she was the cause of their ruin, that she was mixed up with his unspeakable folly. Her ladyship needed to know nothing of Miss Rooth to allude to her as if it were superfluous to give a definite name to the class to which she belonged.

But she gave a name to the group in which Nick had now taken his place, and it made him feel, after the lapse of years, like a small blamed, sorry boy again; for it was so far away he could scarcely remember it (besides there having been but a moment or two of that sort in his happy childhood), the time when his mother had slapped him and called him a little fool. He was a big fool now – a huge, immeasurable one; she repeated the term over and over, with high-pitched passion. The most painful thing in this painful hour was perhaps his glimpse of the strange feminine cynicism that lurked in her fine sense of injury. Where there was such a complexity of revolt it would have been difficult to pick out particular complaints; but Nick could see that to Lady Agnes's imagination he was most a fool for not having kept his relations with the actress, whatever they were, better from Julia's knowledge. He remained indeed freshly surprised at the ardour with which she had rested her hopes on Julia. Julia was certainly a combination – she was fascinating, she was a sort of leading woman and she was rich; but after all (putting aside what she might be to a man in love with her), she was not the keystone of the universe. Yet the form in which the consequences of his apostasy appeared most to come home to Lady Agnes was the loss, for the Dormer family, of the advantages attached to the possession of Mrs Dallow. The larger mortification would round itself later; for the hour the damning thing was that Nick had really made Julia a present of an unforgivable grievance. He had clinched their separation by his letter to his electors; and that above all was the wickedness of the letter. Julia would have got over the other woman, but she would never get over his becoming a nobody.

Lady Agnes challenged him upon this low prospect exactly as if he embraced it with the malignant purpose of making Julia's return impossible. She contradicted her premises and lost her way in her wrath. What had made him suddenly turn round if he had been in good faith before? He had never been in good faith – never, never; he had had from his earliest childhood the nastiest hankerings after a vulgar little daubing, trash-talking life; they were not in him, the grander, nobler aspirations – they never had been – and he had been anything but honest to lead her on, to lead them all on, to think he would do something: the fall and the shame would have been less for them if they had come earlier. Moreover, what need under heaven had he to tell Charles Carteret of his cruel folly on his very death-bed? – as if he mightn't have let it all alone and accepted the benefit the old man was so delighted to confer. No wonder the old man would keep his money for his heirs, if that was the way Nick proposed to repay him; but where was the common sense, where was the common charity, where was the common decency, of tormenting him with such vile news in his last hours? Was he trying what he could invent that would break her heart, that would send her in sorrow down to her grave? Weren't they all miserable enough, and hadn't he a ray of pity for his wretched sisters?

The relation of effect and cause, in regard to his sisters'

wretchedness, was but dimly discernible to Nick, who however easily perceived that his mother genuinely considered that his action had disconnected them all, still more than she held they were already disconnected, from the good things of life. Julia was money, Mr Carteret was money, and everything else was poverty. If these precious people had been primarily money for Nick, it was after all a gracious tribute to his distributive power to have taken for granted that for the rest of the family too the difference would have been so great. For days, for weeks and months afterward the little room on the right of the hall seemed to our young man to vibrate, as if the very walls and windowpanes still suffered, with the most disagreeable ordeal he had ever been through.

35

THAT evening – the evening of his return from Beauclere – Nick was conscious of a keen desire to get away, to go abroad, to leave behind him the little chatter his resignation would be sure to produce in an age of publicity which never discriminated as to the quality of events. Then he felt it was better to stay, to see the business through on the spot. Besides, he would have to meet his constituents (would a parcel of cheese-eating burgesses ever have been 'met' on so queer an occasion?) and when that was over the worst would be over. Nick had an idea that he knew in advance how it would feel to be pointed at as a person who had given up a considerable chance of eventual 'office' to take likenesses at so much a head. He wouldn't attempt, down at Harsh, to touch on the question of motive; for, given the nature of the public mind of Harsh, that would be a strain on his faculty of expression. But as regards the chaff of the political world and of society he had an idea he should find chaff enough for answers. It was true that when his mother chaffed him in her own effective way he had felt rather flattened out; but then one's mother might have a heavier hand than any one else.

He had not thrown up the House of Commons to amuse himself; he had thrown it up to work, to sit quietly down and bend over his task. If he should go abroad his mother might think he had some weak-minded view of joining Julia Dallow and trying, with however little hope, to win her back – an illusion it would be singularly pernicious to encourage. His desire for Julia's society had succumbed, for the present at any rate, to an irresistible interruption – he had become more and more conscious that they spoke a different language. Nick felt like a young man who has gone to the Rhineland to 'get up' his German for an examination – committed to talk, to read, to dream only in the new idiom. Now that he had taken his jump everything was simplified, at the same time that everything was pitched in a higher, more excited key; and he wondered how in the absence of a common dialect he had conversed on the whole so happily with Julia. Then he had after-tastes of understandings tolerably independent of words. He was excited because every fresh responsibility is exciting, and there was no manner of doubt that he had accepted one. No one knew what it was but himself (Gabriel Nash scarcely counted – his whole attitude on the question of responsibility was so wanton), and he would have to ask his dearest friends to take him on trust. Rather, he would ask nothing of any one, but would cultivate independence, mulishness and gaiety and fix his thoughts on a bright if distant morrow. It was disagreeable to have to remember that his task would not be sweetened by a sense of heroism; for if it might be heroic to give up the muses for the strife of great affairs, no romantic glamour worth speaking of would ever gather round an Englishman who in the prime of his strength had given up great or even small affairs for the muses. Such an original might himself privately, perversely regard certain phases of this inferior commerce as a great affair; but who would give him the benefit of that sort of confidence – except indeed a faithful, clever, excited little sister Biddy, if he should have the good luck to have one? Biddy was in fact all ready for heroic flights and eager to think she might fight the battle of the beautiful by her brother's side; so that Nick had really to moderate her and remind her that his actual job was not a crusade, with bugles and banners, but a gray, sedentary grind, whose charm was all at the core. You might have an emotion about it, and an emotion that would be a help, but this was not the sort of thing you could show – the end in view would seem ridiculously small for it. Nick asked Biddy how one could talk to people about the 'responsibility' of what she would see him pottering at in his studio.

Nick therefore didn't talk any more than he was forced to, having moreover a sense that that side of the situation would be plentifully looked after by Gabriel Nash. He left the burden of explanation to others, meeting them on the ground of inexhaustible satire. He saw that he should live for months in a thick cloud of irony, not the finest air of the season, and he adopted the weapon to which a person whose use of tobacco is only occasional resorts when everyone else produces a cigar – he puffed the empirical, defensive cigarette. He accepted the idea of a mystery in his behaviour and abounded so in that sense that his critics were themselves bewildered. Some of them felt that they got, as the phrase is, little out of him – he rose in his good-humour so much higher than the 'rise' they had looked for – on his very first encounter with the world after his scrimmage with his mother. He went to a dinner-party (he had accepted the invitation many days before), having seen his resignation, in the form of a telegram from Harsh, announced in the evening papers. The people he found there had seen it as well, and the most imaginative of them wanted to know what he was going to do. Even the least imaginative asked if it were true he had changed his politics. He gave different answers to different persons, but left most of them under the impression that he had remarkable conscientious scruples. This however was not a formidable occasion, for there happened to be no one present he was particularly fond of. There were old friends whom it would not be so easy to satisfy – Nick was almost sorry, for an hour, that he had so many old friends. If he had had more enemies the case would have been simpler; and he was fully aware that the hardest thing of all would be to be let off too easily. Then he would appear to himself to have been put on his generosity, and his deviation would wear its ugliest face.

When he left the place at which he had been dining he betook himself to Rosedale Road: he saw no reason why he should go down to the House, though he knew he had not done with that yet. He had a dread of behaving as if he supposed he should be expected to make a farewell speech, and was thankful his eminence was not of a nature to create on such an occasion a demand for his oratory. He had in fact nothing whatever to say in public – not a word, not a syllable. Though the hour was late he found Gabriel Nash established in his studio, drawn thither by the fine exhilaration of having seen an evening paper. Trying it late, on the chance, he had been told by Nick's servant that Nick would sleep there that night, and he had come in to wait, he was so eager to congratulate him. Nick submitted with a good grace to his society – he was tired enough to go to bed, but he was restless too – in spite of feeling now, oddly enough, that Nash's congratulations could add little to his fortitude. He had felt a good deal, before, as if he were in Nash's hands; but now that he had made his final choice he seemed to himself to be altogether in his own. Gabriel was wonderful, but no Gabriel could assist him much henceforth.

Gabriel was indeed more wonderful than ever, while he lolled on a divan and emitted a series of reflections which were even more ingenious than opportune. Nick walked up and down the room, and it might have been supposed from his manner that he was impatient for his visitor to withdraw. This idea would have been contradicted however by the fact that subsequently, after Nash had taken leave, he continued to perambulate. He had grown used to Nash – had a sense that he had heard all he had to say. That was one's penalty with persons whose main gift was for talk, however irrigating; talk engendered a sense of sameness much sooner than action. The things a man did were necessarily more different from each other than the things he said, even if he went in for surprising you. Nick felt Nash could never surprise him any more save by doing something.

He talked of his host's future, he talked of Miriam Rooth and of Peter Sherringham, whom he had seen at Miriam Rooth's and whom he described as in a predicament delightful to behold. Nick asked a question or two about Peter's predicament and learned rather to his disappointment that it consisted only of the fact that he was in love with Miriam. He requested his visitor to do better than this; whereupon Nash added the touch that Sherringham wouldn't be able to have her. 'Oh, they have ideas!' he said, when Nick asked him why.

'What ideas? So has he, I suppose.'

'Yes, but they're not the same.'

'Oh well, they'll arrange something,' said Nick.

'You'll have to help them a bit. She's in love with another man,' Nash returned.

'Do you mean with you?'

'Oh, I'm never another man,' said Nash; 'I'm more the wrong one than the man himself. It's you she's after.' And upon Nick's asking him what he meant by this he added: 'While you were engaged in transferring her image to your sensorium you stamped your own upon hers.'

Nick stopped in his walk, staring. 'Ah, what a bore!'

'A bore? Don't you think she's agreeable?'

Nick hesitated. 'I wanted to go on with her – now I can't.'

'My dear fellow, it only makes her handsomer: I wondered what was the matter with her.'

'Oh, that's twaddle,' said Nick, turning away. 'Besides, has she told you?'

'No, but her mother has.'

'Has she told her mother?'

'Mrs Rooth says not. But I have known Mrs Rooth to say that which isn't.'

'Apply that rule then to the information you speak of.'

'Well, since you press me, I know more,' said Nash. 'Miriam knows you are engaged to a certain lady; she told me as much, told me she had seen her here. That was enough to set Miriam off – she likes forbidden fruit.'

'I'm not engaged to any lady. I was, but we've altered our minds.'

'Ah, what a pity!' sighed Nash.

'Mephistopheles!' Nick rejoined, stopping again and looking at his visitor gravely.

'Pray, whom do you call Margaret? May I ask if your failure of interest in the political situation is the cause of this change in your personal one?' Nash went on. Nick signified to him that he might not; whereupon Gabriel added: 'I am not in the least devilish – I only mean it's a pity you've altered your minds, because now perhaps Miriam will alter hers. She goes from one thing to another. However, I won't tell her.'

'I will, then,' said Nick, between jest and earnest.

'Would that really be prudent?' Nash asked, with an intonation that made hilarity prevail.

'At any rate,' Nick resumed, 'nothing would induce me to interfere with Peter Sherringham. That sounds fatuous, but to you I don't mind appearing an ass.'

'The thing would be to get Sherringham – out of spite – to entangle himself with another woman.'

'What good would that do?'

'Oh, Miriam would begin to fancy him then.'

'Spite surely isn't a conceivable motive – for a healthy man.'

'Ah, Sherringham isn't a healthy man. He's too much in love.'

'Then he won't care for another woman.'

'He would try to, and that would produce its effect – its effect on Miriam.'

'You talk like an American novel. Let him try, and God keep us all straight.' Nick thought, in extreme silence, of his poor little Biddy and hoped – he would have to see to it a little – that Peter wouldn't 'try' on her. He changed the subject and before Nash went away took occasion to remark to him – the occasion was offered by some new allusion of the visitor's to the sport he hoped to extract from seeing Nick carry out everything to which he stood committed – that the great comedy would fall very flat, the great incident would pass unnoticed.

'Oh, if you'll simply do your part I'll take care of the rest,' said Nash.

'If you mean by doing my part working like a beaver, it's all right,' Nick replied.

'Ah, you reprobate, you'll become a fashionable painter, a P.R.A.!' his companion groaned, getting up to go.

When he had gone Nick threw himself back on the cushions of the divan and, with his hands locked above his head, sat a long time lost in thought. He had sent his servant to bed; he was unmolested. He gazed before him into the gloom produced by the unheeded burning out of the last candle. The vague outer light came in through the tall studio window, and the painted images, ranged about, looked confused in the dusk. If his mother had seen him she might have thought he was staring at his father's ghost.

36

THE night Peter Sherringham walked away from Balaklava Place with Gabriel Nash the talk of the two men directed itself, as was natural under the circumstances, to the question of Miriam's future renown and the pace, as Nash called it, at which she would go. Critical spirits as they both were, and one of them as dissimulative in passion as the other was paradoxical in the absence of it, they yet took this renown for granted as completely as the simple-minded, a pair of hot spectators in the pit, might have done, and exchanged observations on the assumption that the only uncertain element would be the pace. This was a proof of general subjugation. Peter wished not to show, but he wished to know; and in the restlessness of his anxiety he was ready even to risk exposure, great as the sacrifice might be of the imperturbable, urbane scepticism most appropriate to a secretary of embassy. He was unable to rid himself of the sense that Gabriel Nash had got up earlier than he, had had opportunities in days already distant, the days of Mrs Rooth's hungry foreign rambles. Something of authority and privilege stuck to him from this, and it made Sherringham still more uncomfortable when he was most conscious that at the best even the trained diplomatic mind would never get a grasp of Miriam as a whole. She was constructed to revolve like the terrestrial globe; some part or other of her was always out of sight or in shadow.

Sherringham talked to conceal his feelings and, like every man doing a thing from that sort of intention, did it perhaps too much. They agreed that, putting strange accidents aside, Miriam would go further than any one had gone, in England at least and within the memory of man; and that it was a pity, as regards marking the comparison, that for so long no one had gone any distance worth speaking of. They further agreed that it would naturally seem absurd to any one who didn't know, their prophesying such big things on such small evidence; and they agreed lastly that the absurdity quite vanished as soon as the prophets knew as *they* knew. Their knowledge (they quite recognized this) was simply confidence raised to a high point –

the communication of the girl's own confidence. The conditions were enormous to make, but it was of the very essence of Miriam's confidence that she would make them. The parts, the plays, the theatres, the 'support', the audiences, the critics, the money were all to be found, but she cast a spell which prevented that from seeming a serious hitch. One might not see from one day to the other what she would do or how she would do it, but she would none the less go on. She would have to construct her own road, as it were, but at the worst there would only be delays in making it. These delays would depend on the hardness of the stones she had to break.

As Sherringham had perceived, you never knew where to 'have' Gabriel Nash; a truth exemplified in his unexpected delight at the prospect of Miriam's drawing forth the modernness of the age. You might have thought he would loathe that modernness; but he had a brilliant, amused, amusing vision of it, saw it as something huge and ornamentally vulgar. It's vulgarity would rise to the grand style, like that of a London railway station, and Miriam's publicity would be as big as the globe itself. All the machinery was ready, the platform laid; the facilities, the wires and bells and trumpets, the colossal, deafening newspaperism of the period — its most distinctive sign — were waiting for her, their predestined mistress, to press her foot on the spring and set them all in motion. Gabriel brushed in a large bright picture of her progress through the time and round the world, round it and round it again, from continent to continent and clime to clime; with populations and deputations, reporters and photographers, placards, and interviews and banquets, steamers, railways, dollars, diamonds, speeches and artistic ruin all jumbled into her train. Regardless of expense the spectacle would be thrilling, though somewhat monotonous, the drama — a drama more bustling than any she would put on the stage and a spectacle that would beat everything for scenery. In the end her divine voice would crack, screaming to foreign ears and antipodal barbarians, and her clever manner would lose all quality, simplified to a few unmistakable knock-down dodges. Then she would be at the fine climax of life and glory, still young and insatiate, but already coarse, hard and raddled, with nothing left to do and nothing left to do it with, the remaining years all before her and the *raison d'être* all behind. It would be curious and magnificent and grotesque.

'Oh, she'll have some good years – they'll be worth having,' Sherringham insisted, as they went on. 'Besides, you see her too much as a humbug and too little as a real producer. She has ideas – great ones; she loves the thing for itself. That may keep a woman serious.'

'Her greatest idea must always be to show herself; and fortunately she has a splendid self to show. I think of her absolutely as a real producer, but as a producer whose production is her own person. No "person", even as fine a one as hers, will stand that for more than an hour, so that humbuggery has very soon to lend a hand. However,' Nash continued, 'if she's a fine humbug it will do as well, and perfectly suit the time. We can all be saved by vulgarity; that's the solvent of all difficulties and the blessing of this delightful age. Let no man despair; a new hope has dawned.'

'She'll do her work like any other worker, with the advantage over many that her talent is rare,' Peter replied. 'Compared with the life of many women, that's security and sanity of the highest order. Then she can't help her beauty. You can't vulgarize that.'

'Oh, can't you?' exclaimed Gabriel Nash.

'It will abide with her till the day of her death. It isn't a mere superficial freshness. She's very noble.'

'Yes, that's the pity of it,' said Nash. 'She's a capital girl, and I quite admit that she'll do for a while a lot of good. She will have brightened up the world for a great many people; she will have brought the ideal nearer to them, held it fast for an hour, with its feet on earth and its great wings trembling. That's always something, for blessed is he who has dropped even the smallest coin into the little iron box that contains the precious savings of mankind. Miriam will doubtless have dropped a big gold piece. It will be found, in the general scramble, on the day the race goes bankrupt. And then, for herself, she will have had a great go at life.'

'Oh, yes, she'll have got out of her hole; she won't have vegetated,' said Sherringham. 'That makes her touching to me; it adds to the many good reasons for which one may want to help her. She's tackling a big job, and tackling it by herself; throwing herself upon the world, in good faith, and dealing with it as she can; meeting alone, in her youth, her beauty and her generosity all the embarrassments of notoriety and all the difficulties of a profession of which, if one half is what's called brilliant, the other half is odious.'

'She has great courage, but should you speak of her as solitary, with such a lot of us all round her?' Gabriel asked.

'She's a great thing for you and me, but we're a small thing for her.'

'Well, a good many small things may make up a considerable one,' Nash returned. 'There must always be the man; he's the indispensable element in such a life, and he'll be the last thing she'll ever want for.'

'What man are you talking about?' Sherringham asked, rather confusedly.

'The man of the hour, whoever he is. She'll inspire innumerable devotions.'

'Of course she will, and they will be precisely a part of the insufferable side of her life.'

'Insufferable to whom?' Nash inquired. 'Don't forget that the insufferable side of her life will be just the side she'll thrive on. You can't eat your cake and have it, and you can't make omelettes without breaking eggs. You can't at once sit by the fire and fly about the world, and you can't go round and round the globe without having adventures. You can't be a great actress without quivering nerves. If you haven't them you'll only be a small one. If you have them, your friends will be pretty sure to hear of them. Your nerves and your adventures, your eggs and your cake, are part of the cost of the most expensive of professions. If you do your business at all you should do it handsomely, so that the costs may run up tremendously. You play with human passions, with exaltations and ecstasies and terrors, and if you trade on the fury of the elements you must know how to ride the storm.'

'Those are the fine old commonplaces about the artistic temperament, but I usually find the artist a very meek, decent little person,' said Sherringham.

'You never find the artist – you only find his work, and that's all you need find. When the artist's a woman and the woman's an actress, meekness and decency will doubtless be there in the right proportions,' Nash went on. 'Miriam will represent them for you, if you give her her starting-point, with the utmost charm.'

'Of course she'll have devotions – that's all right,' said Sherringham, impatiently.

'And – don't you see? – they'll mitigate her solitude, they'll even enliven it,' Nash remarked.

'She'll probably box a good many ears: that'll be lively,' Peter rejoined, with some grimness.

'Oh, magnificent! it will be a merry life. Yet with its tragic passages, its distracted or its pathetic hours,' Nash continued. 'In short a little of everything.'

The two men walked on without further speech, till at last Sherringham said: 'The best thing for a woman in her situation is to marry some good fellow.'

'Oh, I dare say she'll do that too!' Nash laughed; a remark in consequence of which Peter again lapsed into silence. Gabriel left him to enjoy his silence for some minutes; after which he added: 'There's a good fellow she'd marry tomorrow.'

Peter hesitated. 'Do you mean her friend Dashwood?'

'No, no, I mean Nick Dormer.'

'She'd marry him?' Sherringham asked.

'I mean her head's full of him. But she'll hardly get the chance.'

'Does she like him so much as that?' Sherringham went on.

'I don't know quite how much you mean, but enough for all practical ends.'

'Marrying a fashionable actress – that's hardly a practical end.'

'Certainly not, but I'm not speaking from his point of view. Moreover I thought you just now said it would be such a good thing for her.'

'To marry Nick Dormer?'

'You said a good fellow, and he's the very best.'

'I wasn't thinking of the man, but of the marriage. It would protect her, make things safe and comfortable for her and keep a lot of cads and blackguards away.'

'She ought to marry the prompter or the box-keeper,' said Nash. 'Then it would be all right. I think indeed they generally do, don't they?'

Sherringham felt for a moment a strong disposition to drop his companion on the spot – to cross to the other side of the street and walk away without him. But there was a different impulse which struggled with this one and, after a minute, overcame it – the impulse which led to his saying presently: 'Has she told you that – that she's in love with Nick?'

'No, no – that's not the way I know it.'

'Has Nick told you, then?'

'On the contrary, I've told him.'

'You've rendered him a questionable service if you've no proof,' said Peter.

'My proof is only that I've seen her with him. She's charming, poor thing.'

'But surely she isn't in love with every man she's charming to.'

'I mean she's charming to me,' Nash replied. 'I see her that way. But judge for yourself – the first time you get a chance.'

'When shall I get a chance? Nick doesn't come near her.'

'Oh, he'll come, he'll come; his picture isn't finished.'

'You mean *he'll* be the box-keeper then?'

'My dear fellow, I shall never allow it,' said Gabriel Nash. 'It would be idiotic and quite unnecessary. He's beautifully arranged, in quite a different line. Fancy his taking that sort of job on his hands! Besides, she would never expect it; she's not such a goose. They're very good friends – it will go on that way. She's an excellent sort of woman for him to know; she'll give him lots of ideas of the plastic kind. He would have been up there before this, but he has been absorbed in this delightful squabble with his constituents. That of course is pure amusement; but when once it's well launched he'll get back to business and his business will be a very different matter from Miriam's. Imagine him writing her advertisements, living on her money, adding up her profits, having rows and recriminations with her agent, carrying her shawl, spending his days in her rouge-pot. The right man for that, if she must have one, will turn up. "*Pour le mariage, non.*" Miriam isn't an idiot; she really, for a woman, quite sees things as they are.'

As Sherringham had not crossed the street and left Gabriel planted, he was obliged to brave the torment of this suggestive flow. But descrying in the dusky vista of the Edgware Road a vague and vigilant hansom, he waved his stick with eagerness and with the abrupt declaration that he was tired, must drive the rest of the way. He offered Nash, as he entered the vehicle, no seat, but this coldness was not reflected in the lucidity with which that master of every subject went on to affirm that there was, of course, a danger – the danger that in given circumstances Miriam would leave the stage.

'Leave it you mean for some man?'

'For the man we're talking about.'

'For Nick Dormer?' Peter asked from his place in the cab, his paleness lighted by its lamps.

'If he should make it a condition. But why should he – why should he make *any* conditions. He's not an ass either. You see it would be a bore,' Nash continued while the hansom waited, 'because if she were to do anything of that sort she would make him pay for the sacrifice.'

'Oh yes, she'd make him pay for the sacrifice,' Sherringham repeated.

'And then, when he had paid, she'd go back to her footlights,' Gabriel added, explicatively, from the curbstone, as Sherringham closed the apron of the cab.

'I see – she'd go back – good-night,' Peter replied. '*Please* go on!' he cried to the driver through the hole in the roof. And when the vehicle rolled away he subjoined to himself: 'Of course she would – and quite right!'

37

'JUDGE for yourself when you get a chance,' Nash had said; and as it turned out Sherringham was able to judge two days later, for he found his cousin in Balaklava Place on the Tuesday following his walk with Gabriel. He had not only stayed away from the theatre on the Monday evening (he regarded this as an achievement of some importance), but had not been near Miriam during the day. He had meant to absent himself from her company on Tuesday as well; a determination confirmed by the fact that the afternoon turned out wet. But when, at ten minutes to five o'clock, he jumped into a hansom and directed its course to St John's Wood, it was precisely upon the weather that he shifted the responsibility of his behaviour.

Miriam had dined when he reached the villa, but she was lying down – she was tired – before going to the theatre. Mrs Rooth was however in the drawing-room with three gentlemen, in two of whom the fourth visitor was not startled to recognize Basil Dashwood and Gabriel Nash. Dashwood appeared to have become Miriam's brother-in-arms and a second child – a fonder one – to Mrs Rooth; it had come to Sherringham's knowledge the last time he was in Balaklava Place that the young actor had finally moved his lodgings into the quarter, making himself a near neighbour for all sorts of convenience. 'Hang his convenience!' Peter thought, perceiving that Mrs Lovick's 'Arty' was now altogether one of the family. Oh, the family – it was a queer one to be connected with; that consciousness was acute in Sherringham's breast today as he entered Mrs Rooth's little circle. The room was filled with cigarette-smoke and there was a messy coffee-service on the piano, whose keys Basil Dashwood lightly touched for his own diversion. Nash, addressing the room, of course, was at one end of a little sofa, with his nose in the air, and Nick Dormer was at the other end, seated much at his ease, with a certain privileged appearance of having been there often before, though Sherringham knew he had not. He looked uncritical and very young, as rosy as a school-boy on a half-holiday. It was past five o'clock in the day, but Mrs Rooth was not dressed; there was however no want of finish in her elegant attitude – the same relaxed grandeur (she seemed to let you understand) for which she used to be distinguished at Castle Nugent when the house was full. She toyed incongruously, in her unbuttoned wrapper, with a large tinsel fan which resembled a theatrical property.

It was one of the discomforts of Sherringham's situation that many of those minor matters which are, superficially at least, most characteristic of the histrionic life had power to displease him, so that he was obliged to make the effort of indulgence. He disliked besmoked drawing-rooms and irregular meals and untidy arrangements; he could suffer from the vulgarity of Mrs Rooth's apartments, the importunate photographs (they gave on his nerves), the barbarous absence of signs of an orderly domestic life, the odd volumes from the circulating library (you could see what they were – the very covers told you – at a glance) tumbled about with cups or glasses on them. He had not waited till now to make the reflection that it was a strange thing fate should have goaded *him* into that sort of contact; but as he stood before Mrs Rooth and her companions he made it perhaps more pointedly than ever. Her companions, somehow, who were not responsible, didn't keep him from making it; which was particularly odd, as they were not, superficially, in the least of Bohemian type. Almost the first thing that struck him, as it happened, in coming into the room, was the essential good looks of his cousin, who was a gentleman to the eye in a different degree from the high-collared Dashwood. Peter didn't hate him for being such a pleasant young Englishman; his consciousness was traversed rather by a fresh wave of annoyance at Julia's failure to get on with him on that substantial basis.

It was Sherringham's first encounter with Nick since his arrival in London: they had been, on one side and the other, so much taken up with their own affairs. Since their last meeting Nick had, as we know, to his kinsman's perception, really taken on a new character: he had done a fine stroke of business in a quiet way. This made him a figure to be counted with, and in just the sense in which Peter desired least to count with him. Poor Sherringham, after his summersault in the blue, was much troubled these last days; he was ravaged by contending passions; he paid, every hour, in a torment of unrest, for what was false in his position, the impossibility of being consistent, the opposition of interest and desire. Nick, his junior and a lighter weight, had settled *his* problem and showed no wounds: there was something impertinent and mystifying in it. He looked too innocently young and happy there, and too careless and modest and amateurish for a rival or for the genius that he was apparently going to try to be – the genius that, the other day in the studio with Biddy, Peter had got a startled glimpse of his capacity for being. Sherringham would have liked to feel that he had grounds of resentment, that Julia had been badly treated or that Nick was fatuous, for in that case he might have regarded him as offensive. But where was the offence of his merely being liked by a woman in respect to whom Peter had definitely denied himself the luxury of pretensions, especially if the offender had taken no action in the matter? It could scarcely be called culpable action to call, casually, on an afternoon when the lady was invisible. Peter, at any rate, was distinctly glad Miriam was invisible; and he proposed to himself to suggest to Nick after a little that they should adjourn together – they had such interesting things to talk about. Meanwhile Nick greeted him with candid tones and pleasant eyes, in which he could read neither confusion nor defiance. Sherringham was reassured against a danger he believed he didn't recognize and puzzled by a mystery he flattered himself he didn't mind. And he was still more ashamed of being reassured than of being puzzled.

It must be recorded that Miriam remained invisible only a few minutes longer. Nick, as Sherringham gathered, had been about a quarter of an hour in the house, which would have given the girl, aroused from her repose, about time to array herself to come down to him. At all events she was in the room, prepared apparently to go to the theatre, very shortly after Sherringham had become sensible of how glad he was she was out of it. Familiarity had never yet cured him of a certain tremor of expectation and even of suspense in regard to her entrances; a flutter caused by the simple circumstance of her infinite variety. To say she was always acting suggests too much that she was often fatiguing; for her changing face affected this particular admirer at least not as a series of masks, but as a response to perceived differences, an intensity of sensibility, or still more as something cleverly constructive, like the shifting of the scene in a play or a room with many windows. Her incarnations were incalculable, but if her present denied her past and declined responsibility for her future, it made a good thing of the hour and kept the actual very actual. This time the actual was a bright, gentle, graceful, smiling young woman in a new dress, eager to go out, drawing on fresh gloves, who looked as if she were about to step into a carriage and (it was Gabriel Nash who thus formulated her physiognomy) do a lot of London things.

The young woman had time to spare however, and she sat down and talked and laughed and presently gave, as it seemed to Sherringham, a finer character to the tawdry little room. It was honourable enough if it belonged to her. She described herself as in a state of nervous bewilderment – exhausted, stupefied, blinded with the rehearsals of the forthcoming piece (the first night was close at hand and it was going to be *d'un mauvais* – they would all see!), but there was no correspondence between this account of the matter and her present kindly gaiety. She sent her mother away – to 'put on some clothes or something' – and, left alone with the visitors, went to a long glass between the windows, talking always to Nick Dormer, and revised and rearranged a little her own attire. She talked to Nick over her shoulder, and to Nick only, as if he were the guest to recognize and the others didn't count. She broke out immediately about his having thrown up his seat, wished to know if the strange story told her by Mr Nash were true – that he had knocked all the hopes of his party into pie.

Nick took it in this way and gave a jocular picture of his party's ruin, the critical condition of public affairs: evidently as yet he remained inaccessible to shame or repentance. Sherringham, before Miriam's entrance, had not, in shaking hands with Nick, made even a roundabout allusion to his odd 'game': there seemed a sort of muddled good taste in being silent about it. He winced a little on seeing how his scruples had been wasted, and was struck with the fine, jocose, direct turn of his kinsman's conversation with the young actress. It was a part of her unexpectedness that she took the heavy literal view of Nick's behaviour; declared frankly, though without ill-nature, that she had no patience with his folly. She was horribly disappointed – she had set her heart on his being a great statesman, one of the rulers of the people and the glories of England. What was so useful, what was so noble? – how it belittled everything else! She had expected him to wear a cordon and a star some day (and to get them very soon), and to come and see her in her *loge*: it would look so well. She talked like a lovely Philistine, except perhaps when she expressed surprise at hearing – she heard it from Gabriel Nash – that in England gentlemen accoutred with those emblems of their sovereign's esteem didn't so far forget themselves as to stray into the dressing-rooms of actresses. She admitted, after a moment, that they were quite right – the dressing-rooms of actresses were nasty places; but she was sorry, for that was the sort of thing she had always figured, in a corner – a distinguished man, slightly bald, in evening dress, with orders, admiring the smallness of a satin shoe and saying witty things. Gabriel Nash was convulsed with hilarity at this – such a vision of the British political hero. Coming back from the glass and making him give her his place on the sofa, she seated herself near Nick and continued to express her regret at his perversity.

'They all say that – all the charming women, but I shouldn't have looked for it from you,' Nick replied. 'I've given you such an example of what I can do in another line.'

'Do you mean my portrait? Oh, I've got it, with your name and "M.P." in the corner, and that's precisely why I'm content. "M.P." in the corner of a picture is delightful, but I want to break the mould: I don't in the least insist on your giving specimens to others. And the artistic life, when you can lead another – if you have any alternative, however, modest – is a very poor business. It comes last in dignity – after everything else. Ain't I up to my eyes in it and don't I know?'

'You talk like my broken-hearted mother,' said Nick.

'Does she hate it so intensely?'

'She has the darkest ideas about it – the wildest theories. I can't imagine where she gets them; partly, I think, from a general conviction that the "aesthetic" – a horrible insidious foreign disease – is eating the healthy core out of English life (dear old English life!) and partly from the charming drawings in *Punch* and the clever satirical articles, pointing at mysterious depths of contamination, in the other weekly papers. She believes there's a dreadful coterie of uncannily clever and desperately refined people, who wear a kind of loose, faded uniform and worship only beauty – which is a fearful thing – that Nash has introduced me to it, that I now spend all my time in it, and that for its sweet sake I have repudiated the most sacred engagements. Poor Nash, who, so far as I can make out, isn't in any sort of society, however bad!'

'But I'm uncannily clever,' Nash interposed, 'and though I

can't afford the uniform (I believe you get it best somewhere in South Audley Street), I do worship beauty. I really think it's me the weekly paper means.'

'Oh, I've read the articles – I know the sort!' said Basil Dashwood.

Miriam looked at him. 'Go and see if the brougham's there – I ordered it early.'

Dashwood, without moving, consulted his watch. 'It isn't time yet – I know more about the brougham than you. I've made a rattling good arrangement for her – it really costs her nothing,' the young actor continued confidentially to Sherringham, near whom he had placed himself.

'Your mother's quite right to be broken-hearted,' Miriam declared, 'and I can imagine exactly what she has been through. I should like to talk with her – I should like to see her.' Nick broke into ringing laughter, reminding her that she had talked to him, while she sat for her portrait, in directly the opposite sense, most suggestively and inspiringly; and Nash explained that she was studying the part of a political duchess and wished to take observations for it, to work herself into the character. Miriam might in fact have been a political duchess as she sat with her head erect and her gloved hands folded, smiling with aristocratic dimness at Nick. She shook her head with stately sadness; she might have been representing Mary Stuart in Schiller's play. 'I've changed since that. I want you to be the grandest thing there is – the counsellor of kings.'

Peter Sherringham wondered if possibly it were not since she had met his sister in Nick's studio that she had changed, if perhaps it had not occurred to her that it would give Julia the sense of being more effectually routed to know that the woman who had thrown the bomb was one who also tried to keep Nick in the straight path. This indeed would involve an assumption that Julia might know, whereas it was perfectly possible that she mightn't and more than possible that if she should she wouldn't care. Miriam's essential fondness for trying different ways was always there as an adequate reason for any particular way; a truth which however sometimes only half prevented the particular way from being vexatious to Sherringham.

'Yet after all who is more aesthetic than you and who goes in more for the beautiful?' Nick asked. 'You're never so beautiful as when you pitch into it.'

'Oh, I'm an inferior creature, of an inferior sex, and I have to earn my bread as I can. I'd give it all up in a moment, my odious trade – for an inducement.'

'And pray what do you mean by an inducement?' Nick demanded.

'My dear fellow, she means you – if you'll give her a permanent engagement to sit for you!' exclaimed Gabriel Nash. 'What crude questions you ask!'

'I like the way she talks,' Basil Dashwood broke in, 'when I gave up the most brilliant prospects, of very much the same kind as Mr Dormer's, expressly to go on the stage.'

'You're an inferior creature too,' said Miriam.

'Miss Rooth is very hard to satisfy,' Sherringham observed. 'A man of distinction, slightly bald, in evening dress, with orders, in the corner of her *loge* – she has such a personage ready made to her hand and she doesn't so much as look at him. Am I not an inducement? Have I not offered you a permanent engagement?'

'Your orders – where are your orders?' Miriam inquired with a sweet smile, getting up.

'I shall be a minister next year and an ambassador before you know it. Then I shall stick on everything that can be had.'

'And they call *us* mountebanks!' cried the girl. 'I've been so glad to see you again – do you want another sitting?' she went on, to Nick, as if to take leave of him.

'As many as you'll give me – I shall be grateful for all,' Nick answered. 'I should like to do you as you are at present. You're totally different from the woman I painted – you're wonderful.'

'The Comic Muse!' laughed Miriam. 'Well, you must wait till our first nights are over – I'm *sur les dents* till then. There's everything to do, and I have to do it all. That fellow's good for nothing – for nothing but domestic life,' and she glanced at Basil Dashwood. 'He hasn't an idea – not one that you'd willingly tell of him, though he's rather useful for the stables. We've got stables now – or we try to look as if we had: Dashwood's ideas are *de cette force*. In ten days I shall have more time.'

'The Comic Muse? Never, never,' Sherringham protested.

'You're not to go smirking through the age and down to posterity – I'd rather see you as Medusa crowned with serpents. That's what you look like when you look best.'

'That's consoling – when I've just bought a new bonnet! I forgot to tell you just now that when you're an ambassador you may propose anything you like,' Miriam went on. 'But excuse me if I make that condition. Seriously speaking, come to me glittering with orders and I shall probably succumb. I can't resist stars and garters. Only you must, as you say, have them all. I *don't* like to hear Mr Dormer talk the slang of the studio – like that phrase just now: it *is* a fall to a lower state. However, when one is low one must crawl, and I'm crawling down to the Strand. Dashwood, see if mamma's ready. If she isn't I decline to wait; you must bring her in a hansom. I'll take Mr Dormer in the brougham; I want to talk with Mr Dormer; he must drive with me to the theatre. His situation is full of interest.' Miriam led the way out of the room as she continued to chatter, and when she reached the house-door, with the four men in her train, the carriage had just drawn up at the garden-gate. It appeared that Mrs Rooth was not ready, and the girl, in spite of a remonstrance from Nick, who had the sense of usurping the old lady's place, repeated her injunction that she should be brought on in a cab. Miriam's companions accompanied her to the gate, and she insisted upon Nick's taking his seat in the brougham and taking it first. Before she entered she put out her hand to Sherringham and, looking up at him, held his own kindly. 'Dear old master, aren't you coming tonight? I miss you when you're not there.'

'Don't go – don't go – it's too much,' Nash interposed.

'She *is* wonderful,' said Basil Dashwood, regarding her admiringly! 'she *has* gone into the rehearsals, tooth and nail. But nothing takes it out of her.'

'Nothing puts it into you, my dear!' Miriam returned. Then she went on, to Sherringham: 'You're the faithful one – you're the one I count on.' He was not looking at her; his eyes travelled into the carriage, where they rested on Nick Dormer, established on the further seat with his face turned toward the further window. He was the one, faithful or no, counted on or no, whom a charming woman had preferred to carry off, and there was a certain triumph for him in that fact; but it pleased Sherringham to imagine that his attitude was a little foolish. Miriam discovered something of this sort in Sherringham's eyes; for she exclaimed abruptly: 'Don't kill him – he doesn't care for me!' With this she passed into the carriage, which rolled away.

Sherringham stood watching it a moment, till he heard Basil Dashwood again beside him. 'You wouldn't believe what I made him do it for – a little fellow I know.'

'Good-bye; take good care of Mrs Rooth,' said Gabriel Nash, waving a cheerful farewell to the young actor. He gave a smiling survey of the heavens and remarked to Sherringham that the rain had stopped. Was he walking, was he driving, should they be going in the same direction? Sherringham cared little about his direction and had little account of it to give; he simply moved away in silence, with Gabriel at his side. Gabriel was partly an affliction to him; indeed the fact that he had assumed a baleful fascination made him only a deeper affliction. Sherringham moreover did him the justice to observe that he could hold his peace occasionally: he had for instance this afternoon taken little part in the conversation in Balaklava Place. Peter greatly disliked to talk to him of Miriam, but he liked Nash to talk of her and he even liked him to say such things as he might contradict. He was not however moved to contradict an assertion dropped by his companion, disconnectedly, at the end of a few minutes, to the effect that she was after all the most goodnatured creature alive. All the same, Nash added, it wouldn't do for her to take possession of an organization like Nick's; and he repeated that for his part he would never allow it. It would be on his conscience to interfere. To which Sherringham replied disingenuously that they might all do as they liked – it didn't matter a button to *him*. And with an effort to carry off that comedy he changed the subject.

38

PETER SHERRINGHAM would not for a moment have admitted that he was jealous of Nick Dormer, but he would almost have liked to be accused of it; for this would have given him an opportunity to declare with plausibility that so uncomfortable a passion had no application to his case. How could a man be jealous when he was not a suitor? how could he pretend to guard a property which was neither his own nor destined to become his own? There could be no question of loss when one had nothing at stake and no question of envy when the responsibility of possession was exactly what one prayed to be delivered from. The measure of one's susceptibility was one's pretensions, and Peter was not only ready to declare over and over again that, thank God, he had none: his spiritual detachment was still more complete – he literally suffered from the fact that the declaration was but little elicited. He connected an idea of virtue and honour with his attitude; for surely it was a high example of conduct to have quenched a personal passion for the sake of the public service. He had gone over the whole question at odd, irrepressible hours; he had returned, spiritually speaking, the buffet administered to him in a moment, that day in Rosedale Road, by the spectacle of the *crânerie* with which Nick could let worldly glories slide. Resolution for resolution he preferred after all another sort, and his own *crânerie* would be shown in the way he should stick to his profession and stand up for British interests. If Nick had leaped over a wall he would leap over a river. The course of his river was already traced and his loins were already girded. Thus he was justified in holding that the measure of a man's susceptibility was a man's attitude: that was the only thing he was bound to give an account of.

He was perpetually giving an account of it to his own soul, in default of other listeners. He was quite angry at having tasted a sweetness in Miriam's assurance, at the carriage door, bestowed indeed with very little solemnity, that Nick didn't care for her. Wherein did it concern him that Nick cared for her or that Nick didn't? Wherein did it signify to him that Gabriel Nash should have taken upon himself to disapprove of a union between the young actress and the young painter and to frustrate an accident that might perhaps be happy? For those had also been cooling words at the hour, though Peter blushed on the morrow to think that he had perceived in them anything but Nash's personal sublimity. He was ashamed of having been refreshed, and refreshed by so sickly a draught, because it was his theory that he was not in a fever. As for keeping an eye on Nick, it would soon become clear to that young man and that young man's charming friend that he had quite other uses for his sharpness. Nick and Miriam and Gabriel Nash could straighten out their complications according to their light. He would never speak to Nick of Miriam; he felt indeed just now as if he should never speak to Nick of anything. He had traced the course of his river, as I say, and the real proof would be the way he should fly through the air. It was a cause for action – for vigorous, unmistakable action. He had done very little since his arrival in London but moon round a *fille de théâtre* who was taken up partly, though she bluffed it off, with another man and partly with arranging new petticoats for a beastly old 'poetic drama'; but this little waste of time should instantly be made up. He had given himself a certain rope and he had danced to the end of his rope, and now he would dance back. That was all right – so right that Sherringham could only express to himself how right it was by whistling gaily.

He whistled as he went to dine with a great personage, the day after his meeting with Nick in Balaklava Place; a great personage to whom he had originally paid his respects – it was high time – the day before that meeting, the Monday previous. The sense of omissions to repair, of a superior line to take, perhaps made him study with more intensity to please the personage, who gave him ten minutes and asked him five questions. A great many doors were successively opened before any palpitating pilgrim who was about to enter the presence of this distinguished man; but they were discreetly closed again behind Sherringham, and I must ask the reader to pause with me at the nearer end of the momentary vista. This particular pilgrim fortunately felt that he could count upon being recognized not only as a faithful if obscure official in the great hierarchy, but as a clever young man who happened to be connected by blood with people his lordship had intimately known. No doubt it was simply as the clever young man that Peter received the next morning from the dispenser of his lordship's hospitality a note asking him to dine on the morrow. He had received such cards before and he always responded to the invitation: he did so however on the present occasion with a sense of unusual intention. In due course his intention was translated into words: before the gentlemen left the dining-room he took the liberty of asking his noble host if during the next few days there would be three minutes more that he might, in his extreme benevolence, bestow upon him.

'What is it you want? Tell me now,' the master of his fate replied, motioning to the rest of the company to pass out and detaining Peter in the dining-room.

Peter's excellent training covered every contingency: he could be concise or diffuse, as the occasion required. Even he himself however was surprised at the quick felicity of the terms in which he was conscious of conveying that if it were compatible with higher conveniences he should peculiarly like to be transferred to duties in a more distant quarter of the globe. Indeed though Sherringham was fond of thinking of himself as a man of emotions controlled by training, it is not impossible that there was a greater candour than he knew in the expression of his face and even the slight tremor of his voice as he presented this petition. He had wished extremely that his manner should be good in doing so, but perhaps the best part of it for his interlocutor was just the part in which it failed – in which it confessed a secret that the highest diplomacy would not have confessed. Sherringham remarked to the minister that he didn't care in the least where the place might be, nor how little coveted a post; the further away the better and the climate didn't matter. He would only prefer of course that there should be really something to do, although he would make the best of it even if there were not. He stopped in time, or at least he thought he did, not to appear to suggest that he covertly sought relief from the misery of a hindered passion in a flight to latitudes unfavourable to human life. His august patron gave him a sharp look which, for the moment, seemed the precursor of a sharper question; but the moment elapsed and the question did not come. This considerate omission, characteristic of a true man of the world and representing quick guesses and still quicker indifferences, made Sherringham from that moment his lordship's ardent partisan. What did come was a good-natured laugh and the exclamation: 'You know there are plenty of swamps and jungles, if you want that sort of thing.' Sherringham replied that it was very much that sort of thing he did want; whereupon his interlocutor continued: 'I'll see – I'll see; if anything turns up, you shall hear.'

Something turned up the very next day: our young man, taken at his word, found himself indebted to the post for a large, stiff, engraved official letter, in which the high position of minister to the smallest of Central American republics was offered to him. The republic, though small, was big enough to be 'shaky', and the position, though high, was not so exalted that there were not much greater altitudes above it to which it was a stepping-stone. Sherringham took one thing with another, rejoiced at his easy triumph, reflected that he must have been even more noticed at headquarters than he had hoped, and, on the spot, consulting nobody and waiting for nothing, signified his unqualified acceptance of the place. Nobody with a grain of sense would have advised him to do anything else. It made him happier than he had supposed he should ever be again; it made him feel professionally in the train, as they said in Paris; it was serious, it was interesting, it was exciting, and Sherringham's imagination, letting itself loose into the future, began once more to scale the crowning heights. It was very simple to hold one's course if one really tried, and he blessed shaky republics. A further communication informed him that he would be expected to return to Paris for a short interval a week later and that he would before that time be advised of the date at which he was to proceed to his remoter duties.

39

THE first thing Peter did now was to go and see Lady Agnes Dormer; it is not unworthy of note that he took on the other hand no step to make his promotion known to Miriam Rooth. To render it more probable he should find her he went at the luncheon-hour; and she was indeed on the point of sitting down to that repast with Grace. Biddy was not at home – Biddy was never at home now, her mother said: she was always at Nick's place, she spent her life there, she ate and drank there, she almost slept there. What she found to do there in so many hours, or what was the irresistible spell, Lady Agnes could not pretend that she had succeeded in discovering. She spoke of this baleful resort only as 'Nick's place', and she spoke of it at first as little as possible. She thought it very probable however that Biddy would come in early that afternoon: there was something or other, some common social duty, that she had condescended to promise she would perform with Grace. Poor Lady Agnes, whom Sherringham found in a very grim yet very tremulous condition (she assured her visitor her nerves were all gone), almost abused her younger daughter for two minutes, having evidently a deep-seated need of abusing some one. I must add however that she didn't wait to meet Grace's eye before recovering, by a rapid gyration, her view of the possibilities of things – those possibilities from which she still might squeeze, as a mother, the drop that would sweeten her cup. 'Dear child,' she had the presence of mind to add, 'her only fault is after all that she adores her brother. She has a capacity for adoration and must always take her gospel from some one.'

Grace declared to Peter that her sister would have stayed at home if she had dreamed he was coming, and Lady Agnes let him know that she had heard all about the hour he had spent with the poor child at Nick's place and about his extraordinary good-nature in taking the two girls to the play. Peter lunched in Calcutta Gardens, spending an hour there which proved at first unexpectedly and, as it seemed to him, unfairly dismal. He knew from his own general perceptions, from what Biddy had told him and from what he had heard Nick say in Balaklava Place, that Lady Agnes would have been wounded by her son's apostasy; but it was not till he saw her that he appreciated the dark difference this young man's behaviour had made in the outlook of his family. Evidently that behaviour had, as he phrased it, pulled the bottom out of innumerable private calculations. These were things that no outsider could measure and they were none of an outsider's business; it was enough that Lady Agnes struck him really as a woman who had received her death-blow. She looked ten years older; she was white and haggard and tragic. Her eyes burned with a strange intermittent fire which made him say to himself that her children had better look out for her. When they were not filled with this unnatural flame they were suffused with comfortless tears; and altogether the afflicted lady was very bad – very bad indeed. It was because he had known she would be very bad that he had in his kindness called upon her in exactly this manner; but he recognized that to undertake to be kind to her in proportion to her need might carry one very far. He was glad he himself had not a wronged, mad mother, and he wondered how Nick Dormer could endure the home he had ruined. Apparently he didn't endure it very much, but had taken definitive and highly convenient refuge in Rosedale Road.

Peter's judgement of his young kinsman was considerably confused, and a sensible element in it was the consciousness that he was perhaps just now not in the best state of mind for judging him at all. At the same time, though he held in general that an intelligent man has a legible warrant for doing the particular thing he prefers, he could scarcely help asking himself whether in the exercise of a virile freedom it had been absolutely indispensable that Nick should work such domestic woe. He admitted indeed that this was an anomalous vision of Nick, as the worker of domestic woe. Then he saw that Lady Agnes's grievance (there came a moment, later, when she asserted as much) was not quite what Nick, in Balaklava Place, had represented it – with questionable taste perhaps – to a mocking actress; was not a mere shocked quarrel with his adoption of a 'low' career, or a horror, the old-fashioned horror, of the strange licences taken by artists under pretext of being conscientious: the day for this was past and English society thought the brush and the fiddle as good as anything else, with two or three exceptions. It was not what he had taken up but what he had put down that made the sorry difference, and the tragedy would have been equally great if he had become a wine-merchant or a horse-dealer. Peter had gathered at first that Lady Agnes would not trust herself to speak directly of her trouble, and he obeyed what he supposed to be the best discretion in making no allusion to it. But a few minutes before they rose from luncheon she broke out, and when he attempted to utter a word of mitigation there was something that went to his heart in the way she returned: 'Oh, you don't know – you don't know!'

He perceived Grace's eyes fixed upon him at this instant with a look of supplication, and he was uncertain as to what she wanted – that he should say something more to console her mother or should hurry away from the subject. Grace looked old and plain and (he had thought, on coming in) rather cross, but she evidently wanted something. 'You don't know,' Lady Agnes repeated, with a trembling voice – 'you don't know.' She had pushed her chair a little away from the table; she held her pocket-handkerchief pressed hard to her mouth, almost stuffed into it, and her eyes were fixed on the floor. She made him feel as if he did know – knew what towering piles of confidence and hope had been dashed to the earth: Then Lady Agnes finished her sentence unexpectedly: 'You don't know what my life with my husband was.' Here, on the other hand, Peter was slightly at fault – he didn't exactly see what her life with her husband had to do with it. What was clear to him however was that they literally had looked for the very greatest things from Nick. It was not quite easy to see why this had been the case – it had not been precisely Sherringham's own prefigurement. Nick appeared to have had the faculty of communicating that sort of faith to women; he had originally given Julia a tremendous dose of it, though she had since shaken off the effects.

'Do you really think he would have done such great things, politically speaking?' Peter inquired. 'Do you consider that the root of the matter was in him?'

Lady Agnes hesitated a moment, looked rather hard at her visitor. 'I only think what all his friends – all his father's friends – have thought. He was his father's son, after all. No young man ever had a finer training, and he gave, from the first, repeated proof of having the highest sort of ability, the highest sort of ambition. See how he got in everywhere. Look at his first seat – look at his second,' Lady Agnes continued. 'Look at what everyone says at this moment.'

'Look at all the papers!' said Grace. 'Did you ever hear him speak?' she asked. And when Peter reminded her that he had spent his life in foreign lands she went on: 'Well, you lost something.'

'It was very charming,' said Lady Agnes quietly.

'Of course he's charming, whatever he does,' Peter conceded. 'He'll be a charming artist.'

'Oh, heaven!' groaned Lady Agnes, rising quickly.

'He won't – that's the worst,' Grace amended. 'It isn't as if he'd do things people would like. I've been to his place and I never saw such a horrid lot of things – not at all clever or pretty.'

'You know nothing whatever about the matter!' Lady Agnes exclaimed, with unexpected asperity. Then she added, to Peter, that, as it happened, her children did have a good deal of artistic taste: Grace was the only one who was totally deficient in it. Biddy was very clever – Biddy really might learn to do pretty things. And anything the poor child could learn was now no more than her duty – there was so little knowing what the future had in store for them all.

'You think too much of the future – you take terribly gloomy views,' said Peter, looking for his hat.

'What other views can one take, when one's son has deliberately thrown away a fortune?'

'Thrown one away? Do you mean through not marrying –?'

'I mean through killing by his perversity the best friend he ever had.'

Sherringham stared a moment; then with laughter: 'Ah, but Julia isn't dead of it?'

'I'm not talking of Julia,' said Lady Agnes, with a good deal of majesty. 'Nick isn't mercenary, and I'm not complaining of that.'

'She means Mr Carteret,' Grace explained. 'He would have done anything if Nick had stayed in the House.'

'But he's not dead?'

'Charles Carteret is dying,' said Lady Agnes – 'his end is very, very near. He has been a sort of providence to us – he was Sir Nicholas's second self. But he won't stand such nonsense, and that chapter's closed.'

'You mean he has dropped Nick out of his will?'

'Cut him off utterly. He has given him notice.'

'The old scoundrel! But Nick will work the better for that – he'll depend on himself.'

'Yes, and whom shall *we* depend on?' Grace demanded.

'Don't be vulgar, for God's sake!' her mother ejaculated with a certain inconsequence.

'Oh, leave Nick alone – he'll make a lot of money,' Peter declared cheerfully, following his two companions into the hall.

'I don't in the least care whether he does or not,' said Lady Agnes. 'You must come upstairs again – I've lots to say to you yet,' she went on, seeing that Peter had taken his hat. 'You must arrange to come and dine with us immediately; it's only because I've been so steeped in misery that I didn't write to you the other day – directly after you called. We don't give parties, as you may imagine, but if you'll come just as we are, for old acquaintance' sake –'

'Just with Nick – if Nick will come – and dear Biddy,' Grace interposed.

'Nick must certainly come, as well as dear Biddy, whom I hoped so much to find,' Peter rejoined. 'Because I'm going away – I don't know when I shall see them again.'

'Wait with mamma. Biddy will come in at any moment,' Grace urged.

'You're going away?' asked Lady Agnes, pausing at the foot of the stairs and turning her white face upon him. Something in the tone of her voice showed that she had been struck by his tone.

'I have had promotion, and you must congratulate me. They are sending me out as minister to a little hot hole in Central America – five thousand miles away. I shall have to go rather soon.'

'Oh, I'm so glad!' Lady Agnes breathed. Still she paused at the foot of the stair and still she gazed.

'How very delightful, because it will lead, straight off, to all sorts of other good things!' Grace exclaimed.

'Oh, I'm crawling up, and I'm an excellency,' Peter laughed.

'Then if you dine with us your excellency must have great people to meet you.'

'Nick and Biddy – they're great enough.'

'Come upstairs – come upstairs,' said Lady Agnes, turning quickly and beginning to ascend.

'Wait for Biddy – I'm going out,' Grace continued, extending her hand to her kinsman. 'I shall see you again – not that you care; but good-bye now. Wait for Biddy,' the girl repeated in a lower tone, fastening her eyes on his with the same urgent, mystifying gleam that he thought he had perceived in them at luncheon.

'Oh, I'll go and see her in Rosedale Road,' he answered.

'Do you mean today – now?'

'I don't know about today, but before I leave England.'

'Well, she'll be in immediately,' said Grace. 'Good-bye to your excellency.'

'Come up, Peter – *please* come up,' called Lady Agnes, from the top of the stairs.

He mounted, and when he found himself in the drawing-room with her, with the door closed, she told him that she was exceedingly interested in his fine prospects, that she wished to hear all about his new position. She rang for coffee and indicated the seat he would find most comfortable: he had for a moment an apprehension that she would tell him he might if he liked light a cigar. For Peter Sherringham had suddenly become restless – too restless to occupy a comfortable chair; he seated himself in it only to jump up again, and he went to the window – while he communicated to his hostess the very little that he knew about his prospective post – on hearing a vehicle drive up to the door. A strong light had just been thrown into his mind, and it seemed to grow stronger when, looking out of the window, he saw Grace Dormer issue from the house in a bonnet and jacket which had all the air of having been assumed with extraordinary speed. Her jacket was unbuttoned, her gloves were dangling from her hand and she was tying her bonnet-strings. The vehicle into which she hastily sprang was a hansom-cab which had been summoned by the butler from the doorstep and which rolled away with her after she had given the cabman an address.

'Where is Grace going in such a hurry?' he asked of Lady Agnes; to which she replied that she had not the least idea – her children, at the pass they had all come to, knocked about as they liked.

Peter sat down again; he stayed a quarter of an hour and then he stayed longer, and during this time his appreciation of what Lady Agnes had in her mind gathered force. She showed him clearly enough what she had in her mind, although she showed it by no clumsy nor reprehensible overtures. It looked out of her sombre, conscious eyes and quavered in her preoccupied, perfunctory tones. She manifested an extravagant interest in his future proceedings, the probable succession of events in his career, the different honours he would be likely to come in for, the salary attached to his actual appointment, the salary attached to the appointments that would follow – they would be sure to, wouldn't they? – and what he might reasonably expect to save. Oh, he must save – Lady Agnes was an advocate of saving; and he must take tremendous pains and get on and be clever and ambitious: he must make himself indispensable and rise to the top. She was urgent and suggestive and sympathetic; she threw herself into the vision of his achievements and emoluments as if to satisfy a little the sore hunger with which Nick's treachery had left her. This was touching to Peter Sherringham, and he did not remain unmoved even at those more importunate moments when, as she fell into silence, fidgeting feverishly with a morsel of fancy-work that she had plucked from a table, her whole presence became an intense repressed appeal to him. What that appeal would have been had it been uttered was: 'Oh, Peter, take little Biddy; oh, my dear young friend, understand your interests at the same time that you understand mine; be kind and reasonable and clever; save me all further anxiety and tribulation and accept my lovely, faultless child from my hands.'

That was what Lady Agnes had always meant, more or less, that was what Grace had meant, and they meant it with singular lucidity on the present occasion. Lady Agnes meant it so much that from one moment to another Peter scarcely knew what she might do; and Grace meant it so much that she had rushed away in a hansom to fetch her sister from the studio. Grace, however, was a fool, for Biddy certainly wouldn't come. The news of his promotion had set them off, adding brightness to their idea of his being an excellent match; bringing home to them sharply the sense that if he were going away to strange countries he must take Biddy with him – that something at all events must be settled about Biddy before he went. They had suddenly begun to throb with the conviction that they had no time to lose.

Strangely enough, the perception of all this had not the effect of throwing Peter on the defensive, or at least of making him wish to bolt. When once he had discovered what was in the air he recognized a propriety, a real felicity in it; could not deny that he was in certain ways a good match, since it was quite probable he would go far; and was even generous enough (as he had no fear of being dragged to the altar) to enter into the conception that he might offer some balm to a mother who had had a horrid disappointment. The feasibility of marrying Biddy was not exactly augmented by the idea that his doing so would be a great offset to what Nick had made Lady Agnes suffer; but at any rate Peter did not dislike his strenuous companion so much as to wish to punish her for being strenuous. He was not afraid of her, whatever she might do; and though he was unable to grasp the practical relevancy of Biddy's being produced on the instant he was willing to linger for half an hour on the chance of her turning up.

There was a certain contagion in Lady Agnes's appeal – it made him appeal sensibly to himself. For indeed, as it is time to say, the glass of our young man's spirit had been polished for that reflection. It was only at this moment that he became really candid with himself. When he made up his mind that his only safety was in flight and took the strong measure of asking for assistance to flee, he was very conscious that another and probably still more effectual safeguard (especially if the two should be conjoined) lay in the hollow of his hand. Julia Dallow's words in Paris had come back to him and had seemed much wiser than when they were spoken: 'She'll save you disappointments; you would know the worst that can happen to you, and it wouldn't be bad.' Julia had put it into a nutshell – Biddy would probably save him disappointments. And then she was – well, she was Biddy. Peter knew better what that was since the hour he had spent with her in Rosedale Road. But he had brushed away the sense of it, though he was aware that in doing so he took only half measures, was even guilty of a sort of fraud upon himself. If he was sincere in wishing to put a gulf between his future and that portion of his past and present which was associated with Miriam Rooth, there was a very simple way to do so. He had dodged that way, dishonestly fixing upon another which, taken alone, was far from being so good; but Lady Agnes brought him back to it. She held him in magnanimous contemplation of it, during which the safety, as Julia had called it, of the remedy became fascinating to his mind, especially as that safety appeared not to exclude a concomitant sweetness. It would be simple and it would swallow up his problems; it would put an end to all alternatives, which, as alternatives were otherwise putting an end to him, would be an excellent thing. It would settle the whole question of his future, and it was high time this should be settled.

Peter took two cups of coffee while he made out his future with Lady Agnes, but though he drank them slowly he had finished them before Biddy turned up. He stayed three-quarters of an hour, saying to himself that she wouldn't come – why should she come? Lady Agnes said nothing about this; she really, in vulgar vocables, said nothing about any part of the business. But she made him fix the next day but one for coming to dinner, and her repeated declaration that there would be no one else, not another creature but themselves, had almost the force of a legal paper. In giving his word that he would come without fail and not write the next day to throw them over for some function that he should choose to dub obligatory, Peter felt quite as if he were putting his name to such a document. He went away at half-past three; Biddy of course hadn't come, and he had been certain she wouldn't. He couldn't imagine what Grace's idea had been, nor what pretext she had put forward to her sister. Whatever it had been, Biddy had seen through it and hated such machinations. Peter could only like her the better for that.

40

LADY AGNES would doubtless have done better, in her own interest or that of her child, to have made sure of Peter's company for the very next evening. This she had indeed attempted, but the application of the idea had failed. Peter had a theory that he was inextricably engaged; moreover her ladyship could not take upon herself to answer for Nick. Of course they must have Nick, though, to tell the truth, the hideous truth, she and her son were scarcely upon terms. Peter insisted on Nick, he wished particularly to see him and he gave his hostess notice that he would make each of them forgive everything to the other. Lady Agnes declared that all her son had to forgive was her loving him more than her life, and she would have challenged Peter, had he allowed it, on the general ground of the comparative dignity of the two arts of painting portraits and governing nations. Peter declined the challenge; the most he did was to intimate that he perhaps saw Nick more vividly as a painter than as a governor. Later he remembered vaguely something Lady Agnes had said about their being a governing family.

He was going, by what he could ascertain, to a very queer climate, and he had many preparations to make. He gave his best attention to these, and for a couple of hours after leaving Lady Agnes he rummaged London for books from which he might extract information about his new habitat. It made apparently no great figure in literature, so that Peter could reflect that he was perhaps destined to find a salutary distraction in filling the void with a volume of impressions. After he had gathered that there were no books he went into the Park. He treated himself to an afternoon or two there when he happened to drop upon London in the summer: it refreshed his sense of the British interests he would have to stand up for. Moreover, he had been hiding more or less, and now all that was changed and this was the simplest way not to hide. He met a host of friends, made his situation as public as possible and accepted on the spot a great many invitations, subject to the mental reservation that he should allow none of them to interfere with his being present the first night of Miriam's new venture. He was going to the equator to get away from her, but to break with the past with some decency of form he must show an affected interest, if he could muster none other, in an occasion that meant so much for her. The least intimate of her associates would do that, and Peter remembered that, at the expense of good manners, he had stayed away from her first appearance on any stage. He would have been shocked if he had found himself obliged to go back to Paris without giving her his personal countenance at the imminent crisis, so good a right had she to expect it.

It was nearly eight o'clock when he went to Great Stanhope Street to dress for dinner and learn that a note which he found on the hall table, and which bore the marks of hasty dispatch, had come in three or four hours before. It exhibited the signature of Miriam Rooth and informed him that she positively expected him at the theatre at eleven o'clock the next morning, for which hour a dress-rehearsal of the revived play had been hurriedly determined upon, the first night being now definitely fixed for the impending Saturday. She counted upon his attendance at both ceremonies, but she had particular reasons for wishing to see him at the rehearsal. 'I want you to see and judge and tell me,' she said, 'for my mind's like a flogged horse – it won't give another kick.' It was for the Saturday he had made Lady Agnes his promise; he had thought of the possibility of the play in doing so, but had rested in the faith that, from valid symptoms, this complication would not occur till the following week. He decided nothing on the spot in relation to the conflict – it was enough to dash off three words to Miriam to the effect that he would sooner perish than fail her on the morrow.

He went to the theatre in the morning, and the episode proved curious and instructive. Though there were twenty people in the stalls it bore little resemblance to those *répétitions générales* to which, in Paris, his love of the drama had often attracted him and which, taking place at night in the theatre closed to the public, are virtually first performances with invited spectators. They were, to his sense, always settled and stately and were rehearsals of the *première* even more than rehearsals of the play. The present occasion was less august; it was not so much a concert as a confusion of sounds, and it took audible and at times disputatious counsel with itself. It was rough and frank and spasmodic, but it was vivid and strong and, in spite of the serious character of the piece, often exceedingly droll; while it gave Sherringham, oddly enough, a livelier sense than he had ever had of bending over the hissing, smoking, sputtering caldron in which a palatable performance is stewed. He looked into the gross darkness that may result from excess of light; that is he understood how knocked up, on the eve of production, every one concerned in the preparation of a play might be, with nerves overstretched and glasses blurred, awaiting the test and the response, the echo to be given back by the big, receptive, artless, stupid, delightful public. Sherringham's interest had been great in advance, and as Miriam, since his arrival, had taken him much into her confidence he knew what she intended to do and had discussed a hundred points with her. They had differed about some of them and she had always said: 'Ah, but wait till you see how I shall do it at the time!' That was usually her principal reason and her most convincing argument. She had made some changes at the last hour – she was going to do several things in another way. But she wanted a touchstone, she wanted a fresh ear and, as she told Sherringham when he went behind after the first act, that was why she had insisted on this private performance, to which a few fresh ears were to be admitted. They didn't want to let her have it – they were a parcel of donkeys; but as to what she meant in general to have she had given them a hint which she flattered herself they wouldn't soon forget.

Miriam spoke as if she had had a great battle with her fellow-workers and had routed them utterly. It was not the first time Sherringham had heard her talk as if such a life as hers could only be a fighting life, so that she frankly recognized the fine uses of a faculty for making a row. She rejoiced that she had this faculty, for she knew what to do with it; and though there might be a certain swagger in taking such a stand in advance, when one had done the infinitely little that she had done, yet she trusted to the future to show how right she should have been in believing that a pack of idiots would never hold out against her, would know that they couldn't afford to. Her assumption of course was that she fought for the light and the right, for the good way and the thorough, for doing a thing properly if one did it at all. What she had really wanted was the theatre closed for a night and the dress-rehearsal put on for a few people, given instead of *Yolande*. That she had not got, but she would have it the next time. She spoke as if her triumphs behind the scenes as well as before would go by leaps and bounds, and Sherringham perfectly believed, for the time, that she would drive her coadjudicators in front of her like sheep. Her tone was the sort of thing that would have struck one as preposterous if one didn't believe in her; but if one did believe in her it only seemed thrown in with the other gifts. How was she going to act that night, and what could be said for such a hateful way of doing things? She asked Sherringham questions that he was quite unable to answer; she abounded in superlatives and tremendously strong objections. He had a sharper vision than usual of the queer fate, for a peaceable man, of being involved in a life of so violent a rhythm: one might as well be hooked to a Catharine-wheel and whiz round in flame and smoke.

It was only for five minutes, in the wing, amid jostling and shuffling and shoving, that they held this conference. Miriam, splendid in a brocaded anachronism, a false dress of the beginning of the century, and excited and appealing, imperious and reckless and good-natured, full of exaggerated propositions, supreme determinations and comical irrelevancies, showed as radiant a young head as the stage had ever seen. Other people quickly surrounded her, and Sherringham saw that though she wanted a fresh ear and a fresh eye she was liable to tell those who possessed these advantages that they didn't know what they were talking about. It was rather hard with her (Basil Dashwood let him into this, wonderfully painted and in a dress even more beautiful than Miriam's – that of a young dandy of the ages of silk): if you were not in the business you were one kind of donkey and if you *were* in the business you were another kind. Sherringham noted with a certain displeasure that Gabriel Nash was not there; he preferred to believe that it was from this observation that his annoyance happened to come when Miriam, after the remark just quoted from Dashwood, laughing and saying that at any rate the thing would do because it would just have to do, thrust vindictively but familiarly into the young actor's face a magnificent feather fan. 'Isn't he too lovely,' she asked, 'and doesn't he know how to do it?' Basil Dashwood had the sense of costume even more than Sherringham supposed, inasmuch as it now appeared that he had gone profoundly into the question of what his clever comrade was to wear. He had drawn patterns and hunted up stuffs, had helped her to try on her clothes, had bristled with ideas and pins. It is not perfectly easy to explain why Sherringham grudged Gabriel Nash the cynicism of his absence; it may even be thought singular that he should have missed him. At any rate he flushed a little when Miriam, of whom he inquired whether she hadn't invited her oldest and dearest friend, exclaimed: 'Oh, he says he doesn't like the kitchen fire – he only wants the pudding!' It would have taken the kitchen fire to account at that moment for the red of Sherringham's cheek; and he was indeed uncomfortably heated by helping to handle, as he phrased it, the saucepans.

This he felt so much after he had returned to his seat, which he forbore to quit again till the curtain had fallen on the last act, that in spite of the high beauty of that part of the performance of which Miriam carried the weight there was a moment when his emancipation led him to give a suppressed gasp of relief, as if he were scrambling up the bank of a torrent after an immersion. The girl herself, at any rate, as was wholly right, was of the incorruptible faith: she had been saturated to good purpose with the great spirit of Madame Carré. That was conspicuous as the play went on and she watched over the detail with weary piety and passion. Sherringham had never liked the piece itself; he thought that as clumsy in form and false in feeling it did little honour to the British theatre; he hated many of the speeches, pitied Miriam for having to utter them and considered that, lighted by that sort of candle, the path of fame might very well lead nowhere.

When the rehearsal was over he went behind again, and in the rose-coloured satin of the *dénouement* the heroine of the occasion said to him: 'Fancy my having to drag through that other stuff tonight – the brutes!' He was vague about the persons designated in this allusion, but he let it pass: he had at the moment a kind of detached foreboding of the way any gentleman familiarly connected with Miriam in the future would probably form the habit of letting objurgations and some other things pass. This had become indeed, now, a frequent state of mind with him; the instant he was before her, near her, next her, he found himself a helpless subject of the spell which, so far at least as he was concerned, she put forth by contact and of which the potency was punctual and absolute: the fit came on, as he said, exactly as some esteemed express-train on a great line bangs at a given moment into the station. At a distance he partly recovered himself – that was the encouragement for going to the shaky republic; but as soon as he entered her presence his life struck him as a thing disconnected from his will. It was as if *he* had been one thing and his behaviour another; he had glimpses of pictures of this difference, drawn, as they might be, from the coming years – little illustrative scenes in which he saw himself in strange attitudes of resignation, always rather sad and still, with a slightly bent head. Such images should not have been inspiring, but it is a fact that they were decidedly fascinating. The gentleman with the bent head had evidently given up something that was dear to him, but it was exactly because he had got his price that he was there. 'Come and see me three or four hours hence,' Miriam said – 'come, that is, about six. I shall rest till then, but I want particularly to talk with you. There will be no one else – not the end of any one's nose. You'll do me good.' So of course Peter drove up to Balaklava Place about six.

41

'I DON'T know – I haven't the least idea – I don't care – don't ask me,' he broke out immediately, in answer to some question that she put to him, with little delay, about his sense of the way she had done certain things at the theatre. Had she not frankly better give up that way and return to their first idea, the one they had talked over so much? Sherringham declared that it was no idea of his; that at any rate he should never have another as long as he lived; and that, so help him heaven, they had talked such things over more than enough.

'You're tired of me – yes, already,' said Miriam, sadly and kindly. They were alone, her mother had not peeped out, and she had prepared herself to return to the theatre. 'However, it doesn't matter, and of course your head is full of other things. You must think me ravenously selfish – perpetually chattering about my little shop. What will you have, when one's a shopgirl? You used to like it, but then you weren't a minister.'

'What do you know about my being a minister?' Sherringham asked, leaning back in his chair and gazing at her from sombre eyes. Sometimes he thought she looked better on the stage than she did off it, and sometimes he thought the exact contrary. The former of these convictions had held his mind in the morning, and it was now punctually followed by the other. As soon as she stepped on the boards a great and special alteration usually took place in her – she was in focus and in her frame; yet there were hours too in which she wore her world's face before the audience, just as there were hours when she wore her stage face in the world. She took up either mask as it suited her humour. Today Sherringham was seeing each in its order, and he thought each the best.

'I should know very little if I waited for you to tell me – that's very certain,' Miriam answered. 'It's in the papers that you've got a high appointment, but I don't read the papers unless there's something in them about myself. Next week I shall devour them and think them drivelling too, no doubt. It was Basil Dashwood told me, this afternoon, of your promotion – he has seen it announced somewhere. I'm delighted if it gives you more money and more advantages, but don't expect me to be glad that you're going away to some distant, disgusting country.'

'The matter has only just been settled and we have each been busy with our own affairs. Even if you hadn't given me these opportunities,' Sherringham went on, 'I should have tried to see you today, to tell you my news and take leave of you.'

'Take leave? Aren't you coming tomorrow?'

'Oh, yes, I shall see you through that. But I shall rush away the very moment it's over.'

'I shall be much better then – really I shall,' the girl said.

'The better you are the worse you are.'

Miriam returned his gaze with a beautiful charity. 'If it would do you any good I would be bad.'

'The worse you are the better you are!' laughed Sherringham. 'You're a kind of devouring demon.'

'Not a bit! It's you.'

'It's I? I like that.'

'It's you who make trouble, who are sore and suspicious and supersubtle, not taking things as they come and for what they are, but twisting them into misery and falsity. Oh, I've watched you enough, my dear friend, and I've been sorry for you – and sorry for myself; for I'm not so taken up with myself as you think. I'm not such a low creature. I'm capable of gratitude, I'm capable of affection. One may live in paint and tinsel, but one isn't absolutely without a soul. Yes, I've got one,' the girl went on, 'though I do paint my face and practise my intonations. If what you are going to do is good for you I'm very glad. If it leads to good things, to honour and fortune and greatness, I'm enchanted. If it means your being away always, forever and ever, of course that's serious. You know it – I needn't tell you – I regard you as I really don't regard any one else. I have a confidence in you – ah, it's a luxury. You're a gentleman, *mon bon* – ah, you're a gentleman! It's just that. And then you see, you understand, and that's a luxury too. You're a luxury altogether, Mr Sherringham. Your being where I shall never see you is not a thing I shall enjoy; I know that from the separation of these last months – after our beautiful life in Paris, the best thing that ever happened to me or that ever will. But if it's your career, if it's your happiness, I can miss you and hold my tongue. I *can* be disinterested – I can!'

'What did you desire me to come for?' Sherringham asked, attentive and motionless. The same impression, the old impression was with him again; the sense that if she was sincere it was sincerity of execution, if she was genuine it was the genuineness of doing it well. She did it so well now that this very fact was charming and touching. When she asked him at the theatre to grant her the hour in the afternoon, she wanted candidly (the more as she had not seen him at home for several days) to go over with him once again, on the eve of the great night (it would be for her second attempt the critics would lie so in wait – the first success might have been a fluke), some of her recurrent doubts: knowing from experience what good ideas he often had, how he could give a worrying alternative its quietus at the last. Then she had heard from Dashwood of the change in his situation, and that had really from one moment to the other made her think sympathetically of his preoccupations – led her openhandedly to drop her own. She was sorry to lose him and eager to let him know how good a friend she was conscious that he had been to her. But the expression of this was already, at the end of a minute, a strange bedevilment: she began to listen to herself, to speak dramatically, to represent. She uttered the things she felt as if they were snatches of old play-books, and really felt them the more because they sounded so well. This however didn't prevent them from being as good feelings as

410

those of anybody else, and at the moment Sherringham, to still a rising emotion – which he knew he shouldn't still – articulated the challenge I have just recorded, she seemed to him to have at any rate the truth of gentleness and generosity.

'There's something the matter with you – you're jealous,' said Miriam. 'You're jealous of Mr Dormer. That's an example of the way you tangle everything up. Lord, he won't hurt you, nor me either!'

'He can't hurt me, my dear, and neither can you; for I have a nice little heart of stone and a smart new breast-plate of iron. The interest I take in you is something quite extraordinary; but the most extraordinary thing in it is that it's perfectly prepared to tolerate the interest of others.'

'The interest of others needn't trouble it much!' Miriam declared. 'If Mr Dormer had broken off his marriage to such an awfully fine woman (for she is that, your swell of a sister), it isn't for a loud wretch like me. He's kind to me because that's his nature, and he notices me because that's his business; but he's away up in the clouds – a thousand miles over my head. He has got something "on", as they say; he's in love with an idea. I think it's a shocking bad one, but that's his own affair. He's quite *exalté*; living on nectar and ambrosia – what he has to spare for us poor crawling things on earth is only a few dry crumbs. I didn't even ask him to come to rehearsal. Besides, he thinks you're in love with me and that it wouldn't be honourable to cut in. He's capable of that – isn't it charming?'

'If he were to relent and give up his scruples, would you marry him?' asked Sherringham.

'Mercy, how you chatter about marrying!' the girl laughed. 'You've all got it on the brain.'

'Why, I put it that way to please you, because you complained to me last year precisely that this was not what seemed generally to be wanted.'

'Oh, last year!' Miriam murmured. Then, differently: 'Yes, it's very tiresome!' she exclaimed.

'You told me moreover in Paris, more than once, that you wouldn't listen to anything but that.'

'Well, I won't, but I shall wait till I find a husband who's bad enough. One who'll beat me and swindle me and spend my money on other women – that's the sort of man for me. Mr Dormer, delightful as he is, doesn't come up to that.'

'You'll marry Basil Dashwood,' Sherringham replied.

'Oh, marry? – call it marry, if you like. That's what poor mother says – she lives in dread of it.'

'To this hour,' said Sherringham, 'I haven't managed to make out what your mother wants. She has so many ideas, as Madame Carré said.'

'She wants me to be a tremendous sort of creature – all her ideas are reducible to that. What makes the muddle is that she isn't clear about the kind of creature she wants most. A great actress or a great lady – sometimes she inclines for one and sometimes for the other; but on the whole she persuades herself that a great actress, if she'll cultivate the right people, may *be* a great lady. When I tell her that won't do and that a great actress can never be anything but a great vagabond, then the dear old thing has tantrums and we have scenes – the most grotesque: they'd make the fortune, for a subject, of some play-writing fellow, if he had the wit to guess them; which, luckily for us perhaps, he never will. She usually winds up by protesting – *devinez un peu quoi!*' Miriam added. And as her companion professed his complete inability to divine: 'By declaring that rather than take it that way I must marry *you*.'

'She's shrewder than I thought. It's the last of vanities to talk about it, but I may mention in passing that if you would marry me you should be the greatest of all possible ladies.'

'Heavens, my dear fellow, what natural capacity have I for that?'

'You're artist enough for anything. I shall be a great diplomatist: my resolution is firmly taken. I'm infinitely cleverer than you have the least idea of, and you shall be a great diplomatist's wife.'

'And the demon, the devil, the devourer and destroyer, that you are so fond of talking about: what, in such a position, do you do with that element of my nature? *Où le fourrez-vous?*'

'I'll look after it, I'll keep it under. Rather perhaps I should say I'll bribe it and lull it – I'll gorge it with earthly grandeurs.'

'That's better,' said Miriam; 'for a demon that's kept under is a shabby little demon. Don't let us be shabby.' Then she added: 'Do you really go away the beginning of next week?'

'Monday night, if possible.'

'That's to Paris. Before you go to your new post they must give you an interval here.'

'I shan't take it – I'm so tremendously keen for my duties. I shall insist on going sooner. Oh, I shall be concentrated now.'

'I'll come and act there,' said Miriam, with her handsome smile. 'I've already forgotten what it was I wanted to discuss with you: it was some trumpery stuff. What I want to say now is only one thing: that it's not in the least true that because my life pitches me in every direction and mixes me up with all sorts of people – or rather with one sort mainly, poor dears! – I haven't a decent character, I haven't common honesty. Your sympathy, your generosity, your patience, your precious suggestions, our dear, sweet days last summer in Paris, I shall never forget. You're the best – you're different from all the others. Think of me as you please and make profane jokes about my matrimonial prospects – I shall think of *you* only in one way. I have a great respect for you. With all my heart I hope you'll be a great diplomatist. God bless you!'

Miriam got up as she spoke and in so doing she glanced at the clock – a movement which somehow only added to the noble gravity of her discourse: it was as if she were considering his time, not her own. Sherringham, at this, rising too, took out his watch and stood a moment with his eyes bent upon it, though without in the least perceiving what the needles marked.

'You'll have to go, to reach the theatre at your usual hour, won't you? Let me not keep you. That is, let me keep you only long enough just to say this, once for all, as I shall never speak of it again. I'm going away to save myself,' Sherringham went on deliberately, standing before her and soliciting her eyes with his own. 'I ought to go, no doubt, in silence, in decorum, in virtuous submission to hard necessity – without asking for credit or sympathy, without provoking any sort of scene or calling attention to my fortitude. But I can't – upon my soul I can't. I can go, I can see it through, but I can't hold my tongue. I want you to know all about it, so that over there, when I'm bored to death, I shall at least have the exasperatingly vain consolation of feeling that you do know – and that it does neither you nor me any good!'

He paused a moment, upon which Miriam asked: 'That I do know what?'

'That I have a consuming passion for you and that it's impossible.'

'Ah, impossible, my friend!' she sighed, but with a quickness in her assent.

'Very good; it interferes, the gratification of it would interfere fatally, with the ambition of each of us. Our ambitions are odious, but we are tied fast to them.'

'Ah, why ain't we simple?' Miriam quavered. 'Why ain't we of the people – *comme tout le monde* – just a man and a girl liking each other?'

Sherringham hesitated a moment; she was so tenderly mocking, so sweetly ambiguous, as she said this. 'Because we are precious asses! However, I'm simple enough, after all, to care for you as I have never cared for any human creature. You have, as it happens, a personal charm for me that no one has ever approached, and from the top of your splendid head to the sole of your theatrical shoe (I could go down on my face – there, abjectly – and kiss it?) every inch of you is dear and delightful to me. Therefore good-bye.'

Miriam stared, at this, with wider eyes: he had put the matter in a way that struck her. For a moment, all the same, he was afraid she would reply as if she had often heard that sort of thing before. But she was too much moved – the pure colour that had risen to her face showed it – to have recourse to this particular facility. She was moved even to the glimmer of tears, though she gave him her hand with a smile. 'I'm so glad you've said all that; for from you I know what it means. Certainly it's better for you to go away. Of course it's all wrong, isn't it? – but that's the only thing it can be: therefore it's all right, isn't it? Some day when we're great people we'll talk these things over; then we shall be quiet, we shall be at peace – let us hope so at least – and better friends than people will know.' She paused a moment, smiling still; then she said while he held her hands: 'Don't, *don't* come tomorrow night.'

With this she attempted to draw her hand away, as if everything were settled and over; but the effect of her movement was that, as he held her hand tight, he was simply drawn toward her and close to her. The effect of this, in turn, was that, releasing her only to possess her more, he seized her in his arms and breathing deeply 'I love you!' clasped her in a long embrace. It was so long that it gave the door of the room time to open before either of them had taken notice. Mrs Rooth, who had not peeped in before, peeped in now, becoming in this matter witness of an incident she could scarcely have expected. The unexpected indeed, for Mrs Rooth, had never been an insuperable element in things; it was her system, in general, to be too much in harmony to be surprised. As the others turned round they saw her standing there and smiling at them, and heard her ejaculate with wise indulgence:

'Oh, you extravagant children!'

Miriam brushed off her tears, quickly but confusedly. 'He's going away – he's bidding us farewell.'

Sherringham – it was perhaps a result of his general agitation – laughed at the 'us' (he had already laughed at the charge of puerility), and Mrs Rooth returned: 'Going away? Ah, then I must have one too!' And she held out both her hands. Sherringham stepped forward and, taking them, kissed her respectfully on each cheek, in the foreign manner, while she continued: 'Our dear old friend – our kind, gallant gentleman!'

'The gallant gentleman has been promoted to a great post – the proper reward of his gallantry,' Miriam said. 'He's going out as minister to some impossible place – where is it?'

'As minister – how very charming! We *are* getting on.' And the old woman gave him a curious little upward interrogative leer.

'Oh, well enough. One must take what one can get,' he answered.

'You'll get everything now, I'm sure, shan't you?' Mrs Rooth asked, with an inflection that called back to him, comically (the source was so different), the very vibrations he had noted the day before in Lady Agnes's voice.

'He's going to glory and he'll forget all about us – forget that he has ever known such people. So we shall never see him again, and it's better so. Good-bye, good-bye,' Miriam repeated; 'the brougham must be there, but I won't take you. I want to talk to mother about you, and we shall say things not fit for you to hear. Oh, I'll let you know what we lose – don't be afraid,' she added to Mrs Rooth. 'He's the rising star of diplomacy.'

'I knew it from the first – I know how things turn out for such people as you!' cried the old woman, gazing fondly at Sherringham. 'But you don't mean to say you're not coming tomorrow night?'

'Don't – don't; it's great folly,' Miriam interposed; 'and it's quite needless, since you saw me today.'

Sherringham stood looking from the mother to the daughter, the former of whom broke out to the latter: 'Oh, you dear rogue, to say one has *seen* you yet! You know how you'll come up to it; you'll be transcendent.'

'Yes, I shall be there – certainly,' said Sherringham, at the door, to Mrs Rooth.

'Oh, you dreadful goose!' Miriam called after him. But he went out without looking round at her.

42

NICK DORMER had for the hour quite taken up his abode at his studio, where Biddy usually arrived after breakfast to give him news of the state of affairs in Calcutta Gardens and where many letters and telegrams were now addressed to him. Among such missives, on the morning of the Saturday on which Peter Sherringham had promised to dine at the other house, was a note from Miriam Rooth, informing Nick that if he should not telegraph to put her off she would turn up about half-past eleven, probably with her mother, for just one more sitting. She added that it was a nervous day for her and that she couldn't keep still, so that it would really be very kind to let her come to him as a refuge. She wished to stay away from the theatre, where everything was now settled (or so much the worse for the others if it wasn't), till the evening, but if she were left to herself should be sure to go there. It would keep her quiet and soothe her to sit – he could keep her quiet (he was such a blessing that way!) at any time. Therefore she would give him two or three hours – or rather she would ask him for them – if he didn't positively turn her from the door.

It had not been definite to Nick that he wanted another sitting at all for the slight work, as he held it to be, that Miriam had already helped him to achieve. He regarded this work as a kind of pictorial *obiter dictum*: he had made what he could of it and would have been at a loss to see how he could make more. If it was not finished, this was because it was not finishable; at any rate he had said all he had to say in that particular phrase. Nick Dormer, as it happened, was not just now in the highest spirits; his imagination had within two or three days become conscious of a check which he tried to explain by the idea of a natural reaction. Any important change, any new selection in one's life was exciting, and exaggerate that importance and one's own as little as one would, there was an inevitable strong emotion in renouncing, in the face of considerable opposition, one sort of responsibility for another sort. That made life not perhaps necessarily joyous, but decidedly thrilling, for the hour; and it was all very well till the thrill abated. When this occurred, as it inevitably would, the romance and the poetry of the thing would be exchanged for the flatness and the prose. It was to these latter elements that Nick Dormer had waked up pretty wide on this particular morning; and the prospect was not appreciably more blooming from the fact that he had warned himself in advance that it would be dull. He had known how dull it would be, but now he would have time to learn that even better. A reaction was a reaction, but it was not after all a catastrophe. A part of its privilege would be to make him ask himself if he had not committed a great mistake; that privilege would doubtless even remain within the limits of its nature in leading him to reply to this question in the affirmative. But he would live to withdraw such a concession – this was the first thing to bear in mind.

He was occupied, even while he dressed, in the effort to get forward mentally with some such retractation when, by the first post, Miriam's note arrived. At first it did little to help him in his effort, for it made him contrast her eagerness with his own want of alacrity and ask himself what the deuce he should do with her. Ambition, with her, was always on the charge, and she was not a person to conceive that others might in bad moments listen for the trumpet in vain. It would never have occurred to her that only the day before he had spent a portion of the afternoon quite at the bottom of the hill. He had in fact turned into the National Gallery and had wandered about there for more than one hour, and it was just while he did so that the immitigable recoil had begun perversely to set in. And the perversity was all the greater from the circumstance that if the experience was depressing it was not because he had been discouraged beyond measure by the sight of the grand things that had been done – things so much grander than any that would ever bear his signature. That variation he was duly acquainted with and should taste in abundance again. What had happened to him, as he passed on this occasion from Titian to Rubens and from Gainsborough to Rembrandt, was that he found himself calling the whole art literally into question. What was it after all, at the best, and why had people given it so high a place? Its weakness, its narrowness appeared to him; tacitly blaspheming he looked at several world-famous performances with a lustreless eye. That is he blasphemed if it were blasphemy to say to himself that, with all respect, they were a poor business, only well enough in their small way. The force that produced them was not one of the greatest forces in human affairs; their place was inferior and their connection with the life of man casual and slight. They represented so inadequately the idea, and it was the idea that won the race, that in the long run came in first. He had incontestably been in much closer relation to the idea a few months before than he was today: it made up a great deal for the bad side of politics that they were after all a clumsy system for applying and propagating the idea. The love of it had really been at certain hours at the bottom of his disposition to follow them up; though this had not been what he used to talk of most with his political comrades or even with Julia. Certainly, political as Julia was, he had not conferred with her much about the idea. However, this might have been his own fault quite as much as hers, and she probably took such an enthusiasm for granted – she took such a tremendous lot of things for granted. On the other hand he had put this enthusiasm forward frequently in his many discussions with Gabriel Nash, with the effect, it is true, of making that worthy scoff transcendentally at what he was pleased to term his hypocrisy. Gabriel maintained precisely that there were more ideas, more of those that man lived by, in a single room of the National Gallery than in all the statutes of Parliament. Nick had replied to this more than once that the determination of what man did live by was required; to which Nash had retorted (and it was very rarely that he quoted Scripture) that it was at any rate not by bread-and-butter alone. The statutes of Parliament gave him bread-and-butter *tout au plus*.

Nick Dormer at present had no pretension of trying this question over again; he reminded himself that his ambiguity was subjective, as the philosophers said; the result of a mood which in due course would be at the mercy of another mood. It made him curse, and cursing, as a finality, was shaky; so he would throw out a platform beyond it. The time far beyond others to do one's work was when it didn't seem worth doing, for then one gave it a brilliant chance, that of resisting the stiffest test of all – the test of striking one as very bad. To do the most when there would be the least to be got by it was to be most in the true spirit of production. One thing at any rate was very certain, Nick reflected: nothing on earth would induce him to change back again; not even if this twilight of the soul should last for the rest of his days. He hardened himself in his posture with a good conscience which, had they had a glimpse of it, would have made him still more diverting to those who already thought him so; but now by good fortune Miriam suddenly knocked together the little bridge that was wanted to carry him over to more elastic ground. If he had made his sketch it was a proof that he had done her, and that he had done her flashed upon him as a sign that she would be still more feasible. He found his platform, as I have called it, and for a moment in his relief he danced upon it. He sent out a telegram to Balaklava Place requesting his beautiful sitter by no manner of means to fail him. When his servant came back it was to usher into the studio Peter Sherringham, whom the man apparently found at the door.

The hour was so early for social intercourse that Nick immediately guessed his visitor had come on some rare errand; but this inference was instantly followed by the reflection that

Peter might after all only wish to make up by present zeal for not having been near him before. He forgot that, as he had subsequently learned from Biddy, their foreign or all but foreign cousin had spent an hour in Rosedale Road, missing him there but pulling out Miriam's portrait, the day of his own hurried visit to Beauclere. These young men were not on a ceremonious footing and it was not in Nick's nature to keep a record of civilities rendered or omitted; nevertheless he had been vaguely conscious that during a stay in London, on Peter's part, which apparently was stretching itself out, he and his kinsman had foregathered less than of yore. It was indeed an absorbing moment in the career of each, but at the same time that he recognized this truth Nick remembered that it was not impossible Peter might have taken upon himself to resent some supposititious failure of consideration for Julia; though this would have been stupid, and the newly-appointed minister (to he had forgotten where) cultivated a finer habit. Nick held that as he had treated Julia with studious generosity she had nothing whatever to reproach him with; so her brother had therefore still less. It was at any rate none of her brother's business. There were only two things that would have made Nick lukewarm about disposing in a few frank words of all this: one of them his general hatred of talking of his private affairs (a reluctance in which he and Peter were well matched); and the other a particular sentiment which would have involved more of a confession and which could not be otherwise described than as a perception that the most definite and even most pleasant consequence of the collapse of his engagement was, as it happened, an extreme consciousness of freedom. Nick Dormer's observation was of a different sort from his cousin's; he noted much less the signs of the hour and kept altogether a looser register of life. Nevertheless, just as one of our young men had during these days in London found the air peopled with personal influences, the concussion of human atoms, so the other, though only asking to live without too many questions and work without too many disasters, to be glad and sorry in short on easy terms, had become aware of a certain social tightness, of the fact that life is crowded and passion is restless, accident frequent and community inevitable. Everybody with whom one had relations had other relations too, and even optimism was a mixture and peace an embroilment. The only chance was to let everything be embroiled but one's temper and everything spoiled but one's work. It must be added that Nick sometimes took precautions against irritation which were in excess of the danger, as departing travellers, about to whiz through foreign countries, study phrase-books for combinations of words they will never use. He was at home in the brightness of things – his longest excursions across the border were short. He had a dim sense that Peter considered that he made him uncomfortable and might have come now to tell him so; in which case he should be sorry for Peter in various ways. But as soon as his visitor began to speak Nick felt suspicion fade into old friendliness, and this in spite of the fact that Peter's speech had a slightly exaggerated promptitude, like the promptitude of business, which might have denoted self-consciousness. To Nick it quickly appeared better to be glad than to be sorry: this simple argument was more than sufficient to make him glad Peter was there.

'My dear Nick, it's an unpardonable hour, isn't it? I wasn't even sure you'd be up, and yet I had to risk it because my hours are numbered. I'm going away tomorrow,' Peter went on; 'I've a thousand things to do. I've had no talk with you this time such as we used to have of old (it's an irreparable loss, but it's your fault, you know), and as I've got to rush about all day I thought I'd just catch you before any one else does.'

'Some one has already caught me, but there's plenty of time,' Nick returned.

Peter stared a moment, as if he were going to ask a question; then he thought better of this and said: 'I see, I see. I'm sorry to say I've only a few minutes at best.'

'Man of crushing responsibilities, you've come to humiliate me!' Nick exclaimed. 'I know all about it.'

'It's more than I do then. That's not what I've come for, but I shall be delighted if I humiliate you a little by the way. I've two things in mind, and I'll mention the most difficult first. I came here the other day – the day after my arrival in town.'

'Ah, yes, so you did; it was very good of you,' Nick interrupted, as if he remembered. 'I ought to have returned your visit, or left a card or written my name or something, in Great

Stanhope Street, oughtn't I? You hadn't got this new thing then, or I would have done so.'

Peter eyed him a moment. 'I say, what's the matter with you? Am I really unforgivable for having taken that liberty?'

'What liberty?' Nick looked now as if there were nothing whatever the matter with him, and indeed his visitor's allusion was not clear to him. He was thinking only for the instant of Biddy, of whom and whose secret inclinations Grace had insisted on talking to him. They were none of his business, and if he would not for the world have let the girl herself suspect that he had violent lights on what was most screened and curtained in her, much less would he have made Peter a clumsy present of this knowledge. Grace had a queer theory that Peter treated Biddy badly – treated them all, somehow, badly; but Grace's zeal (she had plenty of it, though she affected all sorts of fine indifference) almost always took the form of being wrong. Nick wanted to do only what Biddy would thank him for, and he knew very well what she wouldn't. She wished him and Peter to be great friends, and the only obstacle to this was that Peter was too much of a diplomatist. Peter made him for an instant think of her and of the hour they had lately spent together in the studio in his absence – an hour of which Biddy had given him a history full of detail and of omissions; and this in turn brought Nick's imagination back to his visitor's own side of the matter. That complexity of things of which the sense had lately increased with him, and to which it was owing that any thread one might take hold of would probably lead one to something discomfortable, was illustrated by the fact that while poor Biddy was thinking of Peter it was ten to one that poor Peter was thinking of Miriam Rooth. All this danced before Nick's intellectual vision for a space briefer than my too numerous words.

'I pitched into your treasures – I rummaged among your canvases,' Peter said. 'Biddy had nothing whatever to do with it – she maintained an attitude of irreproachable reserve. It has been on my conscience all these days, and I ought to have done penance before. I have been putting it off partly because I am so ashamed of my indiscretion. *Que voulez-vous*, my dear Nick? My provocation was great. I heard you had been painting Miss

Rooth, so that I couldn't restrain my curiosity. I simply went into that corner and struck out there – a trifle wildly, no doubt. I dragged the young lady to the light – your sister turned pale as she saw me. It was a good deal like breaking open one of your letters, wasn't it? However, I assure you it's all right, for I congratulate you both on your style and on your correspondent.'

'You're as clever, as witty, as humorous as ever, old boy,' Nick rejoined, going himself into the corner designated by his companion and laying his hands on the same canvas. 'Your curiosity is the highest possible tribute to my little attempt, and your sympathy sets me right with myself. There is she again,' Nick went on, thrusting the picture into an empty frame; 'you shall see her whether you wish to or not.'

'Right with yourself? You don't mean to say you've been wrong!' Sherringham returned, standing opposite the portrait.

'Oh, I don't know; I've been kicking up such a row; anything is better than a row.'

'She's awfully good – she's awfully true,' said Sherringham. 'You've done more to it, since the other day; you've put in several things.'

'Yes, but I've worked distractedly. I've not altogether conformed to the celebrated recommendation about being off with the old love.'

'With the old love?' Sherringham repeated, looking hard at the picture.

'Before you are on with the new!' Nick had no sooner uttered these words than he coloured; it occurred to him that Peter would probably think he was alluding to Julia. He therefore added quickly: 'It isn't so easy to cease to represent an affectionate constituency. Really, most of my time for a fortnight has been given up to letter-writing. They've all been unexpectedly charming. I should have thought they would have loathed and despised me. But not a bit of it; they cling to me fondly – they struggle with me tenderly. I've been down to talk with them about it, and we've passed the most sociable, delightful hours. I've designated my successor; I've felt a good deal like the Emperor Charles the Fifth when about to retire to the monastery of Yuste. The more I've seen of them in this way the more

I've liked them, and they declare it has been the same with themselves as regards me. We spend our time in assuring each other that we haven't begun to know each other till now. In short, it's all wonderfully jolly, but it isn't business. *C'est magnifique, mais ce n'est pas la guerre.*'

'They're not so charming as they might be if they don't offer to keep you and let you paint.'

'They do, almost; it's fantastic,' said Nick. 'Remember they haven't seen any of my painting yet.'

'Well, I'm sorry for you; we live in too enlightened an age,' Peter declared. 'You can't suffer for art. Your experience is interesting; it seems to show that, at the tremendous pitch of civilization we've reached, you can't suffer from anything but hunger.'

'I shall doubtless do that in abundance.'

'Never, never, when you paint as well as this.'

'Oh, come, you're too good to be true,' Nick replied. 'But where did you learn that one's larder is full in proportion as one's work is fine?'

Peter gave him no satisfaction on this curious point – he only continued to look at the picture; after which, in a moment, he said: 'I'll give you your price for it on the spot.'

'Dear boy, you're so magnanimous that you shall have it for nothing!' Nick exclaimed, passing his arm into his companion's.

Peter was silent at first. 'Why do you call me magnanimous?'

'Oh, bless my soul, it's hers – I forget!' laughed Nick, failing in his turn to answer the other's inquiry. 'But you shall have another.'

'Another? Are you going to do another?'

'This very morning. That is I shall begin it. I've heard from her; she's coming to sit – a short time hence.'

Peter turned away a little at this, releasing himself, and, as if the movement had been an effect of Nick's words, looked at his watch earnestly, to dissipate that appearance. He fell back, to consider the picture from further off. 'The more you do her the better; she has all the qualities of a great model. From that point of view it's a pity she has another trade: she might make so good a thing of this one. But how shall you do her again?' Sherringham continued, ingenuously.

'Oh, I can scarcely say; we'll arrange something; we'll talk it over. It's extraordinary how well she enters into what one wants: she knows more than one does one's self. She isn't the first comer. However, you know all about that, since you invented her, didn't you? That's what she says; she's awfully sweet on you,' Nick pursued. 'What I ought to do is to try something as different as possible from that thing; not the sibyl, the muse, the tremendous creature, but the charming woman, the person one knows, in different gear, as she appears *en ville*, as she calls it. I'll do something really serious and send it to you out there with my respects. It will remind you of home, and perhaps a little even of me. If she knows it's for you she'll throw herself into it in the right spirit. Leave it to us, my dear fellow; we'll turn out something good.'

'It's jolly to hear you; but I shall send you a cheque,' said Peter.

'I suppose it's all right in your position, but you're too proud,' his kinsman answered.

'What do you mean by my position?'

'Your exaltation, your high connection with the country, your treating with sovereign powers as the representative of a sovereign power. Isn't that what they call 'em?'

Sherringham, who had turned again toward his companion, listened to this with his eyes fixed on Nick's face, while at the same time he once more drew forth his watch. 'Brute!' he exclaimed familiarly, at the same time dropping his eyes on the watch. 'At what time did you say you expected your sitter?'

'Oh, we've plenty of time; don't be afraid of letting me see you agitated by her presence.'

'Brute!' Sherringham again ejaculated.

This friendly personal note cleared the air, made the communication between the two men closer. 'Stay with me and talk to me,' said Nick; 'I dare say it's good for me. Heaven knows when I shall see you so independently again.'

'Have you got something more to show me, then – some other work?' Sherringham asked.

'Must I bribe you by setting my signboards in a row? You know what I've done; by which I mean of course you know what I haven't done. My work, as you are so good as to call it, has hitherto been terrible rot. I've had no time, no opportunity, no continuity. I must go and sit down in a corner and learn my alphabet. That thing isn't good; what I shall do for you won't be good. Don't protest, my dear fellow; nothing will be fit to look at for a long time. And think of my ridiculous age. As the populace say (or don't they say it?) it's a rum go. It won't be amusing.'

'Oh, you're so clever you'll get on fast,' Sherringham replied, trying to think how he could most directly disobey his companion's injunction not to protest.

'I mean it won't be amusing for others,' said Nick, unperturbed by this violation. 'They want results, and small blame to them.'

'Well, whatever you do, don't talk like Mr Gabriel Nash,' Peter went on. 'Sometimes I think you're just going to.'

Nick stared a moment. 'Why, he never would have said that. "They want results, the damned fools" – that would have been more in his key.'

'It's the difference of a *nuance*. And are you extraordinarily happy?' Peter added, as Nick now obliged him by arranging half a dozen canvases so that he could look at them.

'Not so much, doubtless, as the artistic life ought to make one: because all one's people are not so infatuated as one's electors. But little by little I'm learning the beauty of obstinacy.'

'Your mother's very bad; I lunched with her the day before yesterday.'

'Yes, I know – I know,' said Nick hastily; 'but it's too late – it's too late. I must just peg away here and not mind. I've after all a great advantage in my life.'

Sherringham hesitated. 'And that would be –?'

'Oh, I mean knowing what I want to do: that's everything, you know.'

'It's an advantage however that you've only just come in for, isn't it?'

'Yes, but having waited only makes me prize it the more. I've got it now; and it makes up, for the present, for the absence of some other things.'

Again Sherringham was silent awhile. 'That sounds a little flat,' he remarked at last.

'It depends upon what you compare it with. It's rather more pointed than the House of Commons.'

'Oh, I never thought I should like that.'

There was another pause, during which Nick moved about the room, turning up old sketches to see if he had anything more to show his visitor, while Sherringham continued to look at the unfinished and in some cases, as it seemed to him, unpromising productions already submitted to his attention. They were much less interesting than the portrait of Miriam Rooth and, it would have appeared, much less significant of ability. For that particular effort Nick's talent had taken an unprecedented spring. This was the reflection that Peter made, as he had made it intensely before; but the words he presently uttered had no visible connection with it. They only consisted of the abrupt inquiry: 'Have you heard anything from Julia?'

'Not a syllable. Have you?'

'Dear, no; she never writes to me.'

'But won't she on the occasion of your promotion?'

'I dare say not,' said Peter: and this was the only reference to Mrs Dallow that passed between her brother and her late intended. It left a slight agitation of the atmosphere, which Sherringham proceeded to allay by an allusion comparatively speaking more relevant. He expressed disappointment that Biddy should not have come in; having had an idea that she was always in Rosedale Road of a morning. That was the other moiety of his present errand – the wish to see her and give her a message for Lady Agnes, upon whom at so early an hour he had not presumed to intrude in Calcutta Gardens. Nick replied that Biddy did in point of fact almost always turn up, and for the most part early; she came to wish him good morning and start him for the day. She was a devoted Electra, laying a cool, healing hand on a distracted Orestes. He reminded Peter however that he would have a chance of seeing her that evening, and of seeing Lady Agnes; for wasn't he to do them the honour of dining in Calcutta Gardens? Biddy, the day before, had arrived full of that excitement. Peter explained that this was exactly the sad subject of his actual *démarche*: the project of the dinner in Calcutta Gardens had, to his exceeding regret, fallen to pieces. The fact was (didn't Nick know it?) the night had been suddenly

427

and perversely fixed for Miriam's *première*, and he was under a definite engagement with her not to stay away from it. To add to the bore of the thing he was obliged to return to Paris the very next morning. He was quite awfully sorry, for he had promised Lady Agnes: he didn't understand then about Miriam's affair, in regard to which he had given a previous pledge. He was more sorry than he could say, but he could never fail Miss Rooth: he had professed from the first an interest in her which he must live up to a little more. This was his last chance – he hadn't been near her at the trying time she first produced herself. And the second night of the play wouldn't do – it must be the first or nothing. Besides, he couldn't wait over till Monday.

While Peter enumerated these complications his companion was occupied in polishing with a cloth a palette that he had just been scraping. 'I see what you mean – I'm very sorry too,' said Nick. 'I'm sorry you can't give my mother this joy – I give her so little.'

'My dear fellow, you might give her a little more. It's rather too much to expect *me* to make up for your omissions!'

Nick looked at Peter with a moment's fixedness while he rubbed his palette; and for that moment he felt the temptation to reply: 'There's a way you could do that, to a considerable extent – I think you guess it! – which wouldn't be intrinsically disagreeable.' But the impulse passed, without expressing itself in speech, and he simply answered: 'You can make this all clear to Biddy when she comes, and she'll make it clear to my mother.'

'Poor little Biddy!' Sherringham mentally exclaimed, thinking of the girl with that job before her; but what he articulated was that this was exactly why he had come to the studio. He had inflicted his company on Lady Agnes on Thursday and had partaken of a meal with her, but he had not seen Biddy, though he had waited for her, hoping she would come in. Now he would wait for her again – she was thoroughly worth it.

'Patience, patience, you've always me,' said Nick; to which he subjoined: 'If it's a question of going to the play I scarcely see why you shouldn't dine at my mother's all the same. People go to the play after dinner.'

'Yes, but it wouldn't be fair, it wouldn't be decent: it's a case when I must be in my seat from the rise of the curtain. I should force your mother to dine an hour earlier than usual, and then, in return for this courtesy, go off to my entertainment at eight o'clock, leaving her and Grace and Biddy languishing there. I wish I had proposed in time that they should go with me,' Peter continued, not very ingenuously.

'You might do that still,' Nick suggested.

'Oh, at this time of day it would be impossible to get a box.'

'I'll speak to Miss Rooth about it, if you like, when she comes,' smiled Nick.

'No, it wouldn't do,' said Peter, turning away and looking once more at his watch. He made tacitly the addition that still less than asking Lady Agnes, for his convenience, to dine early, would *this* be decent, would it be fair. His taking Biddy the night he dined with her and with Miss Tressilian had been something very like a violation of those proprieties. He couldn't say this to Nick, who remarked in a moment that it was all right, for Peter's action left him his own freedom.

'Your own freedom?' Peter echoed interrogatively, turning round.

'Why, you see now I can go to the theatre myself.'

'Certainly; I hadn't thought of that. You would have been going.'

'I gave it up for the prospect of your company.'

'Upon my word, you're too good – I don't deserve such sacrifices,' said Sherringham, who saw from Nick's face that this was not a figure of speech but the absolute truth. 'Didn't it however occur to you that, as it would turn out, I might – that I even naturally would – myself be going?' he added.

Nick broke into a laugh. 'It would have occurred to me if I understood a little better –' And he paused, still laughing.

'If you understood a little better what?' Peter demanded.

'Your situation, simply.'

Peter looked at him a moment. 'Dine with me tonight independently; we'll go to the theatre together, and then you'll understand it.'

'With pleasure, with pleasure: we'll have a jolly evening,' said Nick.

'Call it jolly if you like. When did you say she was coming?' Peter asked.

'Biddy? Oh, probably, as I tell you, at any moment.'

'I mean the great Miriam,' Peter replied.

'The great Miriam, if she's punctual, will be here in about forty minutes.'

'And will she be likely to find your sister?'

'My dear fellow, that will depend on whether my sister remains to see her.'

'Exactly; but the point is whether you'll allow her to remain, isn't it?'

Nick looked slightly mystified. 'Why shouldn't she do as she likes?'

'In that case she'll probably go.'

'Yes, unless she stays.'

'Don't let her,' Peter dropped; 'send her away.' And to explain this he added: 'It doesn't seem exactly the right sort of thing, young girls meeting actresses.' His explanation in turn struck him as requiring another clause; so he went on: 'At least it isn't thought the right sort of thing abroad, and even in England my foreign ideas stick to me.'

Even with this amplification however his proposition evidently still appeared to his companion to have a flaw; which, after he had considered it a moment, Nick exposed in the simple words: 'Why, you originally introduced them, in Paris – Biddy and Miss Rooth. Didn't they meet at your rooms and fraternize, and wasn't that much more abroad than this?'

'So they did, but she didn't like it,' Peter answered, suspecting that for a diplomatist he looked foolish.

'Miss Rooth didn't like it?' Nick persisted.

'That I confess I've forgotten. Besides, she was not an actress then. What I remember is that Biddy wasn't particularly pleased with her.'

'Why, she thought her wonderful – praised her to the skies. I remember too.'

'She didn't like her as a woman; she praised her as an actress.'

'I thought you said she wasn't an actress then,' Nick rejoined.

Peter hesitated. 'Oh, Biddy thought so. She has seen her since, moreover. I took her the other night, and her curiosity's satisfied.'

430

'It's not of any consequence, and if there's a reason for it I'll bundle her off directly. But the great Miriam seems such a kind, good woman.'

'So she is, charming – charming,' said Peter, looking hard at Nick.

'Here comes Biddy now,' this young man went on. 'I hear her at the door; you can warn her yourself.'

'It isn't a question of "warning" – that's not in the least my idea. But I'll take Biddy away,' said Peter.

'That will be still more energetic.'

'Oh, it's simply selfish – I like her company.' Peter had turned as if to go to the door to meet the girl; but he quickly checked himself, lingering in the middle of the room, and the next instant Biddy had come in. When she saw him there she also stopped.

43

'ARRIVE, arrive, my child,' said Nick. 'Peter's weary of waiting for you.'

'Ah, he's come to say he won't dine with us tonight!' Biddy stood with her hand on the latch.

'I leave town tomorrow; I've everything to do; I'm brokenhearted; it's impossible,' Peter pleaded. 'Please make my peace with your mother; I'm ashamed of not having written to her last night.'

Biddy closed the door and came in, while her brother said to her: 'How in the world did you guess it?'

'I saw it in the *Morning Post*,' Biddy answered, looking at Peter.

'In the *Morning Post*?' her cousin repeated.

'I saw there's to be a first night at that theatre, the one you took us to. So I said: "Oh, he'll go there." '

'Yes, I've got to do that too,' Peter admitted.

'She's going to sit to me again this morning, the wonderful actress of that theatre – she has made an appointment: so you see I'm getting on,' Nick announced to Biddy.

'Oh, I'm so glad – she's so splendid!' The girl looked away

from Peter now, but not, though it seemed to fill the place, at the triumphant portrait of Miriam Rooth.

'I'm delighted you've come in. I *have* waited for you,' Peter hastened to declare to Biddy, though he was conscious that this was under the circumstances meagre.

'Aren't you coming to see us again?'

'I'm in despair, but I shall really not have time. Therefore it's a blessing not to have missed you here.'

'I'm very glad,' said Biddy. Then she added: 'And you're going to America – to stay a long time?'

'Till I'm sent to some better place.'

'And will that better place be as far away?'

'Oh Biddy, it wouldn't be better then,' said Peter.

'Do you mean they'll give you something to do at home?'

'Hardly that. But I've got a tremendous lot to do at home today.' For the twentieth time Peter referred to his watch.

Biddy turned to her brother, who murmured to her: 'You might bid me good morning.' She kissed him, and he asked what the news might be in Calcutta Gardens; to which she replied:

'The only news is, of course, that, poor dears! they're making great preparations for Peter. Mamma thinks you must have had such a nasty dinner the other day,' the girl continued, to the guest of that romantic occasion.

'Faithless Peter!' said Nick, beginning to whistle and to arrange a canvas in anticipation of Miriam's arrival.

'Dear Biddy, thank your stars you are not in my horrid profession,' protested the personage thus designated. 'One is bowled about like a cricket-ball, unable to answer for one's freedom or one's comfort from one moment to another.'

'Oh, ours is the true profession – Biddy's and mine,' Nick broke out, setting up his canvas; 'the career of liberty and peace, of charming long mornings, spent in a still north light, in the contemplation, and I may even say in the company, of the amiable and the beautiful.'

'That certainly is the case when Biddy comes to see you,' Peter returned.

Biddy smiled at him. 'I come every day. *Anch' io son pittore!* I encourage Nick awfully.'

'It's a pity I'm not a martyr; she would bravely perish with me,' Nick said.

'You are – you are a martyr – when people say such odious things!' the girl cried. 'They do say them. I've heard many more than I've repeated to you.'

'It's you yourself then, indignant and sympathetic, who are the martyr,' observed Peter, who wanted greatly to be kind to her.

'Oh, I don't care!' she answered, colouring in response to this; and she continued, to Peter: 'Don't you think one can do as much good by painting great works of art as by – as by what papa used to do? Don't you think art is necessary to the happiness, to the greatness of a people? Don't you think it's manly and honourable? Do you think a passion for it is a thing to be ashamed of? Don't you think the artist – the conscientious, the serious one – is as distinguished a member of society as any one else?'

Peter and Nick looked at each other and laughed at the way she had got up her subject, and Nick asked his visitor if she didn't express it all in perfection. 'I delight in general in artists, but I delight still more in their defenders,' Peter jested to Biddy.

'Ah, don't attack me, if you are wise,' Nick said.

'One is tempted to, when it makes Biddy so fine.'

'Well, that's the way she encourages me; it's meat and drink to me,' Nick went on. 'At the same time I'm bound to say there is a little whistling in the dark in it.'

'In the dark?' his sister demanded.

'The obscurity, my dear child, of your own aspirations, your mysterious ambitions and plastic visions. Aren't there some heavyish shadows there?'

'Why, I never cared for politics.'

'No, but you cared for life, you cared for society, and you have chosen the path of solitude and concentration.'

'You horrid boy!' said Biddy.

'Give it up, that arduous step – give it up and come out with me,' Peter interposed.

'Come out with you?'

'Let us walk a little, or even drive a little. Let us at any rate talk a little.'

'I thought you had so much to do,' Biddy candidly objected.

'So I have, but why shouldn't you do a part of it with me? Would there be any harm? I'm going to some tiresome shops – you'll cheer the economical hour.'

The girl hesitated; then she turned to Nick. 'Would there be any harm?'

'Oh, it's none of *his* business!' Peter protested.

'He had better take you home to your mother.'

'I'm going home – I shan't stay here today,' said Biddy. Then to Peter: 'I came in a hansom, but I shall walk back. Come that way with me.'

'With singular pleasure. But I shall not be able to go in,' Sherringham added.

'Oh, that's no matter,' said Biddy. 'Good-bye, Nick.'

'You understand then that we dine together – at seven sharp. Wouldn't a club be best?' Peter, before going, inquired of Nick. He suggested further which club it should be; and his words led Biddy, who had directed her steps towards the door, to turn a moment, as if she were on the point of asking reproachfully whether it was for this Peter had given up Calcutta Gardens. But this impulse, if impulse it was, had no sequel except so far as it was a sequel that Peter spontaneously explained to her, after Nick had assented to his conditions, that her brother too had a desire to go to Miss Rooth's first night and had already promised to accompany him.

'Oh, that's perfect; it will be so good for him – won't it? if he's going to paint her again,' Biddy responded.

'I think there's nothing so good for him as that he happens to have such a sister as you,' Peter observed as they went out. As he spoke he heard outside the sound of a carriage stopping; and before Biddy, who was in front of him, opened the door of the house he had time to say to himself: 'What a bore – there's Miriam!' The opened door showed him that he was right – this young lady was in the act of alighting from the brougham provided by Basil Dashwood's thrifty zeal. Her mother followed her, and both the new visitors exclaimed and rejoiced, in their demonstrative way, as their eyes fell upon their valued friend. The door had closed behind Peter, but he instantly and violently rang, so that they should be admitted with as little delay as possible, while he remained slightly disconcerted by the prompt occurrence of an encounter he had sought to avert. It ministered moreover a little to this particular sensation that Miriam appeared to have come somewhat before her time. The incident promised however to pass off in the happiest way. Before he knew it both the ladies had taken possession of Biddy, who looked at them with comparative coldness, tempered indeed by a faint glow of apprehension, and Miriam had broken out:

'We know you, we know you; we saw you in Paris, and you came to my theatre a short time ago with Mr Sherringham.'

'We know your mother, Lady Agnes Dormer. I hope her ladyship is very well,' said Mrs Rooth, who had never struck Sherringham as a more objectionable old woman.

'You offered to do a head of me, or something or other: didn't you tell me you work in clay? I dare say you've forgotten all about it, but I should be delighted,' Miriam pursued, with the richest urbanity.

Peter was not concerned with her mother's pervasiveness, though he didn't like Biddy to see even that; but he hoped his companion would take the overcharged benevolence of the young actress in the spirit in which, rather to his surprise, it evidently was offered.

'I've sat to your clever brother many times,' said Miriam; 'I'm going to sit again. I dare say you've seen what we've done – he's too delightful. *Si vous saviez comme cela me repose!*' she added, turning for a moment to Sherringham. Then she continued, smiling, to Biddy: 'Only he oughtn't to have thrown up such prospects, you know. I have an idea I wasn't nice to you that day in Paris – I was nervous and scared and perverse. I remember perfectly; I *was* odious. But I'm better now – you'd see if you were to know me. I'm not a bad girl – really I'm not. But you must have your own friends. Happy they – you look so charming! Immensely like Mr Dormer, especially about the eyes; isn't she, mamma?'

'She comes of a beautiful Norman race – the finest, purest strain,' the old woman simpered. 'Mr Dormer is sometimes so good as to come and see us – we are always at home on Sunday; and if some day you were so venturesome as to come with him you might perhaps find it pleasant, though very different of course from the circle in which you habitually move.'

Biddy murmured a vague recognition of these wonderful civilities, and Miriam commented: 'Different, yes; but we're all right, you know. Do come,' she added. Then turning to Sherringham: 'Remember what I told you – I don't expect you tonight.'

'Oh, I understand; I shall come,' Peter answered, growing red.

'It will be idiotic. Keep him, keep him away – don't let him,' Miriam went on, to Biddy; with which, as Nick's portals now were gaping, she drew her mother away.

Peter at this walked off briskly with Biddy, dropping as he did so: 'She's too fantastic!'

'Yes, but so tremendously good-looking. I shall ask Nick to take me there,' the girl continued after a moment.

'Well, she'll do you no harm. They're all right, as she says. It's the world of art – you were standing up so for art just now.'

'Oh, I wasn't thinking so much of that kind,' said Biddy.

'There's only one kind – it's all the same thing. If one sort's good the other is.'

Biddy walked along a moment. 'Is she serious? Is she conscientious?'

'Oh, she has the makings of a great artist,' said Peter.

'I'm glad to hear you think a woman can be one.'

'In that line there has never been any doubt about it.'

'And only in that line?'

'I mean on the stage in general, dramatic or lyric. It's as the actress that the woman produces the most complete and satisfactory artistic results.'

'And only as the actress?'

'Yes, there's another art in which she's not bad.'

'Which one do you mean?' asked Biddy.

'That of being charming and good and indispensable to man.'

'Oh, that isn't an art.'

'Then you leave her only the stage. Take it if you like in the widest sense.'

Biddy appeared to reflect a moment, as if to see what sense this might be. But she found none that was wide enough, for she cried the next minute: 'Do you mean to say there's nothing for a woman but to be an actress?'

'Never in my life. I only say that that's the best thing for a woman to be who finds herself irresistibly carried into the practice of the arts; for there her capacity for them has most application and her incapacity for them least. But at the same time I strongly recommend her not to be an artist if she can possibly help it. It's a devil of a life.'

'Oh, I know; men want women not to be anything.'

'It's a poor little refuge they try to take from the overwhelming consciousness that you are in fact everything.'

'Everything? That's the kind of thing you say to keep us quiet.'

'Dear Biddy, you see how well we succeed!' laughed Sherringham; to which the girl responded by inquiring irrelevantly:

'Why is it so necessary for you to go to the theatre tonight, if Miss Rooth doesn't want you to?'

'My dear child, she does. But that has nothing to do with it.'

'Why then did she say that she doesn't?'

'Oh, because she meant just the contrary.'

'Is she so false then – is she so vulgar?'

'She speaks a special language; practically it isn't false, because it renders her thought, and those who know her understand it.'

'But she doesn't use it only to those who know her, since she asked me, who have so little the honour of her acquaintance, to keep you away tonight. How am I to know that she meant by that that I'm to urge you on to go?'

Sherringham was on the point of replying: 'Because you have my word for it'; but he shrank in fact from giving his word – he had some fine scruples – and endeavoured to get out of his embarrassment by a general tribute. 'Dear Biddy, you're delightfully acute: you're quite as clever as Miss Rooth.' He felt however that this was scarcely adequate, and he continued: 'The truth is, its being important for me to go is a matter quite independent of that young lady's wishing it or not wishing it. There happens to be a definite, intrinsic propriety in it which determines the matter and which it would take me long to explain.'

'I see. But fancy your "explaining" to me: you make me feel so indiscreet!' the girl cried quickly – an exclamation which touched him because he was not aware that, quick as it had been, Biddy had still had time to be struck first (though she wouldn't for the world have expressed it) with the oddity of such a duty at such a time. In fact that oddity, during a silence of some minutes, came back to Peter himself: the note had been forced – it sounded almost ignobly frivolous for a man on the eve of proceeding to a high diplomatic post. The effect of this however was not to make him break out with: 'Hang it, I *will* keep my engagement to your mother!' but to fill him with the wish that he could shorten his actual excursion by taking Biddy the rest of the way in a cab. He was uncomfortable, and there were hansoms about which he looked at wistfully. While he was so occupied his companion took up the talk by an abrupt interrogation.

'Why did she say that Nick oughtn't to have resigned his seat?'

'Oh, I don't know; it struck her so. It doesn't matter much.'

'If she's an artist herself why doesn't she like people to go in for art, especially when Nick has given his time to painting her so beautifully? Why does she come there so often if she disapproves of what he has done?'

'Oh, Miriam's disapproval – it doesn't count; it's a manner of speaking.'

'Of speaking untruths, do you mean? Does she think just the reverse – is that the way she talks about everything?'

'We always admire most what we can do least,' Peter replied; 'and Miriam of course isn't political. She ranks painters more or less with her own profession, about which already, new as she is to it, she has no illusions. They're all artists; it's the same general sort of thing. She prefers men of the world – men of action.'

'Is that the reason she likes you?' Biddy mocked.

'Ah, she doesn't like me – couldn't you see it?'

Biddy said nothing for a moment; then she asked: 'Is that why she lets you call her "Miriam"?'

'Oh, I don't, to her face.'

'Ah, only to mine!' laughed Biddy.

'One says that as one says "Rachel" of her great predecessor.'

'Except that she isn't so great quite yet, is she?'

'Far from it; she's the freshest of novices – she has scarcely been four months on the stage. But no novice has ever been such an adept. She'll go very fast, and I dare say that before long she'll be magnificent.'

'What a pity you'll not see that!' Biddy remarked, after a short interval.

'Not see it?'

'If you're thousands of miles away.'

'It *is* a pity,' Peter said; 'and since you mention it I don't mind frankly telling you – throwing myself on your mercy, as it were – that that's why I make such a point of a rare occasion like tonight. I've a weakness for the drama that, as you perhaps know, I've never concealed, and this impression will probably have to last me, in some barren spot, for many, many years.'

'I understand – I understand. I hope therefore it will be charming.' And Biddy walked faster.

'Just as some other charming impressions will have to last,' Peter added, conscious of a certain effort that he was obliged to make to keep up with her. She seemed almost to be running away from him, a circumstance which led him to suggest, after they had proceeded a little further without more words, that if she were in a hurry they had perhaps better take a cab. Her face was strange and touching to him as she turned it to reply quickly:

'Oh, I'm not in the least in a hurry, and I think, really, I had better walk.'

'We'll walk then, by all means!' Peter declared, with slightly exaggerated gaiety; in pursuance of which they went on a hundred yards. Biddy kept the same pace; yet it was scarcely a surprise to Sherringham that she should suddenly stop with the exclamation:

'After all, though I'm not in a hurry I'm tired! I had better have a cab; please call that one,' she added, looking about her.

They were in a straight, black, ugly street, where the small, cheap, gray-faced houses had no expression save that of a rueful, inconsolable consciousness of its want of identity. They would have constituted a 'terrace' if they could, but they had given it up. Even a hansom which loitered across the end of the vista turned a sceptical back upon it, so that Sherringham had to lift his voice in a loud appeal. He stood with Biddy watching the cab approach them. 'This is one of the charming things you'll remember,' she said, turning her eyes to the general dreariness from the particular figure of the vehicle, which was antiquated and clumsy. Before he could reply she had lightly stepped into the cab; but as he answered: 'Most assuredly it is,' and prepared to follow her, she quickly closed the apron.

'I must go alone; you've lots of things to do – it's all right'; and through the aperture in the roof she gave the driver her address. She had spoken with decision, and Peter recognized that she wished to get away from him. Her eyes betrayed it as well as her voice, in a look – not a hard one however – which as he stood there with his hand on the cab he had time to take from her. 'Good-bye, Peter,' she smiled; and as the cab began to rumble away he uttered the same tepid, ridiculous farewell.

44

WHEN Miriam and her mother went into the studio Nick Dormer had stopped whistling, but he was still gay enough to receive them with every demonstration of sociability. He thought his studio a poor place; ungarnished, untapestried, a mere seat of rude industry, with all its revelations and honours still to come. But both his visitors smiled upon it a good deal in the same way in which they had smiled on Bridget Dormer when they met her at the door: Mrs Rooth because vague, prudent approbation was the habit of her foolish little face – it was ever the least danger; and Miriam because apparently she was genuinely glad to find herself within the walls which she spoke of now as her asylum. She broke out in this strain to her host almost as soon as she had crossed the threshold, commending his circumstances, his conditions of work as infinitely happier than her own. He was quiet, independent, absolute, free to do what he liked as he liked it, shut up in his little temple with his altar and his divinity; not hustled about in a mob of people, having to posture and grin to pit and gallery, to square himself at every step with insufferable conventions and with the ignorance and vanity of others. He was blissfully alone.

'Mercy, how you do abuse your fine profession! I'm sure I never urged you to adopt it!' Mrs Rooth cried, in real bewilderment, to her daughter.

'She was abusing mine still more, the other day,' joked Nick – 'telling me I ought to be ashamed of it and of myself.'

'Oh, I never know from one moment to the other – I live with my heart in my mouth,' sighed the old woman.

'Aren't you quiet about the great thing – about my behaviour?' Miriam smiled. 'My only extravagances are intellectual.'

'I don't know what you call your behaviour.'

'You would very soon if it were not what it is.'

'And I don't know what you call intellectual,' grumbled Mrs Rooth.

'Yes, but I don't see very well how I could make you understand that. At any rate,' Miriam went on, looking at Nick, 'I retract what I said the other day about Mr Dormer. I've no wish to quarrel with him about the way he has determined to dispose of his life, because after all it does suit me very well. It rests me, this little devoted corner; oh, it rests me. It's out of the tussle and the heat, it's deliciously still, and they can't get at me. Ah, when art's like this, *à la bonne heure!*' And she looked round on such a presentment of 'art' with a splendid air that made Nick burst out laughing at its contrast with the humble fact. Miriam smiled at him as if she liked to be the cause of his mirth, and went on appealing to him: 'You'll always let me come here for an hour, won't you, to take breath – to let the whirlwind pass? You needn't trouble yourself about me; I don't mean to impose on you in the least the necessity of painting me, though if that's a manner of helping you to get on you may be sure it will always be open to you. Do what you like with me in that respect; only let me sit here on a high stool, keeping well out of your way, and see what you happen to be doing. I'll tell you my own adventures when you want to hear them.'

'The fewer adventures you have to tell, the better, my dear,' said Mrs Rooth; 'and if Mr Dormer keeps you quiet he will add ten years to my life.'

'This is an interesting comment on Mr Dormer's own quietus, on his independence and sweet solitude,' Nick observed. 'Miss Rooth has to work with others, which is after all only what Mr Dormer has to do when he works with Miss Rooth. What do you make of the inevitable sitter?'

'Oh,' answered Miriam, 'you can say to the sitter: "Hold your tongue, you brute!"'

'Isn't it a good deal in that manner that I've heard you address your comrades at the theatre?' asked Mrs Rooth. 'That's why my heart's in my mouth.'

'Yes, but they hit me back; they reply to me – *comme de raison* – as I should never think of replying to Mr Dormer. It's a great advantage to him that when he's peremptory with his model it only makes her better, adds to her expression of gloomy grandeur.'

'We did the gloomy grandeur in the other picture; suppose therefore we try something different in this,' suggested Nick.

'It *is* serious, it *is* grand,' murmured Mrs Rooth, who had taken up a rapt attitude before the portrait of her daughter. 'It makes one wonder what she's thinking of. Beautiful, commendable things – that's what it seems to say.'

'What can I be thinking of but the tremendous wisdom of my mother?' Miriam inquired. 'I brought her this morning to see that thing – she had only seen it in its earliest stage – and not to presume to advise you about anything else you may be so good as to embark on. She wanted, or she professed that she wanted, terribly to know what you had finally arrived at. She was too impatient to wait till you should send it home.'

'Ah, send it home – send it home; let us have it always with us!' Mrs Rooth urged. 'It will hold us up; it will keep us on the heights, near the stars – be always for us a symbol and a reminder!'

'You see I was right,' Miriam went on; 'for she appreciates thoroughly, in her own way, and understands. But if she worries or distracts you I'll send her directly home – I've kept the carriage there on purpose. I must add that I don't feel quite safe today in letting her out of my sight. She's liable to make dashes at the theatre and play unconscionable tricks there. I shall never again accuse mamma of a want of interest in my profession. Her interest today exceeds even my own. She's all over the place and she has ideas; ah, but ideas! She's capable of turning up at the theatre at five o'clock this afternoon and demanding that the scenery of the third act be repainted. For myself, I've not a word more to say on the subject – I've accepted the situation. Everything is no doubt wrong; but nothing can possibly be right. Let us eat and drink, for tonight we die. If you like, mamma shall go and sit in the carriage, and as there's no means of fastening the doors (is there?) your servant shall keep guard over her.'

'Just as you are now – be so good as to remain so; sitting just that way – leaning back, with a smile in your eyes and one hand on the sofa beside you, supporting you a little. I shall stick a flower into the other hand – let it lie in your lap, just as it is. Keep that thing on your head – it's admirably uncovered: do you call the construction a bonnet? – and let your head fall back a little. There it is – it's found. This time I shall really do something, and it will be as different as you like from that crazy job. *Pazienza!*' It was in these irrelevant but earnest words that Nick responded to his sitter's uttered vagaries, of which her charming tone and countenance diminished the superficial acerbity. He held up his hands a moment, to fix her in her limits, and a few minutes afterwards had a happy sense of having begun to work.

'The smile in her eyes – don't forget the smile in her eyes!' Mrs Rooth exclaimed softly, turning away and creeping about the room. 'That will make it so different from the other picture and show the two sides of her genius, with the wonderful range between them. It will be a magnificent pendant; and though I dare say I shall strike you as greedy, you must let me hope you will send it home too.'

Mrs Rooth explored the place discreetly, on tiptoe, gossiping as she went and bending her head and her eyeglass over various objects with an air of imperfect comprehension which did not prevent Nick from being reminded of the story of her underhand commercial habits told by Gabriel Nash at the exhibition in Paris, the first time her name had fallen on his ear. A queer old woman from whom, if you approached her in the right way, you could buy old pots – it was in this character that she had originally been introduced to him. He had lost sight of it afterwards, but it revived again as his observant eyes, at the same time that they followed his active hand, became aware of her instinctive appraising gestures. There was a moment when he laughed out gaily – there was so little in his poor studio to appraise. Mrs Rooth's vague, polite, disappointed bent back and head made a subject, the subject of a sketch, in an instant: they gave such a sudden pictorial glimpse of the element of race. He found himself seeing the immemorial Jewess in her, holding up a candle in a crammed back-shop. There was no candle indeed, and his studio was not crammed, and it had never occurred to him before that she was of Hebrew strain, except on the general theory, held with pertinacity by several clever people, that most of us are more or less so. The late Rudolf Roth had been, and his daughter was visibly her father's child; so that, flanked by such a pair, good Semitic reasons were surely not wanting to the mother. Receiving Miriam's little satiric shower without shaking her shoulders, she might at any rate have been the descendant of a tribe long persecuted. Her blandness was imperturbable – she professed that she would be as still as a mouse. Miriam, on the other side of the room, in the tranquil beauty of her attitude (it was 'found' indeed, as Nick had said), watched her a little and then exclaimed that she wished she had locked her up at home. Putting aside her humorous account of the dangers to which she was exposed from her mother, it was not whimsical to imagine that, within the limits of that repose from which the Neville-Nugents never wholly departed, Mrs Rooth might indeed be a trifle fidgety and have something on her mind. Nick presently mentioned that it would not be possible for him to 'send home' this second performance; and he added, in the exuberance of having already got a little into relation with his work, that perhaps that didn't matter, inasmuch as – if Miriam would give him his time, to say nothing of her own – a third masterpiece might also some day very well come off. His model rose to this without conditions, assuring him that he might count upon her till she grew too old and too ugly and that nothing would make her so happy as that he should paint her as often as Romney had painted the celebrated Lady Hamilton. 'Ah, Lady Hamilton!' deprecated Mrs Rooth; while Miriam, who had on occasion the candour of a fine acquisitiveness, inquired what particular reason there might be for his not letting them have the picture he was now beginning.

'Why, I've promised it to Peter Sherringham – he has offered me money for it,' Nick replied. 'However, he's welcome to it for nothing, poor fellow, and I shall be delighted to do the best I can for him.'

Mrs Rooth, still prowling, stopped in the middle of the room at this, and Miriam exclaimed: 'He offered you money – just as we came in?'

'You met him, then, at the door, with my sister? I supposed you had – he's taking her home,' said Nick.

'Your sister's a lovely girl – such an aristocratic type!' breathed Mrs Rooth. Then she added: 'I've a tremendous confession to make to you.'

'Mamma's confessions have to be tremendous to correspond with her crimes,' said Miriam. 'She asked Miss Dormer to come and see us – suggested even that you might bring her some Sunday. I don't like the way mamma does such things – too much humility, too many *simagrées*, after all; but I also said what I could to be nice to her. Your sister *is* charming – awfully pretty and modest. If you were to press me I should tell you frankly that it seems to me rather a social muddle, this rubbing shoulders of "nice girls" and *filles de théâtre*; I shouldn't think it would do your young ladies much good. However, it's their own affair, and no doubt there's no more need of their thinking we're worse than we are than of their thinking we're better. The people they live with don't seem to know the difference – I sometimes make my reflections about the public one works for.'

'Ah, if you go in for the public's knowing differences you're far too particular,' Nick laughed. '*D'où tombez-vous?* as you affected French people say. If you have anything at stake on that, you had simply better not play.'

'Dear Mr Dormer, don't encourage her to be so dreadful; for it *is* dreadful the way she talks,' Mrs Rooth broke in. 'One would think we were not respectable – one would think I had never known what I have known and been what I have been.'

'What one would think, beloved mother, is that you are a still greater humbug than you are. It's you, on the contrary, who go down on your knees, who pour forth apologies about our being vagabonds.'

'Vagabonds – listen to her! – after the education I've given her and our magnificent prospects!' wailed Mrs Rooth, sinking, with clasped hands, upon the nearest ottoman.

'Not after our prospects, if prospects they are: a good deal before them. Yes, you've taught me tongues, and I'm greatly obliged to you – they no doubt impart variety, as well as incoherency, to my conversation; and that of people in our line is for the most part notoriously monotonous and shoppy. The gift of tongues is in general the sign of your genuine adventurer. Dear mamma, I've no low standard – that's the last thing,' Miriam went on. 'My weakness is my exalted conception of respectability. Ah, *parlez-moi de ça* and of the way I understand it! Oh, if I were to go in for being respectable you'd see something fine. I'm awfully conservative and I know what respectability is, even when I meet people of society on the accidental middle ground of glowering or smirking. I know also what it isn't – it isn't the sweet union of little girls and actresses. I should carry it much further than any of these people: I should never look at the likes of us! Every hour I live I see that the wisdom of the ages was in the experience of dear old Madame Carré – was in a hundred things she told me. She's founded a rock. After that,' Miriam went on, to her host, 'I can assure you that if you were so good as to bring Miss Dormer to see us we should be angelically careful of her and surround her with every attention and precaution.'

'The likes of us – the likes of us!' Mrs Rooth repeated plaintively, with ineffectual, theoretical resentment. 'I don't know what you are talking about, and I decline to be turned upside down. I have my ideas as well as you, and I repudiate the charge of false humility. I've been through too many troubles to be proud, and a pleasant, polite manner was the rule of my life even in the days when, God knows, I had everything. I've never changed, and if, with God's help, I had a civil tongue then, I have a civil tongue now. It's more than you always have, my poor perverse and passionate child. Once a lady always a lady – all the footlights in the world, turn them up as high as you will, won't make a difference. And I think people know it, people who know anything (if I may use such an expression), and it's because they know it that I'm not afraid to address them courteously. And I must say – and I call Mr Dormer to witness, for if he could reason with you a bit about it he might render several people a service – your conduct to Mr Sherringham simply breaks my heart,' Mrs Rooth concluded, with a jump of several steps in the fine modern avenue of her argument.

Nick was appealed to, but he hesitated a moment, and while he hesitated Miriam remarked: 'Mother is good – mother is very good; but it is only little by little that you discover how good she is.' This seemed to leave Nick free to ask Mrs Rooth, with the preliminary intimation that what she had just said was very striking, what she meant by her daughter's conduct to Peter Sherringham. Before Mrs Rooth could answer this question, however, Miriam interposed irrelevantly with one of her own. 'Do you mind telling me if you made your sister go off with Mr Sherringham because you knew it was about time for me to turn up? Poor Mr Dormer, I get you into trouble, don't I?' she added sympathetically.

'Into trouble?' echoed Nick, looking at her head but not at her eyes.

'Well, we won't talk about that!' Miriam exclaimed, with a rich laugh.

Nick now hastened to say that he had nothing to do with his sister's leaving the studio – she had only come, as it happened, for a moment. She had walked away with Peter Sherringham because they were cousins and old friends: he was to leave England immediately, for a long time, and he had offered her his company going home. Mrs Rooth shook her head very knowingly over the 'long time' that Mr Sherringham would be absent – she plainly had her ideas about that; and she conscientiously related that in the course of the short conversation they had all had at the door of the house her daughter had reminded Miss Dormer of something that had passed between them in Paris in regard to the charming young lady's modelling her head.

'I did it to make the question of our meeting less absurd – to put it on the footing of our both being artists. I don't ask you if she has talent,' said Miriam.

'Then I needn't tell you,' answered Nick.

'I'm sure she has talent and a very refined inspiration. I see something in that corner, covered with a mysterious veil,' Mrs Rooth insinuated; which led Miriam to ask immediately:

'Has she been trying her hand at Mr Sherringham?'

'When should she try her hand, poor dear young lady? He's always sitting with us,' said Mrs Rooth.

'Dear mamma, you exaggerate. He has his moments, when he seems to say his prayers to me; but we've had some success in cutting them down. *Il s'est bien détaché ces-jours-ci*, and I'm very happy for him. Of course it's an impertinent allusion for me to make; but I should be so delighted if I could think of him as a little in love with Miss Dormer,' the girl pursued, addressing Nick.

'He is, I think, just a little – just a tiny bit,' said Nick, working away; while Mrs Rooth ejaculated, to her daughter, simultaneously:

'How can you ask such fantastic questions when you know that he's dying for you?'

'Oh, dying! – he's dying very hard!' cried Miriam. 'Mr Sherringham's a man of whom I can't speak with too much esteem and affection, who may be destined to perish by some horrid fever (which God forbid!) in the unpleasant country he's going to. But he won't have caught his fever from your humble servant.'

'You may kill him even while you remain in perfect health yourself,' said Nick; 'and since we're talking of the matter I don't see the harm in my confessing that he strikes me as bad – oh, as very bad indeed.'

'And yet he's in love with your sister? – *je n'y suis plus.*'

'He tries to be, for he sees that as regards you there are difficulties. He would like to put his hand on some nice girl who would be an antidote to his poison.'

'Difficulties are a mild name for them; poison even is a mild name for the ill he suffers from. The principal difficulty is that he doesn't know what he wants. The next is that I don't either – or what I want myself. I only know what I don't want,' said Miriam brightly, as if she were uttering some happy, beneficent truth. 'I don't want a person who takes things even less simply than I do myself. Mr Sherringham, poor man, must be very uncomfortable, for one side of him is perpetually fighting against the other side. He's trying to serve God and Mammon, and I don't know how God will come off. What I like in you is that you have definitely let Mammon go – it's the only way. That's my earnest conviction, and yet they call us people light. Poor Mr Sherringham has tremendous ambitions – tremendous *riguardi*, as we used to say in Italy. He wants to enjoy every comfort and to save every appearance, and all without making a sacrifice. He expects others – me, for instance – to make all the sacrifices. *Merci*, much as I esteem him and much as I owe him! I don't know how he ever came to stray at all into our bold, bad Bohemia: it was a cruel trick for fortune to play him. He can't keep out of it, he's perpetually making dashes across the border, and yet he's not in the least at home there. There's another in whose position (if I were in it) I wouldn't look at the likes of us!'

'I don't know much about the matter, but I have an idea Peter thinks he has made, or at least is making, sacrifices.'

'So much the better – you must encourage him, you must help him.'

'I don't know what my daughter's talking about – she's much too clever for me,' Mrs Rooth put in. 'But there's one way you can encourage Mr Sherringham – there's one way you can help him; and perhaps it won't make it any worse for a gentleman of your good-nature that it will help me at the same time. Can't I look to you, dear Mr Dormer, to see that he does come to the theatre tonight – that he doesn't feel himself obliged to stay away?'

'What danger is there of his staying away?' Nick asked.

'If he's bent on sacrifices, that's a very good one to begin with,' Miriam observed.

'That's the mad, bad way she talks to him – she has forbidden the dear unhappy gentleman the house!' her mother cried. 'She brought it up to him just now, at the door, before Miss Dormer: such very odd form! She pretends to impose her commands upon him.'

'Oh, he'll be there – we're going to dine together,' said Nick. And when Miriam asked him what that had to do with it he went on: 'Why, we've arranged it; I'm going, and he won't let me go alone.'

'You're going? I sent you no places,' Miriam objected.

'Yes, but I've got one. Why didn't you, after all I've done for you?'

She hesitated a moment. 'Because I'm so good. No matter, she added: 'if Mr Sherringham comes I won't act.'

'Won't you act for me?'

'She'll act like an angel,' Mrs Rooth protested. 'She might do, she might be anything in the world; but she won't take common pains.'

'Of one thing there's no doubt,' said Miriam: 'that compared with the rest of us – poor passionless creatures – mamma does know what she wants.'

'And what is that?' inquired Nick, chalking away.

'She wants everything.'

'Never, never – I'm much more humble,' retorted the old woman; upon which her daughter requested her to give then to Mr Dormer, who was a reasonable man and an excellent judge, a general idea of the scope of her desires.

As, however, Mrs Rooth, sighing and deprecating, was not quick to comply with the injunction, the girl attempted a short cut to the truth with the abrupt inquiry: 'Do you believe for a single moment he'd marry me?'

'Why, he has proposed to you – you've told me yourself – a dozen times.'

'Proposed what to me? I've told you that neither a dozen times nor once, because I've never understood. He has made wonderful speeches, but he has never been serious.'

'You told me he had been in the seventh heaven of devotion, especially that night we went to the foyer of the Français,' Mrs Rooth insisted.

'Do you call the seventh heaven of devotion serious? He's in love with – *je veux bien*; he's so poisoned, as Mr Dormer vividly says, as to require an antidote; but he has never spoken to me as if he really expected me to listen to him, and he's the more of a gentleman from that fact. He knows we haven't a common ground – that a grasshopper can't mate with a fish. So he has taken care to say to me only more than he can possibly mean. That makes it just nothing.'

'Did he say more than he can possibly mean when he took formal leave of you yesterday – forever and ever?'

'Pray don't you call that a sacrifice?' Nick asked.

'Oh, he took it all back, his sacrifice, before he left the house.'

'Then has *that* no meaning?' demanded Mrs Rooth.

'None that I can make out.'

'Oh, I've no patience with you: you can be stupid when you will as well as clever when you will!' the old woman groaned.

'What mamma wishes me to understand and to practise is the particular way to be clever with Mr Sherringham,' said Miriam. 'There are doubtless depths of wisdom and virtue in it. But I can only see one way; namely, to be perfectly honest.'

'I like to hear you talk – it makes you live, brings you out,' Nick mentioned. 'And you sit beautifully still. All I want to say is, please continue to do so; remain exactly as you are – it's rather important – for the next ten minutes.'

'We're washing our dirty linen before you, but it's all right,' Miriam answered, 'because it shows you what sort of people we are, and that's what you need to know. Don't make me vague and arranged and fine, in this new thing,' she continued: 'make me characteristic and real; make life, with all its horrid facts and truths, stick out of me. I wish you could put mother in too; make us live there side by side and tell our little story. "The wonderful actress and her still more wonderful mamma" – don't you think that's an awfully good subject?'

Mrs Rooth, at this, cried shame on her daughter's wanton humours, professing that she herself would never accept so much from Nick's good-nature, and Miriam settled it that at any rate he was some day and in some way to do her mother and sail very near the wind.

'She doesn't believe he wants to marry me, any more than you do,' the girl, taking up her dispute again after a moment, represented to Nick; 'but she believes – how indeed can I tell you what she believes? – that I can work it (that's about it), so that in the fullness of time I shall hold him in a vice. I'm to keep him along for the present, but not to listen to him, for if I listen to him I shall lose him. It's ingenious, it's complicated; but I dare say you follow me.'

'Don't move – don't move,' said Nick. 'Excuse a beginner.'

'No, I shall explain quietly. Somehow (here it's *very* complicated and you mustn't lose the thread), I shall be an actress and make a tremendous lot of money, and somehow, too (I suppose a little later), I shall become an ambassadress and be the favourite of courts. So you see it will all be delightful. Only I shall have to go straight! Mamma reminds me of a story I once heard about the mother of a young lady who was in receipt of much civility from the pretender to a crown, which indeed he, and the young lady too, afterwards more or less wore. The old countess watched the course of events and gave her daughter the cleverest advice: "*Tiens bon, ma fille*, and you shall sit upon a throne." Mamma wishes me to *tenir bon* (she apparently thinks there's a danger I may not), so that if I don't sit upon a throne I shall at least parade at the foot of one. And if before that for ten years I pile up the money they'll forgive me the way I've made it. I should hope so, if I've *tenu bon*! Only ten years is a good while to hold out, isn't it? If it isn't Mr Sherringham it will be some one else. Mr Sherringham has the great merit of being a bird in the hand. I'm going to keep him along, I'm to be still more diplomatic than even he can be.'

Mrs Rooth listened to her daughter with an air of assumed reprobation which melted, before the girl had done, into a diverted, complacent smile – the gratification of finding herself the proprietress of so much wit and irony and grace. Miriam's account of her mother's views was a scene of comedy, and there was instinctive art in the way she added touch to touch and made point upon point. She was so quiet, to oblige her painter, that only her fine lips moved – all her expression was in their charming utterance. Mrs Rooth, after the first flutter of a less cynical spirit, consented to be sacrificed to an effect of an order she had now been educated to recognize; so that she hesitated only for a moment, when Miriam had ceased speaking, before she broke out endearingly with a little titter and '*Comédienne!*' She looked at Nick Dormer as if to say: 'Ain't she fascinating? That's the way she does for you!'

'It's rather cruel, isn't it,' said Miriam, 'to deprive people of the luxury of calling one an actress as they'd call one a liar? I represent, but I represent truly.'

'Mr Sherringham would marry you tomorrow – there's no question of ten years!' cried Mrs Rooth, with a comicality of plainness.

Miriam smiled at Nick, appealing for a sort of pity for her mother. 'Isn't it droll, the way she can't get it out of her head?' Then turning almost coaxingly to the old woman: '*Voyons*, look about you: they don't marry us like that.'

'But they do – *cela se voit tous les jours*. Ask Mr Dormer.'

'Oh, never!' said Miriam: 'it would be as if I asked him to give us a practical illustration.'

'I shall never give any illustration of matrimony; for me that question's over,' said Nick.

Miriam rested kind eyes on him. 'Dear me, how you must hate me!' And before he had time to reply she went on, to her mother: 'People marry them to make them leave the stage; which proves exactly what I say.'

'Ah, they offer them the finest positions,' reasoned Mrs Rooth.

'Do you want me to leave it then?'

'Oh, you can manage if you will!'

'The only managing I know anything about is to do my work. If I manage that, I shall pull through.'

'But, dearest, may our work not be of many sorts?'

'I only know one,' said Miriam.

At this Mrs Rooth got up with a sigh. 'I see you do wish to drive me into the street.'

'Mamma's bewildered – there are so many paths she wants to follow, there are so many bundles of hay. As I told you, she wishes to gobble them all,' Miriam went on. Then she added: 'Yes, go and take the carriage; take a turn round the Park – you always delight in that – and come back for me in an hour.'

'I'm too vexed with you; the air will do me good,' said Mrs Rooth. But before she went she added, to Nick: 'I have your assurance that you will bring him then tonight?'

'Bring Peter? I don't think I shall have to drag him,' said Nick. 'But you must do me the justice to remember that if I should resort to force I should do something that's not particularly in my interest – I should be magnanimous.'

'We must always be that, mustn't we?' moralized Mrs Rooth.

'How could it affect your interest?' Miriam inquired, less abstractedly, of Nick.

'Yes, as you say,' her mother reminded him, 'the question of marriage has ceased to exist for you.'

'Mamma goes straight at it!' laughed the girl, getting up while Nick rubbed his canvas before answering. Miriam went to Mrs Rooth and settled her bonnet and mantle in preparation for her drive; then stood for a moment with a filial arm about her, as if they were waiting for their host's explanation. This however when it came halted visibly.

'Why, you said awhile ago that if Peter was there you wouldn't act.'

'I'll act for *him*,' smiled Miriam, encircling her mother.

'It doesn't matter whom it's for!' Mrs Rooth declared sagaciously.

'Take your drive and relax your mind,' said the girl, kissing her. 'Come for me in an hour; not later, but not sooner.' She went with her to the door, bundled her out, closed it behind her and came back to the position she had quitted. '*This* is the peace I want!' she exclaimed, with relief, as she settled into it.

45

PETER SHERRINGHAM said so little during the performance that his companion was struck by his dumbness, especially as Miriam's acting seemed to Nick Dormer magnificent. He held his breath while she was on the stage – she gave the whole thing, including the spectator's emotion, such a lift. She had not carried out her fantastic menace of not exerting herself, and, as Mrs Rooth had said, it little mattered for whom she acted. Nick was conscious as he watched her that she went through it all for herself, for the idea that possessed her and that she rendered with extraordinary breadth. She could not open the door a part of the way to it and let it simply peep in; if it entered at all it must enter in full procession and occupy the premises in state.

This was what had happened on an occasion which, as Nick noted in his stall, grew larger with each throb of the responsive house; till by the time the play was half over it appeared to stretch out wide arms to the future. Nick had often heard more applause but he had never heard more attention; for they were all charmed and hushed together and success seemed to be sitting down with them. There had been of course plenty of announcement – the newspapers had abounded and the arts of the manager had taken the freest licence; but it was easy to feel a fine universal consensus and to recognize the intrinsic buoyancy of the evening. People snatched their eyes from the stage for an instant to look at each other, and a sense of intelligence deepened and spread. It was a part of the impression that the actress was only now really showing, for this time she had verse to deal with and she made it unexpectedly exquisite. She was beauty, she was music, she was truth; she was passion and persuasion and tenderness. She caught up the obstreperous play in soothing, entwining arms and carried it into the high places of poetry, of style. And she had such tones of nature, such concealments of art, such effusions of life, that the whole scene glowed with the colour she communicated, and the house, as if pervaded with rosy fire, glowed back at the scene. Nick looked round in the intervals; he felt excited and flushed – the night had turned into a feast of fraternity and he expected to see people embrace each other. The crowd, the flutter, the triumph, the surprise, the signals and rumours, the heated air, his associates, near him, pointing out other figures, who presumably were celebrated but whom he had never heard of, all amused him and banished every impulse of criticism. Miriam was as satisfactory as some right sensation – she would feed the memory with the ineffaceable.

One of the things that amused Nick, or at least helped to fill his attention, was Peter's attitude, which apparently did not exclude criticism; rather indeed mainly implied it. Sherringham never took his eyes off the actress, but he made no remark about her and he never stirred out of his chair. Nick had from the first a plan of going round to speak to her, but as his companion evidently meant not to move he felt a delicacy in regard to being more forward. During their brief dinner together (they made a rigid point of not being late), Peter had been silent and irremediably serious, but also, his kinsman judged, full of the wish to make it plain that he was calm. In his seat he was calmer than ever; had an air even of trying to suggest to Nick that his attendance, preoccupied as he was with deeper solemnities, was slightly mechanical, the result of a conception of duty, a habit of courtesy. When during a scene in the second act – a scene from which Miriam was absent – Nick observed to him that from his inexpressiveness one might gather he was not pleased, he replied after a moment: 'I've been looking for her mistakes.' And when Nick rejoined to this that he certainly wouldn't find them he said again, in an odd tone: 'No, I shan't find them – I shan't find them.' It might have seemed that since the girl's performance was a dazzling success he regarded his evening as rather a failure.

After the third act Nick said candidly: 'My dear fellow, how can you sit here? Aren't you going to speak to her?'

To which Peter replied inscrutably: 'Lord, no, never again; I bade her good-bye yesterday. She knows what I think of her form. It's very good, but she carries it a little too far. Besides, she didn't want me to come, and it's therefore more discreet to keep away from her.'

'Surely it isn't an hour for discretion!' cried Nick. 'Excuse me, at any rate, for five minutes.'

He went behind and reappeared only as the curtain was rising on the fourth act; and in the interval between the fourth and fifth he went again for a shorter time. Peter was personally detached, but he consented to listen to his companion's vivid account of the state of things on the stage, where the elation of victory had made every one merry. The strain was over, the ship was in port, and they were all wiping their faces and grinning. Miriam – yes, positively – was grinning too, and she hadn't asked a question about Peter nor sent him a message. They were kissing all round and dancing for joy. They were on the eve, worse luck, of a tremendous run. Peter groaned, irrepressibly, at this; it was, save for a slight manifestation a moment later, the only sign of emotion that Nick's report elicited from him. There was but one voice of regret that they hadn't put on the piece earlier, as the end of the season would interrupt the run. There was but one voice too about the fourth act – it was believed that all London would rush to see the fourth act. There was a wonderful lot of people, and Miriam was charming; she was receiving there, in the ugly place, like a kind of royalty, with a smile and a word for each. She was like a young queen on her accession. When she saw him, Nick, she had kissed her hand to him over the heads of the courtiers. Nick's artless comment on this was that she had such pretty manners. It made Sherringham laugh, apparently at his companion's conception of the manners of a young queen. Mrs Rooth, with a dozen shawls on her arm, was as red as a turkey; but you couldn't tell whether Miriam was red or pale: she was so cleverly, awfully cleverly, painted – perhaps a little too much. Dashwood of course was greatly to the fore, but you didn't have to mention his own performance to him: he was magnanimous and would use nothing but the feminine pronoun. He didn't say much, indeed, but he evidently had ideas; he nodded significant things and whistled inimitable sounds – 'heuh, heuh!' He was perfectly satisfied; moreover he looked further ahead than any one.

It was on coming back to his place after the fourth act that Nick put in, for Sherringham's benefit, most of these touches in his sketch of the situation. If Peter had continued to look for Miriam's mistakes he had not yet found them: the fourth act, bristling with dangers, putting a premium on every sort of cheap effect, had rounded itself without a flaw. Sitting there alone while Nick was away he had leisure to meditate on the wonder of this – on the art with which the girl had separated passion from violence, filling the whole place and never screaming; for it had seemed to him in London sometimes of old that the yell of theatrical emotion rang through the shrinking night like a fatal warning. Miriam had never been more present to him than at this hour; but she was inextricably transmuted – present essentially as the romantic heroine she represented. His state of mind was of the strangest, and he was conscious of its strangeness; just as he was conscious, in his person, of a cessation of resistance which identified itself absurdly with liberation. He felt weak at the same time that he felt excited, and he felt excited at the same time that he knew or believed he knew that his face was a blank. He saw things as a shining confusion, and yet somehow something monstrously definite kept surging out of them. Miriam was a beautiful, actual, fictive, impossible young woman, of a past age and undiscoverable country, who spoke in blank verse and overflowed with metaphor, who was exalted and heroic beyond all human convenience, and who yet was irresistibly real and related to one's own affairs. But that reality was a part of her spectator's joy, and she was not changed back to the common by his perception of the magnificent trick of art with which it was connected. Before Nick Dormer rejoined him Sherringham, taking a visiting-card from his pocket, wrote on it in pencil a few words in a foreign tongue; but as at that moment he saw Nick coming in he immediately put it out of view.

The last thing before the curtain rose on the fifth act Nick mentioned that he had brought him a message from Basil Dashwood, who hoped they both, on leaving the theatre, would come to supper with him, in company with Miriam and her mother and several others: he had prepared a little informal banquet in honour of so famous a night. At this, while the curtain was rising, Peter immediately took out his card again and added something – he wrote the finest small hand you could see. Nick asked him what he was doing, and after an hesitation he replied:

'It's a word to say I can't come.'

'To Dashwood? Oh, I shall go,' said Nick.

'Well, I hope you'll enjoy it!' his companion replied in a tone which came back to him afterwards.

When the curtain fell on the last act the people stayed, standing up in their places for the most part. The applause shook the house – the recall became a clamour, the relief from a long tension. This was a moment, in any performance, that Sherringham detested, but he stood for an instant beside Nick, who clapped like a school-boy. There was a veritable roar and the curtain drew back at the side most removed from them. Sherringham could see that Basil Dashwood was holding it, making a passage for the male 'juvenile lead', who had Miriam in tow. Nick redoubled his efforts; heard the plaudits swell; saw the bows of the leading gentleman, who was hot and fat; saw Miriam, personally conducted and closer to the footlights, grow brighter and bigger and more swaying; and then became aware that Sherringham had with extreme agility slipped out of the stalls. Nick had already lost sight of him – he had apparently taken but a minute to escape from the house. Nick wondered at his quitting him without a farewell, if he was to leave England on the morrow and they were not to meet at the hospitable Dashwood's.

He wondered even what Peter was 'up to', since as he had assured him, there was no question of his going round to Miriam. He waited to see this young lady reappear three times, dragging Dashwood behind her at the second with a friendly arm, to whom, in turn, was hooked Miss Fanny Rover, the actress entrusted, in the piece, with the inevitable comic relief. He went out slowly, with the crowd, and at the door looked again for Peter, who struck him as deficient for once in finish. He couldn't know that, in another direction and while he was helping the house to 'rise' at Miriam, his kinsman had been particularly explicit.

On reaching the lobby Sherringham had pounced upon a small boy in buttons, who appeared to be superfluously connected with a desolate refreshment-room and was peeping, on tiptoe, at the stage, through the glazed hole in the door of a box. Into one of the child's hands he thrust the card he had drawn again from his waistcoat, and into the other the largest silver coin he could find in the same receptacle, while he bent over him with words of adjuration — words which the little page tried to help himself to apprehend by instantly attempting to peruse the other words written on the card.

'That's no use — it's Italian,' said Peter; 'only carry it round to Miss Rooth without a minute's delay. Place it in her hand and she'll give you some object — a bracelet, a glove or a flower — to bring me back as a sign that she has received it. I shall be outside; bring me there what she gives you and you shall have another shilling — only fly!'

Sherringham's small messenger sounded him a moment with the sharp face of London wage-earning, and still more of London tip-earning, infancy, and vanished as swiftly as a slave of the Arabian Nights. While his patron waited in the lobby the audience began to pour out, and before the urchin had come back to him Peter was clapped on the shoulder by Nick Dormer.

'I'm glad I haven't lost you,' said Nick; 'but why didn't you stay to give her a hand?'

'Give her a hand? I hated it.'

'My dear fellow, I don't follow you,' Nick rejoined. 'If you won't come to Dashwood's supper I fear our ways don't lie together.'

'Thank him very much; say I have to get up at an unnatural hour.' To this Peter added: 'I think I ought to tell you she may not be there.'

'Miss Rooth? Why, it's for her.'

'I'm waiting for a word from her – she may change her mind.'

Nick stared at his companion. 'For you? Why, what have you proposed?'

'I've proposed marriage,' said Peter in a strange voice.

'I say –!' Nick broke out; and at the same moment Peter's messenger squeezed through the press and stood before him.

'She has given me nothing, sir,' the boy announced; 'but she says I'm to say "All right!"'

Nick marvelled a moment. 'You've proposed through *him*?'

'Aye, and she accepts. Good-night!' Peter exclaimed; and, turning away, he bounded into a hansom. He said something to the driver through the roof, and Nick's eyes followed the cab as it started off. Nick was mystified, was even amused; especially when the youth in buttons, planted there and wondering too, remarked to him:

'Please, sir, he told me he'd give me a shilling, and he've forgot it.'

'Oh, I can't pay you for *that*!' Nick laughed. He was vexed about the supper.

46

PETER SHERRINGHAM rolled away through the summer night to St John's Wood. He had put the pressure of strong words upon Miriam, entreating her to drive home immediately, without any one, without even her mother. He wished to see her alone, for a purpose that he would fully and satisfactorily explain – couldn't she trust him? He implored her to remember his own situation and throw over her supper, throw over everything. He would wait for her with unspeakable impatience in Balaklava Place.

He did so when he got there, but it took half an hour. Interminable seemed his lonely vigil in Miss Lumley's drawing-room, where the character of the original proprietress came out to him more than before in a kind of after-glow of old sociabilities, a vulgar ghostly vibration. The numerous candles had been lighted for him, and Mrs Rooth's familiar fictions were lying about; but his nerves forbade him the solace of taking a chair and a book. He walked up and down, thinking and listening, and as the long window, the balmy air permitting, stood open into the garden he passed several times in and out. A carriage appeared to stop at the gate – then there was nothing; he heard the rare rattle of wheels and the far-off hum of London. His impatience was unreasonable, and though he knew this it persisted; it would have been no easy matter for Miriam to break away from the flock of her felicitators. Still less simple was it doubtless for her to leave poor Dashwood with his supper on his hands. Perhaps she would bring Dashwood with her to time her; she was capable of playing him – that is playing Sherringham – or even playing them both, that trick. Perhaps the little wretch in buttons (Peter remembered now the omitted shilling) had only pretended to go round with his card, had come back with an invented answer. But how could he know, since presumably he couldn't read Italian, that his answer would fit the message? Peter was sorry now that he himself had not gone round, not snatched Miriam bodily away, made sure of her and of what he wanted of her.

When half an hour had elapsed he regarded it as proved that she wouldn't come, and, asking himself what he should do, determined to drive off again and seize her at Basil Dashwood's feast. Then he remembered Nick had mentioned that this entertainment was not to be held at the young actor's lodgings, but at some tavern or restaurant, the name of which he had not heeded. Suddenly however Sherringham became aware with joy that this name didn't matter, for there was something at the garden-door at last. He rushed out before Miriam had had time to ring, and saw as she stepped out of the carriage that she was alone. Now that she was there, that he had this evidence she had listened to him and trusted him, all his impatience and exasperation melted away and a flood of pleading tenderness came out in the first words he spoke to her. It was far 'dearer' of her than he had any right to dream, but she was the best and kindest creature – this showed it – as well as the most wonderful. He was really not off his head with his contradictory ways; no, before heaven, he wasn't, and he would explain, he would make everything clear. Everything was changed.

Miriam stopped short, in the little dusky garden, looking at him in the light of the open window. Then she called back to the coachman – they had left the garden-door open: 'Wait for me, mind; I shall want you again.'

'What's the matter – won't you stay?' Peter asked. 'Are you going out again at this absurd hour? I won't hurt you,' he urged gently. And he went back and closed the garden-door. He wanted to say to the coachman: 'It's no matter; please drive away.' At the same time he wouldn't for the world have done anything offensive to Miriam.

'I've come because I thought it better tonight, as things have turned out, to do the thing you ask me, whatever it may be. That's probably what you calculated I would think eh? What this evening has been you've seen, and I must allow that your hand's in it. That you know for yourself – that you doubtless felt as you sat there. But I confess I don't imagine what you want of me here, now,' Miriam added. She had remained standing in the path.

Peter felt the irony of her 'now', and how it made a fool of him, but he had been prepared for it and for much worse. He had begged her not to think him a fool, but in truth at present he cared little if she did. Very likely he was, in spite of his plea that everything was changed – he cared little even himself. However, he spoke in the tone of intense reason and of the fullest disposition to satisfy her. This lucidity only took still more from the dignity of his tergiversation: his separation from her the day before had had such pretensions to being lucid. But the explanation, the justification, were in the very fact, and the fact had complete possession of him. He named it when he replied to Miriam: 'I've simply overrated my strength.'

'Oh, I knew – I knew! That's why I entreated you not to come!' she groaned. She turned away impatiently, and for a moment he thought she would retreat to her carriage. But he passed his hand into her arm, to draw her forward, and after an instant he felt her yield.

'The fact is we must have this thing out,' he said. Then he

added, as he made her go into the house, bending over her: 'The failure of my strength – that was just the reason of my coming.'

She burst out laughing at these words, as she entered the drawing-room, and her laugh made them sound pompous in their false wisdom. She flung off, as a good-natured tribute to the image of their having the thing out, a white shawl that had been wrapped round her. She was still painted and bedizened, in the splendid dress of her fifth act, so that she seemed protected and alienated by the character she had been representing. 'Whatever it is you want (when I understand), you'll be very brief, won't you? Do you know I've given up a charming supper for you? Mamma has gone there. I've promised to go back to them.'

'You're an angel not to have let her come with you. I'm sure she wanted to,' said Sherringham.

'Oh, she's all right, but she's nervous,' Miriam rejoined. Then she added quickly: 'Couldn't she keep you away after all?'

'Whom are you talking about?' Biddy Dormer was as absent from Sherringham's mind as if she had never existed.

'The charming girl you were with this morning. Is she so afraid of obliging me? Oh, she'd be so good for you!'

'Don't speak of that,' said Peter, gravely. 'I was in perfect good faith yesterday when I took leave of you. I was – I was. But I can't – I can't: you are too unutterably dear to me.'

'Oh, don't – please don't,' moaned Miriam. She stood before the fireless chimney-piece with one of her hands upon it. 'If it's only to say that, don't you know, what's the use?'

'It isn't only to say that. I've a plan, a perfect plan: the whole thing lies clear before me.'

'And what is the whole thing?'

He hesitated a moment. 'You say your mother's nervous. Ah, if you knew how nervous I am!'

'Well, I'm not. Go on.'

'Give it up – give it up!' stammered Sherringham.

'Give it up?' Miriam fixed him like a mild Medusa.

'I'll marry you tomorrow if you'll renounce; and in return for the sacrifice you make for me I'll do more for you than ever was done for a woman before.'

'Renounce, after tonight? Do you call that a plan?' asked Miriam. 'Those are old words and very foolish ones: you wanted something of the sort a year ago.'

'Oh, I fluttered round the idea then; we were talking in the air. I didn't really believe I could make you see it then, and certainly you didn't see it. My own future moreover wasn't definite to me. I didn't know what I could offer you. But these last months have made a difference, and I do know now. Now what I say is deliberate, it's deeply meditated. I simply can't live without you, and I hold that together we may do great things.'

'What sort of things?' Miriam inquired.

'The things of my profession – of my life – the things one does for one's country, the responsibility and the honour of great affairs; deeply fascinating when one's immersed in them, and more exciting than the excitements of the theatre. Care for me only a little and you'll see what they are, they'll take hold of you. Believe me, believe me,' Sherringham pleaded, 'every fibre of my being trembles in what I say to you.'

'You admitted yesterday it wouldn't do,' said Miriam. 'Where were the fibres of your being then?'

'They trembled even more than now, and I was trying, like an ass, not to feel them. Where was this evening, yesterday – where were the maddening hours I've just spent? Ah, you're the perfection of perfections, and as I sat there tonight you taught me what I really want.'

'The perfection of perfections?' the girl repeated interrogatively, with the strangest smile.

'I needn't try to tell you: you must have felt tonight, with such rapture, what you are, what you can do. How can I give that up?' Sherringham asked.

'How can *I*, my poor friend? I like your plans and your responsibilities and your great affairs, as you call them. *Voyons*, they're infantile. I've just shown that I'm a perfection of perfections: therefore it's just the moment to renounce, as you gracefully say? Oh, I was sure, I was sure!' And Miriam paused, resting solicitous, pitying eyes upon her visitor, as if she were trying to think of some arrangement that would help him out of his absurdity. 'I was sure, I mean, that if you did come your poor dear doating brain would be quite addled,' she presently went on. 'I can't be a muff in public just for you, *pourtant*. Dear me, why do you like us so much?'

'Like you? I loathe you!'

'*Je le vois parbleu bien!* I mean, why do you feel us, judge us, understand us so well! I please you because you see, because you know; and because I please you, you must adapt me to your convenience, you must take me over, as they say. You admire me as an artist and therefore you wish to put me into a box in which the artist will breathe her last. Ah, be reasonable; you must let her live!'

'Let her live? As if I could prevent her living!' Peter cried, with unmistakable conviction. 'Even if I wanted, how could I prevent a spirit like yours from expressing itself? Don't talk about my putting you in a box, for, dearest child, I'm taking you out of one. The artist is irrepressible, eternal; she'll be in everything you are and in everything you do, and you'll go about with her triumphantly, exerting your powers, charming the world, carrying everything before you.'

Miriam's colour rose, through her paint, at this vivid picture, and she asked whimsically: 'Shall you like that?'

'Like my wife to be the most brilliant woman in Europe? I think I can do with it.'

'Aren't you afraid of me?'

'Not a bit.'

'Bravely said. How little you know me after all!' sighed the girl.

'I tell the truth,' Peter went on; 'and you must do me the justice to admit that I have taken the time to dig deep into my feelings. I'm not an infatuated boy; I've lived, I've had experience, I've observed; in short I know what I'm about. It isn't a thing to reason about; it's simply a need that consumes me. I've put it on starvation diet, but it's no use – really it's no use, Miriam,' poor Sherringham pursued, with a soft quaver that betrayed all his sincerity. 'It isn't a question of my trusting you; it's simply a question of your trusting me. You're all right, as I've heard you say yourself; you're frank, spontaneous, generous; you're a magnificent creature. Just quietly marry me, and I'll manage you.'

'Manage me?' The girl's inflection was droll; it made Sherringham change colour.

'I mean I'll give you a larger life than the largest you can get in any other way. The stage is great, no doubt, but the world is greater. It's a bigger theatre than any of those places in the Strand. We'll go in for realities instead of fables, and you'll do them far better than you do the fables.'

Miriam had listened to him attentively, but her face showed her despair at his perverted ingenuity. 'Excuse me for saying so, after your delightful tributes to my worth,' she returned in a moment, 'but I've never listened to such a flood of determined sophistry. You think so well of me that humility itself ought to keep me silent; nevertheless I *must* utter a few shabby words of sense. I'm a magnificent creature on the stage – well and good; it's what I want to be and it's charming to see such evidence that I succeed. But off the stage – come, come: I should lose all my advantages. The fact is so patent that it seems to me I'm very good-natured even to discuss it with you.'

'Are you on the stage now, pray? Ah, Miriam, if it were not for the respect I owe you!' her companion murmured.

'If it were not for that I shouldn't have come here to meet you. My talent is the thing that takes you: could there be a better proof than that it's tonight's exhibition of it that has settled you? It's indeed a misfortune that you're so sensitive to this particular kind of talent, since it plays such tricks with your power to see things as they are. Without it I should be a dull, ignorant, third-rate woman, and yet that's the fate you ask me to face and insanely pretend you are ready to face yourself.'

'Without it – without it?' Sherringham cried. 'Your own sophistry is infinitely worse than mine. I should like to see you without it for the fiftieth part of a second. What I ask you to give up is the dusty boards of the playhouse and the flaring footlights, but not the very essence of your being. Your talent is yourself, and it's because it's yourself that I yearn for you. If it had been a thing you could leave behind by the easy dodge of stepping off the stage I would never have looked at you a second time. Don't talk to me as if I were a simpleton, with your false simplifications! You were made to charm and console, to represent beauty and harmony and variety to miserable human beings; and the daily life of man is the theatre for that – not a vulgar shop with a turnstile, that's open only once in the twenty-four hours. Without it, verily?' Sherringham went on, with rising scorn and exasperated passion. 'Please let me know the first time you're without your face, without your voice, your step, your exquisite spirit, the turn of your head and the wonder of your eye!'

Miriam, at this, moved away from him with a port that resembled what she sometimes showed on the stage when she turned her young back upon the footlights and then, after a few steps, grandly swept round again. This evolution she performed (it was over in an instant) on the present occasion; even to stopping short with her eyes upon him and her head erect. 'Surely it's strange,' she said, 'the way the other solution never occurs to you.'

'The other solution?'

'That *you* should stay on the stage.'

'I don't understand you,' Sherringham confessed.

'Stay on *my* stage; come off your own.'

Sherringham hesitated a moment. 'You mean that if I'll do that you'll have me?'

'I mean that if it were to occur to you to offer me a little sacrifice on your own side, it might place the matter in a slightly more attractive light.'

'Continue to let you act – as my wife?' Sherringham demanded. 'Is it a real condition? Am I to understand that those are your terms?'

'I may say so without fear, because you'll never accept them.'

'Would *you* accept them, from me – accept the sacrifice, see me throw up my work, my prospects (of course I should have to do that), and simply become your appendage?'

'My dear fellow, you invite me with the best conscience in the world to become yours.'

'The cases are not equal. You would make of me the husband of an actress. I should make of you the wife of an ambassador.'

'The husband of an actress, *c'est bientôt dit*, in that tone of scorn! If you're consistent,' said Miriam, 'it ought to be a proud position for you.'

'What do you mean, if I'm consistent?'

'Haven't you always insisted on the beauty and interest of our art and the greatness of our mission? Haven't you almost come to blows with poor Gabriel Nash about it? What did all that mean if you won't face the first consequences of your theory? Either it was an enlightened conviction or it was an empty pretence. If it was heartless humbug I'm glad to know it,' Miriam rolled out, with a darkening eye. 'The better the cause, it seems to me, the better the deed; and if the theatre *is* important to the "human spirit", as you used to say so charmingly, and if into the bargain you have the pull of being so fond of me, I don't see why it should be monstrous to give us your services in an intelligent indirect way. Of course if you're not serious we needn't talk at all; but if you are, with your conception of what the actor can do, why is it so base to come to the actor's aid, taking one devotion with another? If I'm so fine I'm worth looking after a bit, and the place where I'm finest is the place to look after me!'

'You were never finer than at this minute, in the deepest domesticity of private life,' Sherringham returned. 'I have no conception whatever of what the actor can do, and no theory whatever about the importance of the theatre. Any infatuation of that sort has completely quitted me, and for all I care the theatre may go to the dogs.'

'You're dishonest, you're ungrateful, you're false!' Miriam flashed. 'It was the theatre that brought you here; if it hadn't been for the theatre I never would have looked at you. It was in the name of the theatre you first made love to me; it is to the theatre that you owe every advantage that, so far as I'm concerned, you possess.'

'I seem to possess a great many!' groaned Sherringham.

'You might certainly make more of those you have! You make me angry, but I want to be fair,' said the glowing girl, 'and I can't be unless you will. You are not fair, nor candid, nor honourable, when you swallow your words and abjure your faith, when you throw over old friends and old memories for a selfish purpose.'

' "Selfish purpose" is in your own convenient idiom, *bientôt dit*,' Sherringham answered. 'I suppose you consider that if I truly esteemed you I should be ashamed to deprive the world of the light of your genius. Perhaps my esteem isn't of the right quality: there are different kinds, aren't there? At any rate I've explained that I propose to deprive the world of nothing at all. You shall be celebrated, *allez*!'

'Rubbish – rubbish!' Miriam mocked, turning away again. 'I know of course,' she added quickly, 'that to befool yourself with such platitudes you must be pretty bad.'

'Yes, I'm pretty bad,' Sherringham admitted, looking at her dismally. 'What do you do with the declaration you made me the other day – the day I found my cousin here – that you'd take me if I should come to you as one who had risen high?'

Miriam reflected a moment. 'I remember – the chaff about the orders, the stars and garters. My poor dear friend, don't be so painfully literal. Don't you know a joke when you see it? It was to worry your cousin, wasn't it? But it didn't in the least succeed.'

'Why should you wish to worry my cousin?'

'Because he's so provoking. And surely I had my freedom no less than I have it now. Pray, what explanations should I have owed you and in what fear of you should I have gone? However, that has nothing to do with it. Say I did tell you that we might arrange it on the day that you should come to me covered with glory in the shape of little tinkling medals: why should you anticipate that transaction by so many years and knock me down such a long time in advance? Where is the glory, please, and where are the medals?'

'Dearest girl, am I not going to America (a capital promotion) next month,' Sherringham argued, 'and can't you trust me enough to believe that I speak with a real appreciation of the facts – that I'm not lying to you, in short – when I tell you that I've my foot in the stirrup? The glory's dawning. *I'm* all right too.'

'What you propose to me then is to accompany you *tout bonnement* to your new post.'

'You put it in a nutshell,' smiled Sherringham.

'You're touching; it has its charm. But you can't get anything in America, you know. I'm assured there are no medals to be picked up there. That's why the diplomatic body hate it.'

'It's on the way – it's on the way,' Sherringham hammered, feverishly. 'They don't keep us long in disagreeable places unless we want to stay. There's one thing you can get anywhere if you're clever, and nowhere if you're not, and in the disagreeable places generally more than in the others: and that (since it's the element of the question we're discussing) is simply success. It's odious to be put on one's swagger, but I protest against being treated as if I had nothing to offer – to offer to a person who has such glories of her own. I'm not a little presumptuous ass; I'm a man accomplished and determined, and the omens are on my side,' Peter faltered a moment, and then with a queer expression he went on: 'Remember, after all, that, strictly speaking, your glories are also still in the future.' An exclamation, at these words, burst from Miriam's lips, but her companion resumed quickly: 'Ask my official superiors, ask any of my colleagues if they consider that I've nothing to offer.'

Peter Sherringham had an idea, as he ceased speaking, that Miriam was on the point of breaking out with some strong word of resentment at his allusion to the contingent nature of her prospects. But it only twisted the weapon in his wound to hear her saying with extraordinary mildness: 'It's perfectly true that my glories are still to come, that I may fizzle out and that my little success of today is perhaps a mere flash in the pan. Stranger things have been – something of that sort happens every day. But don't we talk too much of that part of it?' she asked, with a weary tolerance that was noble in its effect. 'Surely it's vulgar to consider only the noise one's going to make; especially when one remembers how unintelligent nine-tenths of it will be. It isn't to my glories that I cling; it's simply to my idea, even if it's destined to sink me into obscurity. I like it better than anything else – a thousand times better (I'm sorry to have to put it in such a way) than tossing up my head as the fine lady of a little coterie.'

'A little coterie? I don't know what you're talking about!' Peter retorted, with considerable heat.

'A big coterie, then! It's only that, at the best. A nasty, prim "official" woman, who is perched on her little local pedestal and thinks she's a queen for ever because she's ridiculous for an hour! Oh, you needn't tell me. I've seen them abroad, I could imitate them here. I could do one for you on the spot if I were not so tired. It's scarcely worth mentioning perhaps, but I'm ready to drop.' Miriam picked up the white mantle she had tossed off, flinging it round her with her usual amplitude of gesture. 'They're waiting for me, and I confess I'm hungry. If I don't hurry they'll eat up all the nice things. Don't say I haven't been obliging, and come back when you're better. Good-night.'

'I quite agree with you that we've talked too much about the vulgar side of our question,' Peter responded, walking round to get between her and the French window, by which she apparently had a view of leaving the room. 'That's because I've wanted to bribe you. Bribery is almost always vulgar.'

'Yes, you should do better. *Merci!* There's a cab; some of them have come for me. I must go,' Miriam added, listening for a sound that reached her from the road.

Sherringham listened too, making out no cab. 'Believe me, it isn't wise to turn your back on such an affection as mine and on such a confidence,' he went on, speaking almost in a warning tone (there was a touch of superior sternness in it, as of a rebuke for real folly, but it was meant to be tender), and stopping her within a few feet of the window. 'Such things are the most precious that life has to give us,' he added, all but didactically.

Miriam had listened again for a moment; then she appeared to give up the idea of the cab. The reader need hardly be told, at this stage of her youthful history, that the right way for her lover to soothe her was not to represent himself as acting for her highest good. 'I like your calling it confidence,' she presently said; and the deep note of the few words had something of the distant mutter of thunder.

'What is it, then, when I offer you everything I am, everything I have, everything I shall achieve?'

She seemed to measure him for a moment, as if she were thinking whether she should try to pass him. But she remained where she was and she returned: 'I'm sorry for you, yes, but I'm also rather ashamed of you.'

'Ashamed of me?'

'A brave offer to see me through – that's what I should call confidence. You say today that you hate the theatre; and do you know what has made you do it? The fact that it has too large a place in your mind to let you repudiate it and throw it over with a good conscience. It has a deep fascination for you, and yet you're not strong enough to make the concession of taking up with it publicly, in my person. You're ashamed of yourself for that, as all your constant high claims for it are on record; so you blaspheme against it, to try and cover your retreat and your treachery and straighten out your personal situation. But it won't do, my dear fellow – it won't do at all,' Miriam proceeded, with a triumphant, almost judicial lucidity which made her companion stare; 'you haven't the smallest excuse of stupidity, and your perversity is no excuse at all. Leave her alone altogether – a poor girl who's making her way – or else come frankly to help her, to give her the benefit of your wisdom. Don't lock her up for life under the pretence of doing her good. What does one most good is to see a little honesty. You're the best judge, the best critic, the best observer, the best *believer*, that I've ever come across; you're committed to it by everything you've said to me for a twelvemonth, by the whole turn of your mind, by the way you've followed up this business of ours. If an art is noble and beneficent, one shouldn't be afraid to offer it one's arm. Your cousin isn't: he can make sacrifices.'

'My cousin?' shouted Peter. 'Why, wasn't it only the other day that you were throwing his sacrifices in his teeth?'

Under this imputation upon her consistency Miriam flinched but for an instant. 'I did that to worry *you*,' she smiled.

'Why should you wish to worry me if you care so little about me?'

'Care little about you? Haven't I told you often, didn't I tell you yesterday, how much I care? Ain't I showing it now by spending half the night here with you (giving myself away to all those cynics), taking all this trouble to persuade you to hold up your head and have the courage of your opinions?'

'You invent my opinions for your convenience,' said Peter. 'As long ago as the night I introduced you, in Paris, to Mademoiselle Voisin, you accused me of looking down on those who practise your art. I remember you almost scratched my eyes out because I didn't kootoo enough to your friend Dashwood. Perhaps I didn't; but if already at that time I was so wide of the mark you can scarcely accuse me of treachery now.'

'I don't remember, but I dare say you're right,' Miriam meditated. 'What I accused you of then was probably simply what I reproach you with now: the germ at least of your deplorable weakness. You consider that we do awfully valuable work, and yet you wouldn't for the world let people suppose that you really take our side. If your position was even at that time so false, so much the worse for you, that's all. Oh, it's refreshing,' the girl exclaimed, after a pause during which Sherringham seemed to himself to taste the full bitterness of despair, so baffled and derided he felt – 'oh, it's refreshing to see a man burn his ships in a cause that appeals to him, give up something for it and break with hideous timidities and snobberies! It's the most beautiful sight in the world.'

Sherringham, sore as he was and angry and exasperated, nevertheless burst out laughing at this. 'You're magnificent, you give me at this moment the finest possible illustration of what you mean by burning one's ships. Verily, verily, there's no one like you: talk of timidity, talk of refreshment! If I had any talent for it I'd go on the stage tomorrow, to spend my life with you the better.'

'If you'll do that I'll be your wife the day after your first appearance. That would be really respectable,' said Miriam.

'Unfortunately I've no talent.'

'That would only make it the more respectable.'

'You're just like Nick,' Peter rejoined: 'you've taken to imitating Gabriel Nash. Don't you see that it's only if it were a question of my going on the stage myself that there would be a certain fitness in your contrasting me invidiously with Nick Dormer and in my giving up one career for another? But simply to stand in the wing and hold your shawl and your smelling-bottle!' Peter concluded mournfully, as if he had ceased to debate.

'Holding my shawl and my smelling-bottle is a mere detail, representing a very small part of the various precious services, the protection and encouragement for which a woman in my position might be indebted to a man interested in her work and accomplished and determined, as you very justly describe yourself.'

'And would it be your idea that such a man should live on

the money earned by an exhibition of the person of his still more accomplished and still more determined wife?'

'Why not, if they work together – if there's something of his spirit and his support in everything she does?' Miriam demanded. '*Je vous attendais*, with the famous "person"; of course that's the great stick they beat us with. Yes, we show it for money, those of us who have anything to show, and some no doubt who haven't, which is the real scandal. What will you have? It's only the envelope of the idea, it's only our machinery, which ought to be conceded to us; and in proportion as the idea takes hold of us do we become unconscious of the clumsy body. Poor old "person" – if you knew what *we* think of it! If you don't forget it, that's your own affair: it shows that you're dense before the idea.'

'That I'm dense?'

'I mean the public is – the public who pays us. After all, they expect us to look at them too, who are not half so well worth it. If you should see some of the creatures who have the face to plant themselves there in the stalls, before one, for three mortal hours! I dare say it would be simpler to have no bodies, but we're all in the same box, and it would be a great injustice to the idea, and we're all showing ourselves all the while; only some of us are not worth paying.'

'You're extraordinarily droll, but somehow I can't laugh at you,' said Peter, his handsome face lengthened to a point that sufficiently attested the fact. 'Do you remember the second time I ever saw you – the day you recited at my place?' he abruptly inquired, a good deal as if he were drawing from his quiver an arrow which, if it was the last, was also one of the most pointed.

'Perfectly, and what an idiot I was, though it was only yesterday!'

'You expressed to me then a deep detestation of the sort of self-exposure to which the profession you were seeking to enter would commit you. If you compared yourself to a contortionist at a country fair I'm only taking my cue from you.'

'I don't know what I may have said then,' replied Miriam, whose steady flight was not arrested by this ineffectual bolt; 'I was, no doubt, already wonderful for talking of things I know nothing about. I was only on the brink of the stream and I perhaps thought the water colder than it is. One warms it a bit one's self, when once one is in. Of course I'm a contortionist and of course there's a hateful side; but don't you see how that very fact puts a price on every compensation, on the help of those who are ready to insist on the *other* side, the grand one, and especially on the sympathy of the person who is ready to insist most and to keep before us the great thing, the element that makes up for everything?'

'The element?' Peter questioned, with a vagueness which was pardonably exaggerated. 'Do you mean your success?'

'I mean what you've so often been eloquent about,' the girl returned, with an indulgent shrug – 'the way we simply stir people's souls. Ah, there's where life can help us,' she broke out, with a change of tone, 'there's where human relations and affections can help us; love and faith and joy and suffering and experience – I don't know what to call 'em! They suggest things, they light them up and sanctify them, as you may say; they make them appear worth doing.' She became radiant for a moment, as if with a splendid vision; then melting into still another accent, which seemed all nature and harmony, she proceeded: 'I must tell you that in the matter of what we can do for each other I have a tremendously high ideal. I go in for closeness of union, for identity of interest. A true marriage, as they call it, must do one a lot of good!'

Sherringham stood there looking at her a minute, during which her eyes sustained the rummage of his gaze without a relenting gleam of the sense of cruelty or of paradox. With a passionate but inarticulate ejaculation he turned away from her and remained, on the edge of the window, his hands in his pockets, gazing defeatedly, doggedly, into the featureless night, into the little black garden which had nothing to give him but a familiar smell of damp. The warm darkness had no relief for him, and Miriam's histrionic hardness flung him back against a fifth-rate world, against a bedimmed, star-punctured nature which had no consolation – the bleared, irresponsive eyes of the London heaven. For the brief space that he glared at these things he dumbly and helplessly raged. What he wanted was something that was not in *that* thick prospect. What was the meaning of this sudden offensive importunity of 'art', this senseless mocking catch, like some irritating chorus of conspirators in a bad opera, in which Miriam's voice was so incongruously conjoined with Nick's and in which Biddy's sweet little pipe had not scrupled still more bewilderingly to mingle? Art be damned: what commission, after all, had he ever given it to better him or bother him? If the pointless groan in which Peter exhaled a part of his humiliation had been translated into words, these words would have been as heavily charged with the genuine British mistrust of the bothersome principle as if the poor fellow speaking them had never quitted his island. Several acquired perceptions had struck a deep root in him, but there was an immemorial compact formation which lay deeper still. He tried at the present hour to rest upon it spiritually, but found it inelastic; and at the very moment when he was most conscious of this absence of the rebound or of any tolerable ease his vision was solicited by an object which, as he immediately guessed, could only add to the complication of things.

An undefined shape hovered before him in the garden, halfway between the gate and the house; it remained outside of the broad shaft of lamplight projected from the window. It wavered for a moment after it had become aware of Peter's observation, and then whisked round the corner of the little villa. This characteristic movement so effectually dispelled the mystery (it could only be Mrs Rooth who resorted to such conspicuous secrecies) that, to feel that the game was up and his interview over, Sherringham had no need of seeing the figure reappear on second thoughts and dodge about in the dusk with a vexatious sportive imbecility. Evidently Miriam's warning of a few minutes before had been founded: a cab had deposited her anxious mother at the garden-door. Mrs Rooth had entered with precautions; she had approached the house and retreated; she had effaced herself – had peered and waited and listened. Maternal solicitude and muddled calculations had drawn her away from a festival as yet only imperfectly commemorative. The heroine of the occasion of course had been intolerably missed, so that the old woman had both obliged the company and quieted her own nerves by jumping insistently into a hansom and rattling up to St John's Wood to reclaim the absentee. But if she had wished to be in time she had also desired not to be abrupt; she would have been still more embarrassed to say what she aspired to promote than to phrase what she had proposed to hinder. She wanted to abstain tastefully, to interfere felicitously, and, more generally and justifiably (the small hours had come), to see what her young charges were doing. She would probably have gathered that they were quarrelling, and she appeared now to be telegraphing to Sherringham to know if it were over. He took no notice of her signals, if signals they were; he only felt that before he made way for the odious old woman there was one faint little spark he might strike from Miriam's flint.

Without letting her guess that her mother was on the premises he turned again to his companion, half expecting that she would have taken her chance to regard their discussion as more than terminated and by the other egress flit away from him in silence. But she was still there; she was in the act of approaching him with a manifest intention of kindness, and she looked indeed, to his surprise, like an angel of mercy.

'Don't let us part so disagreeably,' she said, 'with your trying to make me feel as if I were merely disobliging. It's no use talking – we only hurt each other. Let us hold our tongues like decent people and go about our business. It isn't as if you hadn't any cure – when you have such a capital one. Try it, try it, my dear friend – you'll see! I wish you the highest promotion and the quickest – every success and every reward. When you've got them all, some day, and I've become a great swell too, we'll meet on that solid basis and you'll be glad I've been nasty now.'

'Surely before I leave you I've a right to ask you this,' Sherringham answered, holding fast in both his own the cool hand of farewell that she had finally tormented him with. 'Are you ready to follow up by a definite promise your implied assurance that I have a remedy?'

'A definite promise?' Miriam benignly gazed, with a perfection of evasion. 'I don't "imply" that you have a remedy. I declare it on the housetops. That delightful girl –'

'I'm not talking of any delightful girl but you!' Peter broke in with a voice which, as he afterwards learned, struck Mrs Rooth's ears, in the garden, with affright. 'I simply hold you, under pain of being convicted of the grossest prevarication, to the strict sense of what you said a quarter of an hour ago.'

'Ah, I've said so many things; one has to do that to get rid of you. You rather hurt my hand,' she added, jerking it away in a manner that showed that if she was an angel of mercy her mercy was partly for herself.

'As I understand you, then, I may have some hope if I do renounce my profession?' Peter pursued. 'If I break with everything, my prospects, my studies, my training, my emoluments, my past and my future, the service of my country and the ambition of my life, and engage to take up instead the business of watching your interests so far as I may learn how and ministering to your triumphs so far as may in me lie – if after further reflection I decide to go through these preliminaries, have I your word that I may definitely look to you to reward me with your precious hand?'

'I don't think you have the right to put the question to me now,' said Miriam, with a promptitude partly produced perhaps by the clear-cut form Peter's solemn speech had given (it was a charm to hear it) to each item of his enumeration. 'The case is so very contingent, so dependent on what you ingeniously call your further reflection. While you reserve yourself you ask me to commit myself. If it's a question of further reflection, why did you drag me up here? And then,' she added, 'I'm so far from wishing you to take any such monstrous step.'

'Monstrous, you call it? Just now you said it would be sublime.'

'Sublime if it's done with spontaneity, with passion; ridiculous if it's done after further reflection. As you said perfectly a while ago, it isn't a thing to reason about.'

'Ah, what a help you'd be to me in diplomacy!' Sherringham cried. 'Will you give me a year to consider?'

'Would you trust me for a year?'

'Why not, if I'm ready to trust you for life?'

'Oh, I shouldn't be free then, worse luck. And how much you seem to take for granted one must like you!'

'Remember that you've made a great point of your liking me. Wouldn't you do so still more if I were heroic?'

Miriam looked at him a moment. 'I think I should pity you, in such a cause. Give it all to *her*; don't throw away a real happiness!'

'Ah, you can't back out of your position with a few vague and even rather impertinent words!' Sherringham declared. 'You accuse me of swallowing my protestations, but you swallow yours. You've painted in heavenly colours the sacrifice I'm talking of, and now you must take the consequences.'

'The consequences.'

'Why, my coming back in a year to square you.'

'Ah, you're tiresome!' cried Miriam. 'Come back when you like. I don't wonder you've grown desperate, but fancy *me* then!' she added, looking past him at a new interlocutor.

'Yes, but if he'll square you!' Peter heard Mrs Rooth's voice respond conciliatingly behind him. She had stolen up to the window now, she had passed the threshold, she was in the room, but her daughter had not been startled. 'What is it he wants to do, dear?' she continued, to Miriam.

'To induce me to marry him if he'll go upon the stage. He'll practise over there, where he's going, and then he'll come back and appear. Isn't it too dreadful? Talk him out of it, stay with him, soothe him!' the girl hurried on. 'You'll find some *bibite* and some biscuits in the cupboard – keep him with you, pacify him, give him *his* little supper. Meanwhile I'll go to mine; I'll take the brougham; don't follow!'

With these words Miriam bounded into the garden, and her white drapery shone for an instant in the darkness before she disappeared. Peter looked about him, to pick up his hat, and while he did so he heard the bang of the gate and the quick carriage getting into motion. Mrs Rooth appeared to sway excitedly for a moment in opposed directions: that of the impulse to rush after Miriam and that of the extraordinary possibility to which the young lady had alluded. She seemed in doubt, but at a venture, detaining him with a maternal touch, she twinkled up at their visitor like an insinuating glow-worm.

'I'm so glad you came.'

'I'm not. I've got nothing by it,' he said, finding his hat.

'Oh, it was so beautiful!' she coaxed.

'The play – yes, wonderful. I'm afraid it's too late for me to avail myself of the privilege your daughter offers me. Goodnight.'

'Oh, it's a pity; won't you take *anything*?' asked Mrs Rooth.

'When I heard your voice so high I was scared and I hung back.' But before he could reply she added: 'Are you really thinking of the stage?'

'It comes to the same thing.'

'Do you mean you've proposed?'

'Oh, unmistakably.'

'And what does she say?'

'Why, you heard: she says I'm an ass.'

'Ah, the little rascal!' laughed Mrs Rooth. 'Leave her to me. I'll help you. But you *are* mad. Give up nothing – least of all your advantages.'

'I won't give up your daughter,' said Peter, reflecting that if this was cheap it was at any rate good enough for Mrs Rooth. He mended it a little indeed by adding darkly: 'But you can't make her take me.'

'I can prevent her taking any one else.'

'Oh, can you!' Peter ejaculated, with more scepticism than ceremony.

'You'll see – you'll see.' He passed into the garden, but, after she had blown out the candles and drawn the window to, Mrs Rooth went with him. 'All you've got to do is to be yourself – to be true to your fine position,' she explained as they proceeded. 'Trust me with the rest – trust me and be quiet.'

'How can one be quiet after this magnificent evening?'

'Yes, but it's just that!' panted the eager old woman. 'It has launched her so, on this sea of dangers, that to make up for the loss of the old security (don't you know?) we must take a still firmer hold.'

'Ay, of what?' asked Sherringham, as Mrs Rooth's comfort became vague while she stopped with him at the garden-door.

'Ah, you know: of the *real* life, of the true anchor!' Her hansom was waiting for her, and she added: 'I kept it, you see; but a little extravagance, on the night one's fortune has come!'

Peter stared. Yes, there were people whose fortune had come; but he managed to stammer: 'Are you following her again?'

'For you – for you!' And Mrs Rooth clambered into the vehicle. From the seat, enticingly, she offered him the place beside her. 'Won't you come too? I know he asked you.' Peter declined with a quick gesture, and as he turned away he heard her call after him, to cheer him on his lonely walk: 'I shall keep this up; I shall never lose sight of her!'

47

WHEN Mrs Dallow returned to London, just before London broke up, the fact was immediately known in Calcutta Gardens and was promptly communicated to Nick Dormer by his sister Bridget. He had learnt it no other way – he had had no correspondence with Julia during her absence. He gathered that his mother and sisters were not ignorant of her whereabouts (he never mentioned her name to them); but as to this he was not sure whether the source of their information was the *Morning Post* or a casual letter received by the inscrutable Biddy. He knew that Biddy had some epistolary commerce with Julia, and he had an impression that Grace occasionally exchanged letters with Mrs Gresham. Biddy, however, who, as he was well aware, was always studying what he would like, forbore to talk to him about the absent mistress of Harsh, beyond once dropping the remark that she had gone from Florence to Venice and was enjoying gondolas and sunsets too much to leave them. Nick's comment on this was that she was a happy woman to have such a go at Titian and Tintoret: as he spoke and for some time afterwards the sense of how he himself should enjoy a similar 'go' made him ache with ineffectual longing.

He had forbidden himself for the present to think of absence, not only because it would be inconvenient and expensive, but because it would be a kind of retreat from the enemy, a concession to difficulty. The enemy was no particular person and no particular body of persons: not his mother; not Mr Carteret, who, as Nick heard from the doctor at Beauclere, lingered on, sinking and sinking till his vitality appeared to have the vertical depth of a gold-mine; not his pacified constituents, who had found a healthy diversion in returning another Liberal, wholly without Mrs Dallow's aid (she had not participated even to the extent of a responsive telegram in the election); not his late colleagues in the House, nor the biting satirists of the newspapers, nor the brilliant women he took down at dinner-parties (there was only one sense in which he ever took them down), nor his friends, nor his foes, nor his private thoughts, nor the periodical phantom of his shocked father: it was simply the general awkwardness of his situation. This awkwardness was connected with the sense of responsibility that Gabriel Nash so greatly deprecated – ceasing to roam, of late, on purpose to miss as few scenes as possible of the drama, rapidly growing dull, alas, of his friend's destiny; but that compromising relation scarcely drew the soreness from it. The public flurry produced by Nick's collapse had only been large enough to mark the flatness of his position when it was over. To have had a few jokes cracked audibly at one's expense was not an ordeal worth talking of; the hardest thing about it was merely that there had not been enough of them to yield a proportion of good ones. Nick had felt, in short, the benefit of living in an age and in a society where number and pressure have, for the individual figure, especially when it's a zero, compensations almost equal to their cruelties.

No, the pinch, for our young man's conscience, after a few weeks had passed, was simply an acute mistrust of the superficiality of performance into which the desire to justify himself might hurry him. That desire was passionate as regards Julia Dallow; it was ardent also as regards his mother; and, to make it absolutely uncomfortable it was complicated with the conviction that neither of them would recognize his justification even when they should see it. They probably couldn't if they would, and very likely they wouldn't if they could. He assured himself however that this limitation wouldn't matter; it was their affair – his own was simply to have the right sort of thing to show. The work he was now attempting was not the right sort of thing; though doubtless Julia for instance would dislike it almost as much as if it were. The two portraits of Miriam, after the first exhilaration of his finding himself at large, filled him with no private glee: they were not in the direction in which for the present he wished really to move. There were moments when he felt almost angry, though of course he held his tongue, when by the few persons who saw them they were pronounced wonderfully clever. That they were wonderfully clever was just the detestable thing in them, so active had that cleverness been in making them seem better than they were. There were people to whom he would have been ashamed to show them, and these were the people whom it would give him most pleasure some day to please. Not only had he many an hour of disgust with his actual work, but he thought he saw, as in an ugly revelation, that nature had cursed him with an odious facility and that the lesson of his life, the sternest and wholesomest, would be to keep out of the trap it had laid for him. He had fallen into this trap on the threshold and he had only scrambled out with his honour. He had a talent for appearance, and that was the fatal thing; he had a damnable suppleness and a gift of immediate response, a readiness to oblige, that made him seem to take up causes which he really left lying, enabled him to learn enough about them in an hour to have all the air of having made them his own. Many people call them their own who had taken them in much less. He was too clever by half, since this pernicious overflow had been at the bottom of deep disappointments and heart-burnings. He had assumed a virtue and enjoyed assuming it, and the assumption had cheated his father and his mother and his affianced wife and his rich benefactor and the candid burgesses of Harsh and the cynical reporters of the newspapers. His enthusiasms had been but young curiosity, his speeches had been young agility, his professions and adhesions had been like postage-stamps without glue: the head was all right, but they wouldn't stick. He stood ready now to wring the neck of the irrepressible vice which certainly would like nothing better than to get him into further trouble. His only real justification would be to turn patience (his own, of course) inside out; yet if there should be a way to misread that recipe his humbugging genius could be trusted infallibly to discover it. Cheap and easy results would dangle before him, little amateurish conspicuities, helped by his history, at exhibitions; putting it in his power to triumph with a quick 'What do you say to that?' over those he had wounded. The fear of this danger was corrosive; it poisoned even legitimate joys. If he should have a striking picture at the Academy next year it wouldn't be a crime: yet he couldn't help suspecting any conditions that would enable him to be striking so soon. In this way he felt quite enough how Gabriel Nash 'had' him whenever he railed at his fever for proof, and how inferior as a productive force the desire to win over the ill-disposed might be to the principle of quiet growth. Nash had a foreign manner of lifting up his finger and waving it before him, as if to put an end to everything, whenever it became, in conversation or discussion, to any extent a question whether any one would like anything.

It was presumably, in some degree at least, a due respect for the principle of quiet growth that kept Nick on the spot at present, made him stick fast to Rosedale Road and Calcutta Gardens and deny himself the simplifications of absence. Do what he would he could not despoil himself of the impression that the disagreeable was somehow connected with the salutary and the 'quiet' with the disagreeable, when stubbornly borne; so he resisted a hundred impulses to run away to Paris or to Florence and the temptation to persuade himself by material motion that he was launched. He stayed in London because it seemed to him that he was more conscious of what he had undertaken, and he had a horror of shirking that consciousness. One element in it indeed was the perception that he would have found no great convenience in a foreign journey, even had his judgement approved such a subterfuge. The stoppage of his supplies from Beauclere had now become an historic fact, with something of the majesty of its class about it: he had had time to see what a difference this would make in his life. His means were small and he had several old debts, the number of which, as he believed, loomed large to his mother's imagination. He could never tell her that she exaggerated, because he told her nothing of that sort now: they had no intimate talk, for an impenetrable partition, a tall bristling hedge of untrimmed misconceptions had sprung up between them. Poor Biddy had made a hole in it, through which she squeezed, from side to side, to keep up communications, at the cost of many rents and scratches; but Lady Agnes walked straight and stiff, never turning her head, never stopping to pluck the least little daisy of consolation. It was in this manner she wished to signify that she had accepted her wrongs. She draped herself in them as in a kind of Roman mantle, and she had never looked so proud and wasted and handsome as now that her eyes rested only upon ruins.

Nick was extremely sorry for her, though he thought there was a dreadful want of grace in her never setting a foot in Rosedale Road (she mentioned his studio no more than if it had been a private gambling-house, or something worse); sorry because he was well aware that for the hour everything must appear to her to have crumbled. The luxury of Broadwood would have to crumble: his mind was very clear about that. Biddy's prospects had withered to the finest, dreariest dust, and Biddy indeed, taking a lesson from her brother's perversities, seemed little disposed to better a bad business. She professed the most peacemaking sentiments, but when it came really to doing something to brighten up the scene she showed herself portentously corrupted. After Peter Sherringham's heartless flight she had wantonly slighted an excellent opportunity to repair her misfortune. Lady Agnes had reason to know, about the end of June, that young Mr Grindon, the only son (the other children were girls) of an immensely rich industrial and political baronet in the north, was literally waiting for the faintest sign. This reason she promptly imparted to her younger daughter, whose intelligence had to take it in but who had shown it no other consideration. Biddy had set her charming face as a stone; she would have nothing to do with signs, and she, practically speaking, wilfully, wickedly refused a magnificent offer, so that the young man carried his noble expectations elsewhere. How much in earnest he had been was proved by the fact that before Goodwood had come and gone he was captured by Lady Muriel Macpherson. It was superfluous to insist on the frantic determination to get married revealed by such an accident as that. Nick knew of this episode only through Grace, and he deplored its having occurred in the midst of other disasters.

He knew or he suspected something more as well – something about his brother Percival which, if it should come to light, no season would be genial enough to gloss over. It had usually been supposed that Percy's store of comfort against the ills of life was confined to the infallibility of his rifle. He was not sensitive, but he had always the consolation of killing something. It had suddenly come to Nick's ears however that he had another resource as well, in the person of a robust countrywoman, housed in an ivied corner of Warwickshire, in whom he had long been interested and whom, without any flourish of magnanimity, he had ended by making his wife. The situation of the latest born of the pledges of this affection, a blooming boy (there had been two or three previously), was therefore perfectly regular and of a nature to make a difference in the worldly position, as the phrase is, of his moneyless uncle. If there be degrees in the absolute and Percy had an heir (others, moreover, would supposably come), Nick would have to regard himself as still more moneyless than before. His brother's last step was doubtless, under the circumstances, to be commended; but such discoveries were enlivening only when they were made in other families, and Lady Agnes would scarcely enjoy learning to what tune she had become a grandmother.

Nick forbore from delicacy to intimate to Biddy that he thought it a pity she couldn't care for Mr Grindon; but he had a private sense that if she had been capable of such an achievement it would have lightened a little the weight he himself had to carry. He bore her a slight grudge, which lasted until Julia Dallow came back; when the circumstance of the girl's being summoned immediately down to Harsh created a diversion that was perhaps after all only fanciful. Biddy, as we know, entertained a theory, which Nick had found occasion to combat, that Mrs Dallow had not treated him perfectly well; therefore in going to Harsh the very first time Julia held out a hand to her, so jealous a little sister must have recognized a special inducement. The inducement might have been that Julia had comfort for her, that she was acting by the direct advice of this acute lady, that they were still in close communion on the question of the offers Biddy was not to accept, that in short Peter Sherringham's sister had taken upon herself to see that Biddy should remain free until the day of the fugitive's inevitable return. Once or twice indeed Nick wondered whether Mrs Dallow herself was visited, in a larger sense, by the thought of retracing her steps – whether she wished to draw out her young friend's opinion as to how she might do so gracefully. During the few days she was in town Nick had seen her twice, in Great Stanhope Street, but not alone. She had said to him on one of these occasions, in her odd, explosive way: 'I should have thought you'd have gone away somewhere – it must be such a bore.' Of course she firmly believed he was staying for Miriam, which he really was not; and probably she had written this false impression off to Peter, who still more probably would prefer to regard it as just. Nick was staying for Miriam only in the sense that he should be very glad of the money he might receive for the portrait he was engaged in painting. That money would be a great convenience to him, in spite of the obstructive ground Miriam had taken in pretending (she had blown half a gale about it) that he had no right to dispose of such a production without her consent. His answer to this was simply that the purchaser was so little of a stranger that it didn't go, as it were, out of the family, out of hers. It didn't matter that Miriam should protest that if Mr Sherringham had formerly been no stranger he was now utterly one, so that there could be nothing less soothing to him than to see her hated image on his wall. He would back out of the bargain, and Nick would be left with his work on his hands. Nick jeered at this shallow theory and, when she came to sit, the question served as well as another to sprinkle their familiar silences with chaff. Nick already knew something, as we have seen, of the conditions in which his distracted kinsman had left England; and this connected itself in casual meditation with some of the calculations that he had attributed to Julia and Biddy. There had naturally been a sequel to the queer behaviour in which Peter had indulged, at the theatre, on the eve of his departure – a sequel embodied in a remark dropped by Miriam in the course of the first sitting she gave Nick after her great night. 'Fancy' – so this observation ran – 'fancy the dear man finding time, in the press of all his last duties, to ask me to marry him!'

'He told me you had found time, in the press of all yours, to say you would,' Nick replied. And this was pretty much all that had passed on the subject between them, save of course, that Miriam immediately made it clear that Peter had grossly misinformed him. What had happened was that she had said she would do nothing of the sort. She professed a desire not to be confronted again with this trying theme, and Nick easily fell in with it, from a definite preference he now had not to handle that kind of subject with her. If Julia had false ideas about him, and if Peter had them too, his part of the business was to take the simplest course to establish that falsity. There were difficulties indeed attached even to the simplest course, but there would be a difficulty the less if, in conversation, one should forbear to meddle with the general suggestive topic of intimate unions. It is certain that in these days Nick cultivated the practice of forbearances for which he did not receive, for which perhaps he never would receive, due credit.

He had been convinced for some time that one of the next things he should hear would be that Mrs Dallow had arranged to marry Mr Macgeorge or some such leader of multitudes. He could think of that now, he found – think of it with resignation, even when Julia was before his eyes, looking so handsomely forgetful that her air had to be taken as referring still more to their original intimacy than to his comparatively superficial offence. What made this accomplishment of his own remarkable was that there was something else he thought of quite as much – the fact that he had only to see her again to feel by how great a charm she had in the old days taken possession of him. This charm operated apparently in a very direct, primitive way: her presence diffused it and fully established it, but her absence left comparatively little of it behind. It dwelt in the very facts of her person – it was something that she happened physically to be; yet (considering that the question was of something very like loveliness) its envelope of associations, of memories and recurrences, had no great density. She packed it up and took it away with her, as if she had been a woman who had come to sell a set of laces. The laces were as wonderful as ever when they were taken out of the box, but to get another look at them you had to send for the woman. What was above all remarkable was that Miriam Rooth was much less irresistible to our young man than Mrs Dallow could be when Mrs Dallow was on the spot. He could paint Miriam, day after day, without any agitating blur of vision; in fact the more he saw of her the clearer grew the atmosphere through which she blazed, the more her richness became one with that of the flowering picture. There are reciprocities and special sympathies in such relations; mysterious affinities they used to be called, divinations of private congruity. Nick had an unexpressed conviction that if, as he had often wanted and proposed, he had embarked with Mrs Dallow in this particular quest of a great prize, disaster would have overtaken them on the deep waters. Even with the limited risk, indeed, disaster had come; but it was of a different kind and it had the advantage for him that now she couldn't reproach and accuse him as the cause of it – couldn't do so at least on any ground he was obliged to recognize. She would never know how much he had cared for her, how much he cared for her still; inasmuch as the conclusive proof, for himself, was his conscious reluctance to care for another woman, which she positively misread. Some day he would doubtless try to do that; but such a day seemed as yet far off, and he had no spite, no vindictive impulse to help him. The soreness that was mingled with his liberation, the sense of indignity even, as of a full cup suddenly dashed by a blundering hand from his lips, demanded certainly a balm; but it found it for a time in another passion, not in a rancorous exercise of the same – a passion strong enough to make him forget what a pity it was that he was not made to care for two women at once.

As soon as Mrs Dallow returned to England he broke ground to his mother on the subject of her making Julia understand that she and the girls now regarded their occupancy of Broadwood as absolutely terminated. He had already, several weeks before, picked a little at this arid tract, but in the interval the soil appeared to have formed again. It was disagreeable to him to impose such a renunciation on Lady Agnes, and it was especially disagreeable to have to phrase it and discuss it and perhaps insist upon it. He would have liked the whole business to be tacit – a little triumph of silent delicacy. But he found reasons to suspect that what in fact would be most tacit was Julia's certain endurance of any chance *in*delicacy. Lady Agnes had a theory that they had virtually – 'practically', as she said – given up the place, so that there was no need of making a splash about it; but Nick discovered, in the course of a conversation with Biddy more rigorous perhaps than any to which he had ever subjected her, that none of their property had been removed from her delightful house – none of the things (there were ever so many things) that Lady Agnes had caused to be conveyed there when they took possession. Her ladyship was the proprietor of innumerable articles of furniture, relics and survivals of her former greatness, and moved about the world with a train of heterogeneous baggage; so that her quiet overflow into the spaciousness of Broadwood had had all the luxury of a final subsidence. What Nick had to propose to her now was a dreadful combination, a relapse into all the things she most hated – seaside lodgings, bald storehouses in the Marylebone Road, little London rooms crammed with things that caught the dirt and made them stuffy. He was afraid he should really finish her, and he himself was surprised in a degree at his insistence. He wouldn't have supposed that he should have cared so much, but he found he did care intensely. He cared enough – it says everything – to explain to his mother that practically her retention of Broadwood would be the violation of an agreement. Julia had given them the place on the understanding that he was to marry her, and since he was not to marry her they had no right to keep the place. 'Yes, you make the mess and *we* pay the penalty!' Lady Agnes flashed out; but this was the only overt protest that she made, except indeed to contend that their withdrawal would be an act ungracious and offensive to Julia. She looked as she had looked during the months that succeeded his father's death, but she gave a general grim assent to the proposition that, let Julia take it as she would, their own duty was unmistakably clear.

It was Grace who was the principal representative of the idea that Julia would be outraged by such a step; she never ceased to repeat that she had never heard of anything so 'nasty'. Nick would have expected this of Grace, but he felt rather deserted and betrayed when Biddy murmured to him that *she* knew – that there was really no need of their sacrificing their mother's comfort to a mere fancy. She intimated that if Nick would only consent to their going on with Broadwood as if nothing had happened (or rather as if everything had happened), she would answer for Julia. For almost the first time in his life Nick disliked what Biddy said to him, and he gave her a sharp rejoinder, embodying the general opinion that they all had enough to do to answer for themselves. He remembered afterwards the way she looked at him, startled, even frightened, with rising tears, before turning away. He held that it would be time enough to judge how Julia would take it after they had thrown up the place; and he made it his duty to see that his mother should address to Mrs Dallow by letter a formal notification of their retirement. Mrs Dallow could protest then if she liked. Nick was aware that in general he was not practical; he could imagine why, from his early years, people should have joked him about it. But this time he was determined that his behaviour should be founded on a rigid view of things as they were. He didn't see his mother's letter sent to Julia, but he knew that it went. He thought she would have been more loyal if she had shown it to him, though of course there could be but little question of loyalty now. That it had really been written however, very much on the lines he dictated, was clear to him from the subsequent surprise which Lady Agnes's blankness did not prevent him from divining.

Julia answered her letter, but in unexpected terms: she had apparently neither resisted nor protested; she had simply been very glad to get her house back again and had not accused any of them of nastiness. Nick saw no more of her letter than he had seen of his mother's, but he was able to say to Grace (to Lady Agnes he was studiously mute): 'My poor child, you see, after all, that we haven't kicked up such a row.' Grace shook her head and looked gloomy and deeply wise, replying that he had no cause to triumph – they were so far from having seen the end of it yet. Then he guessed that his mother had complied with his wish on the calculation that it would be a mere form, that Julia would entreat them not to be so fantastic, and that he would then, in the presence of her wounded surprise, consent to a quiet continuance, so much in the interest (the air of Broadwood had a purity!) of the health of all of them. But since Julia jumped at their relinquishment he had no chance to be mollified: he had only to persist in having been right.

At bottom, probably, he himself was a little surprised at her eagerness. Literally speaking, it was not perfectly graceful. He was sorry his mother had been so deceived, but he was sorrier still for Biddy's mistake – it showed she might be mistaken about other things. Nothing was left now but for Lady Agnes to say, as she did, substantially, whenever she saw him: 'We are to prepare to spend the autumn at Worthing then, or some other horrible place? I don't know their names: it's the only thing we can afford.' There was an implication in this that if he expected her to drag her girls about to country-houses, in a continuance of the fidgety effort to work them off, he must understand at once that she was now too weary and too sad and too sick. She had done her best for them, and it had all been vain and cruel, and now the poor creatures must look out for themselves. To the grossness of Biddy's misconduct she needn't refer, nor to the golden opportunity this young lady had forfeited by her odious treatment of Mr Grindon. It was clear that this time Lady Agnes was incurably discouraged; so much so as to fail to glean the dimmest light from the fact that the girl was really making a long stay at Harsh. Biddy went to and fro two or three times and then, in August, fairly settled there; and what her mother mainly saw in her absence was the desire to keep out of the way of household reminders of her depravity. In fact, as it turned out, Lady Agnes and Grace, in the first days of August gathered themselves together for another visit to the old lady who had been Sir Nicholas's godmother; after which they went somewhere else, so that the question of Worthing had not to be immediately faced.

Nick stayed on in London with a passion of work fairly humming in his ears; he was conscious with joy that for three or four months, in the empty Babylon, he would have generous days. But toward the end of August he got a letter from Grace in which she spoke of her situation, and her mother's, in a manner that made him feel he ought to do something tactful. They were paying a third visit (he knew that in Calcutta Gardens lady's-maids had been to and fro with boxes, replenishments of wardrobes), and yet somehow the outlook for the autumn was dark. Grace didn't say it in so many words, but what he read between the lines was that they had no more invitations. What therefore was to become of them? People liked them well enough when Biddy was with them, but they didn't care for her mother and her, *tout pur*, and Biddy was cooped up indefinitely with Julia. This was not the manner in which Grace used to allude to her sister's happy visits to Mrs Dallow, and the change of tone made Nick wince with a sense of all that had collapsed. Biddy was a little fish worth landing, in short, scantily as she seemed disposed to bite, and Grace's rude probity could admit that she herself was not.

Nick had an inspiration: by way of doing something tactful he went down to Brighton and took lodgings for the three ladies, for several weeks, the quietest and sunniest he could find. This he intended as a kindly surprise, a reminder of how he had his mother's comfort at heart, how he could exert himself and save her trouble. But he had no sooner concluded his bargain (it was a more costly one than he had at first calculated) than he was bewildered, as he privately phrased it quite 'stumped', at learning that the three ladies were to pass the autumn at Broadwood with Julia. Mrs Dallow had taken the place into familiar use again and she was now correcting their former surprise at her crude concurrence (this was infinitely characteristic of Julia) by inviting them to share it with her. Nick wondered vaguely what she was 'up to'; but when his mother treated herself to the fine irony of addressing him an elaborately humble inquiry as to whether he would consent to their accepting the merciful refuge (she repeated this expression three times), he replied that she might do exactly as she liked: he would only mention that he should not feel himself at liberty to come and see her at Broadwood. This condition proved, apparently, to Lady Agnes's mind, no hindrance, and she and her daughters were presently re-installed in the very apartments they had learned to love. This time it was even better than before; they had still fewer expenses. The expenses were Nick's: he had to pay a forfeit to the landlady at Brighton for backing out of his contract. He said nothing to his mother about this bungled business – he was literally afraid; but an event that befell at the same moment reminded him afresh that it was not the time to choose to squander money. Mr Carteret drew his last breath; quite painlessly it seemed, as the closing scene was described at Beauclere when our young man went down to the funeral. Two or three weeks afterwards the contents of his will were made public in the *Illustrated London News*, where it definitely appeared that he had left a very large fortune, not a penny of which was to go to Nick. The provision for Mr Chayter's declining years was very handsome.

48

MIRIAM had mounted at a bound, in her new part, several steps in the ladder of fame, and at the climax of the London season this fact was brought home to her from hour to hour. It produced a thousand solicitations and entanglements, so that she rapidly learned that to be celebrated takes up almost as much of one's own time as of other people's. Even though, as she boasted, she had reduced to a science the practice of 'working' her mother (she made use of the good lady socially to the utmost, pushing her perpetually into the breach), there were many occasions on which it was represented to her that she could not be disobliging without damaging her cause. She made almost an income out of the photographers (their appreciation of her as a subject knew no bounds), and she supplied the newspapers with columns of irreducible copy. To the gentlemen who sought speech of her on behalf of these organs she poured forth, vindictively, floods of unscrupulous romance; she told them all different tales, and as her mother told them others yet more marvellous publicity was cleverly caught by rival versions, surpassing each other in authenticity. The whole case was remarkable, was unique; for if the girl was advertised by the bewilderment of her readers, she seemed to every sceptic, when he went to see her, as fine as if he had discovered her for himself. She was still accommodating enough however, from time to time, to find an hour to come and sit to Nick Dormer, and he helped himself further by going to her theatre whenever he could. He was conscious that Julia Dallow would probably hear of that and triumph with a fresh sense of how right she had been; but this reflection only made him sigh resignedly, so true it struck him as being that there are some things explanation can never better, can never touch.

Miriam brought Basil Dashwood once to see her portrait, and Basil, who commended it in general, directed his criticism mainly to two points – its not yet being finished and its not having gone into that year's Academy. The young actor was visibly fidgety: he felt the contagion of Miriam's rapid pace, the quick beat of her success, and, looking at everything now from the standpoint of that speculation, could scarcely contain his impatience at the painter's clumsy slowness. He thought the second picture much better than the other one, but somehow it ought by that time to be before the public. Having a great deal of familiar proverbial wisdom, he put forth with vehemence the idea that in every great crisis there is nothing like striking while the iron is hot. He even betrayed a sort of impression that with a little good-will Nick might wind up the job and still get the Academy people to take him in. Basil knew some of them; he all but offered to speak to them – the case was so exceptional; he had no doubt he could get something done. Against the approbation of the work by Peter Sherringham he explicitly and loudly protested, in spite of the homeliest recommendations of silence from Miriam; and it was indeed easy to guess how such an arrangement would interfere with his own conception of the eventual right place for the two portraits – the vestibule of the theatre, where every one going in and out would see them, suspended face to face and surrounded by photographs, artistically disposed, of the young actress in a variety of characters. Dashwood showed a largeness of view in the way he jumped to the conviction that in this position the pictures would really help to draw. Considering the virtue he attributed to Miriam the idea was exempt from narrow prejudice.

Moreover, though a trifle feverish, he was really genial; he repeated more than once: 'Yes, my dear sir, you've done it this time.' This was a favourite formula with him; when some allusion was made to the girl's success he greeted it also with a comfortable 'This time she *has* done it.' There was a hint of knowledge and far calculation in his tone. It appeared before he went that this time even he himself had done it – he had taken up something that would really answer. He told Nick more about Miriam, more about her affairs at that moment at least, than she herself had communicated, contributing strongly to our young man's impression that one by one every element of a great destiny was being dropped into her cup. Nick himself tasted of success vicariously for the hour. Miriam let Dashwood talk only to contradict him, and contradicted him only to show how indifferently she could do it. She treated him as if she had nothing more to learn about his folly, but as if it had taken intimate friendship to reveal to her the full extent of it. Nick didn't mind her intimate friendships, but he ended by disliking Dashwood, who irritated him – a circumstance in which poor Julia, if it had come to her knowledge, would doubtless have found a damning eloquence. Miriam was more pleased with herself than ever: she now made no scruple of admitting that she enjoyed all her advantages. She was beginning to have a fuller vision of how successful success could be; she took everything as it came – dined out every Sunday, and even went into the country till the Monday morning; she had a hundred distinguished names on her lips and wonderful tales about the people who were making up to her. She struck Nick as less serious than she had been hitherto, as making even an aggressive show of frivolity; but he was conscious of no obligation to reprehend her for it – the less as he had a dim vision that some effect of that sort, some irritation of his curiosity, was what she desired to produce. She would perhaps have liked, for reasons best known to herself, to look as if she were throwing herself away, not being able to do anything else. He couldn't talk to her as if he took an immense interest in her career, because in fact he didn't; she remained to him primarily and essentially a pictorial subject, with the nature of whose vicissitudes he was concerned (putting common charity and his personal good-nature of course aside) only so far as they had something to say in her face. How could he know in advance what twist of her life would say most? so possible was it even that complete failure or some incalculable perversion would only make her, for this particular purpose, more magnificent.

After she had left him, at any rate, the day she came with Basil Dashwood, and still more on a later occasion, as he turned back to his work when he had put her into her carriage, the last time, for that year, that he saw her – after she had left him it occurred to him, in the light of her quick distinction, that there were mighty differences in the famous artistic life. Miriam was already in the glow of a glory which moreover was probably but a faint spark in relation to the blaze to come; and as he closed the door upon her and took up his palette to rub it with a dirty cloth the little room in which his own battle was practically to be fought looked woefully cold and gray and mean. It was lonely, and yet it peopled with unfriendly shadows (so thick he saw them gathering in winter twilights to come) the duller conditions, the longer patiences, the less immediate and less personal joys. His late beginning was there, and his wasted youth, the mistakes that would still bring forth children after their image, the sedentary solitude, the clumsy obscurity, the poor explanations, the foolishness that he foresaw in having to ask people to wait, and wait longer, and wait again, for a fruition which, to their sense at least, would be an anti-climax. He cared enough for it, whatever it would be, to feel that his pertinacity might enter into comparison even with such a productive force as Miriam's. This was, after all, in his bare studio, the most collective dim presence, the one that was most sociable to him as he sat there and that made it the right place however wrong it was – the sense that it was to the thing in itself he was attached. This was Miriam's case, but the contrast, which she showed him she also felt, was in the number of other things that she got with the thing in itself.

I hasten to add that our young man had hours when this fine substance struck him as requiring, for a complete appeal, no adjunct whatever – as being in its own splendour a summary of all adjuncts and apologies. I have related that the great collections, the National Gallery and the Museum were sometimes rather a series of dead surfaces to him; but the sketch I have attempted of him will have been inadequate if it fails to suggest that there were other days when, as he strolled through them, he plucked right and left perfect nosegays of reassurance. Bent as he was on working in the modern, which spoke to him with a thousand voices, he judged it better, for long periods, not to haunt the earlier masters, whose conditions had been so different (later he came to see that it didn't matter much, especially if one didn't go); but he was liable to accidental deflections from this theory – liable in particular to want to take a look at one of the great portraits of the past. These were the things that were the most inspiring, in the sense that they were the things that, while generations, while worlds had come and gone, seemed most to survive and testify. As he stood before them sometimes the perfection of their survival struck him as the supreme eloquence, the reason that included all others, thanks to the language of art, the richest and most universal. Empires and systems and conquests had rolled over the globe and every kind of greatness had risen and passed away; but the beauty of the great pictures had known nothing of death or change, and the ages had only sweetened their freshness. The same faces, the same figures looked out at different centuries, knowing a deal the century didn't, and when they joined hands they made the indestructible thread on which the pearls of history were strung.

Miriam notified her artist that her theatre was to close on the 10th of August, immediately after which she was to start, with the company, on a tremendous tour of the provinces. They were to make a lot of money, but they were to have no holiday and she didn't want one; she only wanted to keep at it and make the most of her limited opportunities for practice; inasmuch as, at that rate, playing but two parts a year (and such parts – she despised them!) she shouldn't have mastered the rudiments of her trade before decrepitude would compel her to lay it by. The first time she came to the studio after her visit with Dashwood she sprang up abruptly, at the end of half an hour, saying she could sit no more – she had had enough of it. She was visibly restless and preoccupied, and though Nick had not waited till now to discover that she had more moods than he had tints on his palette, he had never yet seen her fitfulness at this particular angle. It was a trifle unbecoming and he was ready to let her go. She looked round the place as if she were suddenly tired of it, and then she said mechanically, in a heartless London way, while she smoothed down her gloves: 'So you're just going to stay on?' After he had confessed that this was his dark purpose she continued in the same casual, talk-making manner: 'I dare say it's the best thing for you. You're just going to grind, eh?'

'I see before me an eternity of grinding.'

'All alone, by yourself, in this dull little hole? You *will* be conscientious, you *will* be virtuous.'

'Oh, my solitude will be mitigated – I shall have models and people.'

'What people – what models?' Miriam asked, before the glass, arranging her hat.

'Well, no one so good as you.'

'That's a prospect!' the girl laughed; 'for all the good you've got out of me!'

'You're no judge of that quantity,' said Nick, 'and even I can't measure it just yet. Have I been rather a brute? I can easily believe it; I haven't talked to you – I haven't amused you as I might. The truth is, painting people is a very absorbing, exclusive occupation. You can't do much to them besides.'

'Yes, it's a cruel honour.'

'Cruel – that's too much,' Nick objected.

'I mean it's one you shouldn't confer on people you like, for when it's over it's over: it kills your interest in them and after you've finished them you don't like them any more.'

'Surely I like you,' Nick returned, sitting tilted back, before his picture, with his hands in his pockets.

'We've done very well: it's something not to have quarrelled,' said Miriam, smiling at him now and seeming more in it. 'I wouldn't have had you slight your work – I wouldn't have had you do it badly. But there's no fear of that for you,' she went on. 'You're the real thing and the rare bird. I haven't lived with you this way without seeing that: you're the sincere artist so much more than I. No, no, don't protest,' she added, with one of her sudden fine transitions to a deeper tone. 'You'll do things that will hand on your name when my screeching is happily over. Only you do seem to me, I confess, rather high and dry here – I speak from the point of view of your comfort and of my personal interest in you. You strike me as kind of lonely, as the Americans say – rather cut off and isolated in your grandeur. Haven't you any *confrères* – fellow-artists and people of that sort? Don't they come near you?'

'I don't know them much, I've always been afraid of them, and how can they take me seriously?'

'Well, I've got *confrères*, and sometimes I wish I hadn't! But does your sister never come near you any more, or is it only the fear of meeting me?'

Nick was aware that his mother had a theory that Biddy was constantly bundled home from Rosedale Road at the approach of improper persons: she was as angry at this as if she wouldn't have been more so if the child had been suffered to stay; but the explanation he gave his present visitor was nearer the truth. He reminded Miriam that he had already told her (he had been careful to do this, so as not to let it appear she was avoided) that his sister was now most of the time in the country, staying with an hospitable relation.

'Oh, yes,' the girl rejoined to this, 'with Mr Sherringham's sister, Mrs – what's her name? I always forget it.' And when Nick had pronounced the word with a reluctance he doubtless failed sufficiently to conceal (he hated to talk about Mrs Dallow; he didn't know what business Miriam had with her), she exclaimed: 'That's the one – the beauty, the wonderful beauty. I shall never forget how handsome she looked the day she found me here. I don't in the least resemble her, but I should like to have a try at that type, some day, in a comedy of manners. But who will write me a comedy of manners? There it is! The trouble would be, no doubt, that I should push her *à la charge*.'

Nick listened to these remarks in silence, saying to himself that if Miriam should have the bad taste (she seemed trembling on the brink of it) to make an allusion to what had passed between the lady in question and himself, he should dislike her utterly. It would show him she was a coarse creature after all. Her good genius interposed however, as against this hard penalty, and she quickly, for the moment at least, whisked away from the topic, demanding, apropos of comrades and visitors, what had become of Gabriel Nash, whom she had not encountered for so many days.

'I think he's tired of me,' said Nick; 'he hasn't been near me, either. But, after all, it's natural – he has seen me through.'

'Seen you through? Why, you've only just begun.'

'Precisely, and at bottom he doesn't like to see me begin. He's afraid I'll do something.'

'Do you mean he's jealous?'

'Not in the least, for from the moment one does anything one ceases to compete with him. It leaves him the field more clear. But that's just the discomfort, for him – he feels, as you said just now, kind of lonely; he feels rather abandoned and even, I think, a little betrayed. So far from being jealous he yearns for me and regrets me. The only thing he really takes seriously is to speculate and understand, to talk about the reasons and the essence of things: the people who do that are the highest. The applications, the consequences, the vulgar little effects belong to a lower plane, to which one must doubtless be tolerant and indulgent, but which is after all an affair of comparative accidents and trifles. Indeed he'll probably tell me frankly, the next time I see him, that he can't but feel that to come down to the little questions of action – the little prudences and compromises and simplifications of practice – is for the superior person a really fatal descent. One may be inoffensive and even commendable after it, but one can scarcely pretend to be interesting. *Il en faut comme ça*, but one doesn't haunt them. He'll do his best for me; he'll come back again, but he'll come back sad, and finally he'll fade away altogether. He'll go off to Granada, or somewhere.'

'The simplifications of practice?' cried Miriam. 'Why, they are just precisely the most blessed things on earth. What should we do without them?'

'What indeed?' Nick echoed. 'But if we need them it's because we're not superior persons. We're awful Philistines.'

'I'll be one with *you*,' the girl smiled. 'Poor Nash isn't worth talking about. What was it but a little question of action when he preached to you, as I know he did, to give up your seat?'

'Yes, he has a weakness for giving up – he'll go with you as far as that. But I'm not giving up any more, you see. I'm pegging away, and that's gross.'

'He's an idiot – *n'en parlons plus!*' Miriam dropped, gathering up her parasol, but lingering.

'Ah, never for me! He helped me at a difficult time.'

'You ought to be ashamed to confess it.'

'Oh, you *are* a Philistine,' said Nick.

'Certainly I am,' Miriam returned going toward the door, 'if it makes me one to be sorry, awfully sorry and even rather angry, that I haven't before me a period of the same sort of unsociable pegging away that you have. For want of it I shall never really be good. However, if you don't tell people I've said so, they'll never know. Your conditions are far better than mine and far more respectable: you can do as many things as you like, in patient obscurity, while I'm pitchforked into the *mêlée*, and into the most improbable fame, upon the back of a solitary *cheval de bataille*, a poor broken-winded screw. I foresee that I shall be condemned for the greater part of the rest of my days (do you see that?) to play the stuff I'm acting now. I'm studying Juliet and I want awfully to do her, but really I'm mortally afraid lest, if I should succeed, I should find myself in such a box. Perhaps they'd want Juliet for ever, instead of my present part. You see amid what delightful alternatives one moves.

What I want most I never shall have had – five quiet years of hard, all-round work, in a perfect company, with a manager more perfect still, playing five hundred parts and never being heard of. I may be too particular, but that's what I should have liked. I think I'm disgusting, with my successful crudities. It's discouraging; it makes one not care much what happens. What's the use, in such an age, of being good?'

'Good? Your haughty claim is that you're bad.'

'I mean *good*, you know – there are other ways. Don't be stupid.' And Nick's visitor tapped him – he was at the door with her – with her parasol.

'I scarcely know what to say to you, for certainly it's your fault if you get on so fast.'

'I'm too clever – I'm a humbug.'

'That's the way I used to be,' said Nick.

Miriam rested her wonderful eyes on him; then she turned them over the room, slowly, after which she attached them again, kindly, musingly, on his own. 'Ah, the pride of that – the sense of purification! He "used" to be, forsooth! Poor me! Of course you'll say: "Look at the sort of thing I've undertaken to produce, compared with what you have." So it's all right. Become great in the proper way and don't expose me.' She glanced back once more into the studio, as if she were leaving it for ever, and gave another last look at the unfinished canvas on the easel. She shook her head sadly. 'Poor Mr Sherringham – with *that*!' she murmured.

'Oh, I'll finish it – it will be very decent,' said Nick.

'Finish it by yourself?'

'Not necessarily. You'll come back and sit when you return to London.'

'Never, never, never again.'

Nick stared. 'Why, you've made me the most profuse offers and promises.'

'Yes, but they were made in ignorance, and I've backed out of them. I'm capricious too – *faites la part de ça*. I see it wouldn't do – I didn't know it then. We're too far apart – I *am*, as you say, a Philistine.' And as Nick protested with vehemence against this unscrupulous bad faith, she added: 'You'll find other models; paint Gabriel Nash.'

'Gabriel Nash – as a substitute for you?'

'It will be a good way to get rid of him. Paint Mrs Dallow too,' Miriam went on as she passed out of the door which Nick had opened for her – 'paint Mrs Dallow if you wish to eradicate the last possibility of a throb.'

It was strange that since only a moment before Nick had been in a state of mind to which the superfluity of this reference would have been the clearest thing about it, he should now have been moved to receive it, quickly, naturally, irreflectively, with the question: 'The last possibility? Do you mean in her or in me?'

'Oh, in you. I don't know anything about her.'

'But that wouldn't be the effect,' rejoined Nick, with the same supervening candour. 'I believe that if she were to sit to me the usual law would be reversed.'

'The usual law?'

'Which you cited awhile since and of which I recognize the general truth. In the case you speak of I should probably make a frightful picture.'

'And fall in love with her again? Then, for God's sake, risk the daub!' Miriam laughed out, floating away to her victoria.

49

MIRIAM had guessed happily in saying to Nick that to offer to paint Gabriel Nash would be the way to get rid of him. It was with no such invidious purpose indeed that our young man proposed to his intermittent friend to sit; rather, as August was dusty in the London streets, he had too little hope that Nash would remain in town at such a time to oblige him. Nick had no wish to get rid of his private philosopher; he liked his philosophy, and though of course premeditated paradox was the light to read him by, yet he had frequently, in detail, an inspired unexpectedness. He remained, in Rosedale Road, the man in the world who had most the quality of company. All the other men of Nick's acquaintance, all his political friends, represented, often very communicatively, their own affairs, and their own affairs alone; which, when they did it well, was the most their host could ask them. But Nash had the rare distinction that he seemed somehow to stand for *his* affairs, the said host's, with an interest in them unaffected by the ordinary social limitations of capacity. This relegated him to the class of high luxuries, and Nick was well aware that we hold our luxuries by a fitful and precarious tenure. If a friend without personal eagerness was one of the greatest of these it would be evident to the simplest mind that by the law of distribution of earthly boons such a convenience should be expected to forfeit in duration what it displayed in intensity. He had never been without a suspicion that Nash was too good to last, though for that matter nothing had happened to confirm a vague apprehension that the particular way he would break up, or break down, would be by wishing to put Nick in relation with his other disciples.

That would practically amount to a catastrophe, Nick felt; for it was odd that one could both have a great kindness for him and not in the least, when it came to the point, yearn for a view of his belongings. His originality had always been that he appeared to have none; and if in the first instance he had introduced Nick to Miriam and her mother, that was an exception for which Peter Sherringham's interference had been in a great measure responsible. All the same however it was some time before Nick ceased to think it might eventually very well happen that to complete his education, as it were, Gabriel would wish him to forgather a little with minds formed by the same mystical influence. Nick had an instinct, in which there was no consciousness of detriment to Nash, that the pupils, perhaps even the imitators of such a genius would be, as he mentally phrased it, something awful. He could be sure, even Gabriel himself could be sure, of his own reservations, but how could either of them be sure of those of others? Imitation is a fortunate homage only in proportion as it is delicate, and there was an indefinable something in Nash's doctrine that would have been discredited by exaggeration or by zeal. Providence happily appeared to have spared it this probation; so that, after months, Nick had to remind himself that his friend had never pressed upon his attention the least little group of fellow-mystics, nor offered to produce them for his edification. It scarcely mattered now that he was just the man to whom the superficial would attribute that sort of tail; it would probably have been hard, for example, to persuade Lady Agnes, or Julia Dallow, or Peter Sherringham, that he was not most at home in some dusky, untidy, dimly-imagined suburb of 'culture', peopled by unpleasant phrasemongers who thought him a gentleman and who had no human use but to be held up in the comic press, which was probably restrained by decorum from touching upon the worst of their aberrations.

Nick, at any rate, never discovered his academy, nor the suburb in question; never caught, from the impenetrable background of his life, the least reverberation of flitting or flirting, the smallest aesthetic ululation. There were moments when he was even moved to a degree of pity by the silence that poor Gabriel's own faculty of sound made around him – when at least it qualified with thinness the mystery he could never wholly dissociate from him, the sense of the transient and occasional, the likeness to vapour or murmuring wind or shifting light. It was for instance a symbol of this unclassified condition, the lack of all position as a name in well-kept books, that Nick in point of fact had no idea where he lived, would not have known how to go and see him or send him a doctor if he had heard he was ill. He had never walked with him to any door of Gabriel's own, even to pause at the threshold, though indeed Nash had a club, the Anonymous, in some improbable square, of which Nick suspected him of being the only member – he had never heard of another – where it was vaguely understood that letters would some day or other find him. Fortunately it was not necessary to worry about him, so comfortably his whole aspect seemed to imply that he could never be ill. And this was not perhaps because his bloom was healthy, but because it was morbid, as if he had been universally inoculated.

He turned up in Rosedale Road one day after Miriam had left London; he had just come back from a fortnight in Brittany, where he had drawn unusual refreshment from the subtle sadness of the landscape. He was on his way somewhere else; he was going abroad for the autumn, but he was not particular what he did, professing that he had returned to London on purpose to take one last superintending look at Nick. 'It's very nice, it's very nice; yes, yes, I see,' he remarked, giving a little general assenting sigh as his eyes wandered over the simple scene – a sigh which, to a suspicious ear, would have testified to an insidious reaction.

Nick's ear, as we know, was already suspicious; a fact which would sufficiently account for the expectant smile (it indicated the pleasant apprehension of a theory confirmed) with which he inquired: 'Do you mean my pictures are nice?'

'Yes, yes, your pictures and the whole thing.'

'The whole thing?'

'Your existence here, in this little remote independent corner of the great city. The disinterestedness of your attitude, the persistence of your effort, the piety, the beauty, in short the example, of the whole spectacle.'

Nick broke into a laugh. 'How near to having had enough of me you must be when you talk of my example!' Nash changed colour slightly at this; it was the first time in Nick's remembrance that he had given a sign of embarrassment. '*Vous allez me lâcher*, I see it coming; and who can blame you? – for I've ceased to be in the least spectacular. I had my little hour; it was a great deal, for some people don't even have that. I've given you your curious case and I've been generous; I made the drama last for you as long as I could. You'll "slope", my dear fellow – you'll quietly slope; and it will be all right and inevitable, though I shall miss you greatly at first. Who knows whether, without you, I shouldn't still have been representing Harsh, heaven help me? You rescued me; you converted me from a representative into an example – that's a shade better. But don't I know where you must be when you're reduced to praising my piety?'

'Don't turn me away,' said Nash plaintively; 'give me a cigarette.'

'I shall never dream of turning you away; I shall cherish you till the latest possible hour. I'm only trying to keep myself in tune with the logic of things. The proof of how I cling is that, precisely, I want you to sit to me.'

'To sit to you?' Nick thought his visitor looked a little blank.

'Certainly, for after all it isn't much to ask. Here we are, and the hour is peculiarly propitious – long light days, with no one coming near me, so that I have plenty of time. I had a hope I should have some orders: my younger sister, whom you know and who is a great optimist, plied me with that vision. In fact we invented together a charming sordid little theory that there might be rather a "run" on me, from the chatter (such as it was) produced by my taking up this line. My sister struck out the idea that a good many of the pretty ladies would think me interesting, would want to be done. Perhaps they do, but they've controlled themselves, for I can't say the run has commenced. They haven't even come to look, but I dare say they don't yet quite take it in. Of course it's a bad time, with every one out of town; though you know they might send for me to come and do them at home. Perhaps they will, when they settle down. A portrait-tour of a dozen country-houses, for the autumn and winter – what do you say to that for a superior programme? I know I excruciate you,' Nick added, 'but don't you see how it's my interest to try how much you'll still stand?'

Gabriel puffed his cigarette with a serenity so perfect that it might have been assumed to falsify Nick's words. 'Mrs Dallow will send for you – *vous allez voir ça*,' he said in a moment, brushing aside all vagueness.

'She'll send for me?'

'To paint her portrait; she'll recapture you on that basis. She'll get you down to one of the country-houses, and it will all go off as charmingly – with sketching in the morning, on days you can't hunt, and anything you like in the afternoon, and fifteen courses in the evening; there'll be bishops and ambassadors staying – as if you were a "well-known" awfully clever amateur. Take care, take care, for, fickle as you may think me, I can read the future: don't imagine you've come to the end of me yet. Mrs Dallow and your sister, of both of whom I speak with the greatest respect, are capable of hatching together the most conscientious, delightful plan for you. Your differences with the beautiful lady will be patched up, and you'll each come round a little and meet the other half-way. Mrs Dallow will swallow your profession if you'll swallow hers. She'll put up with the palette if you'll put up with the country-house. It will be a very unusual one in which you won't find a good north room where you can paint. You'll go about with her and do all her friends, all the bishops and ambassadors, and you'll eat your cake and have it, and every one, beginning with your wife, will forget there is anything queer about you, and everything will be for the best in the best of worlds; so that, together – you and she – you'll become a great social institution, and everyone will think she has a delightful husband; to say nothing of course of your having a delightful wife. Ah, my dear fellow, you turn pale, and with reason!' Nash went on: 'that's to pay you for having tried to make me let you have it. You have it, then, there! I may be a bore' – the emphasis of this, though a mere shade, testified to the first personal resentment Nick had ever heard his visitor express – 'I may be a bore, but once in a while I strike a light, I make things out. Then I venture to repeat, "Take care, take care." If, as I say, I respect those ladies infinitely, it is because they will be acting according to the highest wisdom of their sex. That's the sort of thing women invent when they're exceptionally good and clever. When they're not, they don't do so well; but it's not for want of trying. There's only one thing in the world that's better than their charm: it's their conscience. That indeed is a part of their charm. And when they club together, when they earnestly consider, as in the case we're supposing,' Nash continued, 'then the whole thing takes a lift; for it's no longer the conscience of the individual, it's that of the sex.'

'You're so remarkable that, more than ever, I must paint you,' Nick returned, 'though I'm so agitated by your prophetic words that my hand trembles and I shall doubtless scarcely be able to hold my brush. Look how I rattle my easel trying to put it into position. I see it all there, just as you say it. Yes, it will be a droll day, and more modern than anything else, when the conscience of women perceives objections to men's being in love with them. You talk of their goodness and cleverness, and it's much to the point. I don't know what else they themselves might do with these things, but I don't see what men can do with them but be fond of them.'

'Oh, you'll do it – you'll do it!' cried Nash, brightly jubilant.

'What is it I shall do?'

'Exactly what I just said; if not next year, then the year after, or the year after that. You'll go half-way to meet her, and she'll drag you about and pass you off. You'll paint the bishops and become a social institution. That is, you will if you don't take great care.'

'I shall, no doubt, and that's why I cling to you. You must still look after me; don't melt away into a mere improbable reminiscence, a delightful symbolic fable – don't if you can possibly help it. The trouble is, you see that you can't really keep hold very tight, because at bottom it will amuse you much more to see me in another pickle than to find me simply jogging down the vista of the years on the straight course. Let me at any rate have some sort of sketch of you, as a kind of feather from the angel's wing, or a photograph of the ghost, to prove to me in the future that you were once a solid, sociable fact, that I didn't utterly fabricate you. Of course I shall be able to say to myself that you can't have been a fable – otherwise you would have had a moral; but that won't be enough, because I'm not sure you won't have had one. Some day you'll peep in here languidly and find me in such an attitude of piety – presenting my bent back to you as I niggle over some interminable botch – that I shall give cruelly on your nerves, and you'll draw away, closing the door softly (for you'll be gentle and considerate about it and spare me – you won't even make me look round), and steal off on tiptoe, never, never to return.'

Gabriel consented to sit; he professed he should enjoy it and be glad to give up for it his immediate Continental projects, so vague to Nick, so definite apparently to himself; and he came back three times for the purpose. Nick promised himself a great deal of interest from this experiment; for from the first hour he began to feel that really, as yet, compared to the scrutiny to which he now subjected him, he had never with any intensity looked at his friend. His impression had been that Nash had a head quite fine enough to be a challenge, and that as he sat there, day by day, all sorts of pleasant and paintable things would come out in his face. This impression was not falsified, but the whole problem became more complicated. It struck our young man that he had never *seen* his subject before, and yet somehow this revelation was not produced by the sense of actually seeing it. What was revealed was the difficulty – what he saw was the indefinite and the elusive. He had taken things for granted which literally were not there, and he found things there (except that he couldn't catch them) which he had not hitherto counted in. This baffling effect, being eminently in Nash's line, might have been the result of his whimsical volition, had it not appeared to Nick, after a few hours of the job, that his sitter was not the one who enjoyed it most. He was uncomfortable, at first vaguely and then definitely so – silent, restless, gloomy, dim, as if, when it came to the test, it proved less of a pleasure to him than he could have had an idea of in advance to be infinitely examined and handled, sounded and sifted. He had been willing to try it, in good faith; but frankly he didn't like it. He was not cross, but he was clearly unhappy, and Nick had never heard him say so little, seen him give so little.

Nick felt, accordingly, as if he had laid a trap for him: he asked himself if it were really fair. At the same time there was something fascinating in the oddity of such a relation between the subject and the artist, and Nick was disposed to go on until he should have to stop for very pity. He caught eventually a glimmer of the truth that lay at the bottom of this anomaly; guessed that what made his friend uncomfortable was simply the reversal, in such a combination, of his usual terms of intercourse. He was so accustomed to living upon irony and the interpretation of things that it was strange to him to be himself interpreted, and (as a gentleman who sits for his portrait is always liable to be) interpreted ironically. From being outside of the universe he was suddenly brought into it, and from the position of a free commentator and critic, a sort of amateurish editor of the whole affair, reduced to that of humble ingredient and contributor. It occurred afterwards to Nick that he had perhaps brought on a catastrophe by having happened to say to his companion, in the course of their disjointed pauses, and not only without any cruel intention but with an impulse of genuine solicitude: 'But, my dear fellow, what will you do when you're old?'

'Old? What do you call old?' Nash had replied bravely enough, but with another perceptible tinge of irritation. 'Must I really inform you, at this time of day, that that term has no application to such a condition as mine? It only belongs to you wretched people who have the incurable superstition of "doing": it's the ignoble collapse you prepare for yourselves when you cease to be able to do. For me there'll be no collapse, no transition, no clumsy readjustment of attitude; for I shall only *be*, more and more, with all the accumulations of experience, the longer I live.'

'Oh, I'm not particular about the term,' said Nick. 'If you don't call it old, the ultimate state, call it weary – call it exhausted. The accumulations of experience are practically accumulative of fatigue.'

'I don't know anything about weariness. I live easily – it doesn't fatigue me.'

'Then you need never die,' rejoined Nick.

'Certainly; I dare say I'm eternal.'

Nick laughed out at this – it would be such fine news to some people. But it was uttered with perfect gravity, and it might very well have been in the spirit of that gravity that Nash failed to observe his agreement to sit again the next day. The next, and the next, and the next passed, but he never came back.

True enough, punctuality was not important for a man who felt that he had the command of all time. Nevertheless, his disappearance, 'without a trace', like a personage in a fairy-tale or a melodrama, made a considerable impression on his friend, as the months went on; so that, though he had never before had the least difficulty about entering into the play of Gabriel's humour, Nick now recalled with a certain fanciful awe the unusual seriousness with which he had ranked himself among imperishable things. He wondered a little whether he had at last gone quite mad. He had never before had such a literal air, and he would have had to be mad to be so commonplace. Perhaps indeed he was acting only more than usual in his customary spirit – thoughtfully contributing, for Nick's enlivenment, a mystery to an horizon now grown unromantic. The mystery at any rate remained; another too came near being added to it. Nick had the prospect, for the future, of the harmless excitement of waiting to see when Nash would turn up, if ever, and the further diversion (it almost consoled him for the annoyance of being left with a second unfinished portrait on his hands) of imagining that the picture he had begun had a singular air of gradually fading from the canvas. He couldn't catch it in the act, but he could have a suspicion, when he glanced at it, that the hand of time was rubbing it away little by little (for all the world as in some delicate Hawthorne tale), making the surface indistinct and bare – bare of all resemblance to the model. Of course the moral of the Hawthorne tale would be that this personage would come back on the day when the last adumbration should have vanished.

50

ONE day, toward the end of March of the following year, or in other words more than six months after the incident I have last had occasion to narrate, Bridget Dormer came into her brother's studio and greeted him with the effusion that accompanies a return from an absence. She had been staying at Broadwood – she had been staying at Harsh. She had various things to tell him about these episodes, about his mother, about Grace, about herself, and about Percy's having come, just before, over to Broadwood for two days; the longest visit with which, almost since they could remember, the head of the family had honoured Lady Agnes. Nick noted however that it had apparently been taken as a great favour, and Biddy loyally testified to the fact that her elder brother was awfully jolly and that his presence had been a pretext for tremendous fun. Nick asked her what had passed about his marriage – what their mother had said to him.

'Oh, nothing,' Biddy replied; and he had said nothing to Lady Agnes and not a word to herself. This partly explained, for Nick, the awful jollity and the tremendous fun – none but cheerful topics had been produced; but he questioned his sister further, to a point which led her to say: 'Oh, I dare say that before long she'll write to her.'

'Who will write to whom?'

'Mamma'll write to his wife. I'm sure he'd like it. Of course we shall end by going to see her. He was awfully disappointed at what he found in Spain – he didn't find anything.'

Biddy spoke of his disappointment almost with commiseration, for she was evidently inclined this morning to a fresh and kindly view of things. Nick could share her feeling only so far as was permitted by a recognition merely general of what his brother must have looked for. It might have been snipe and it might have been bristling boars. Biddy was indeed brief, at first, about everything, in spite of the two months that had intervened since their last meeting; for he saw in a few minutes that she had something behind – something that made her gay and that she wanted to come to quickly. Nick was vaguely vexed at her being, fresh from Broadwood, so gay as that; for (it was impossible to shut one's eyes to it) what had come to pass, in practice, in regard to that rural retreat, was exactly what he had desired to avert. All winter, while it had been taken for granted that his mother and sisters were doing what he wished, they had been doing the precise contrary. He held Biddy perhaps least responsible, and there was no one he could exclusively blame. He washed his hands of the matter and succeeded fairly well for the most part in forgetting that he was not pleased. Julia Dallow herself in fact appeared to have been the most active member of the little group united to make light of his scruples. There had been a formal restitution of the place, but the three ladies were there more than ever, with the slight difference that they were mainly there with its mistress. Mahomet had declined to go any more to the mountain, so the mountain had virtually come to Mahomet.

After their long visit in the autumn Lady Agnes and her girls had come back to town; but they had gone down again for Christmas, and Julia had taken this occasion to write to Nick that she hoped very much he wouldn't refuse them all his own company for just a little scrap of the supremely sociable time. Nick, after reflection, judged it best not to refuse, and he spent three days under Mrs Dallow's roof. The 'all' proved a great many people, for she had taken care to fill the house. She was a magnificent entertainer, and Nick had never seen her so splendid, so free-handed, so gracefully practical. She was a perfect mistress of the revels; she had organized something festive for every day and for every night. The Dormers were so much in it, as the phrase was, that after all their discomfiture their fortune seemed in an hour to have come back. There had been a moment when, in extemporized charades, Lady Agnes, an elderly figure being required, appeared on the point of undertaking the part of the housekeeper at a castle, who, dropping her *h*'s, showed sheeplike tourists about; but she waived the opportunity in favour of her daughter Grace. Even Grace had a great success. Nick of course was in the charades and in everything, but Julia was not; she only invented, directed, led the applause. When nothing else was going on Nick 'sketched' the whole company: they followed him about, they waylaid him on staircases, clamouring to be allowed to sit. He obliged them, so far as he could, all save Julia, who didn't clamour; and, growing rather red, he thought of Gabriel Nash, while he bent over the paper. Early in the new year he went abroad for six weeks, but only as far as Paris. It was a new Paris for him then: a Paris of the Rue Bonaparte and three or four professional friends (he had more of these there than in London); a Paris of studios and studies and models, of researches and revelations, comparisons and contrasts, of strong impressions and long discussions and rather uncomfortable economies, small cafés and bad fires and the general sense of being twenty again.

While he was away his mother and sisters (Lady Agnes now sometimes wrote to him) returned to London for a month, and before he was again established in Rosedale Road they went back for a third period to Broadwood. After they had been there five days – and this was the salt of the whole dish – Julia took herself off to Harsh, leaving them in undisturbed possession. They had remained so; they would not come up to town till after Easter. The trick was played, and Biddy, as I have mentioned, was now very content. Her brother presently learned however that the reason of this was not wholly the success of the trick; unless indeed her further ground was only a continuation of it. She was not in London as a forerunner of her mother; she was not even as yet in Calcutta Gardens. She had come to spend a week with Florence Tressilian, who had lately taken the dearest little flat in a charming new place, just put up, on the other side of the Park, with all kinds of lifts and tubes and electricities. Florence had been awfully nice to her – she had been with them ever so long at Broadwood, while the flat was being painted and prepared – and mamma had then let her, let Biddy, promise to come to her, when everything was ready, so that they might have a kind of old maids' house-warming together. If Florence could do without a chaperon now (she had two latch-keys and went alone on the top of omnibuses, and her name was in the Red Book), she was enough of a duenna for another girl. Biddy alluded, with sweet and cynical eyes, to the fine, happy stride, she had thus taken in the direction of enlightened spinsterhood; and Nick hung his head, somewhat abashed and humiliated, for, modern as he had supposed himself, there were evidently currents more modern yet.

It so happened on this particular morning Nick had drawn out of a corner his interrupted study of Gabriel Nash; for no purpose more definite (he had only been looking round the room in a rummaging spirit) than to see curiously how much or how little of it remained. It had become to his apprehension such a shadowy affair (he was sure of this, and it made him laugh) that it didn't seem worth putting away, and he left it leaning against a table, as if it had been a blank canvas or a 'preparation' to be painted over. In this attitude it attracted Biddy's attention, for to her, on a second glance, it had distinguishable features. She had not seen it before and she asked whom it might represent, remarking also that she could almost guess, but not quite: she had known the original, but she couldn't name him.

'Six months ago, for a few days, it represented Gabriel Nash,' Nick replied. 'But it doesn't represent anybody or anything now.'

'Six months ago? What's the matter with it and why don't you go on?'

'What's the matter with it is more than I can tell you. But I can't go on, because I've lost my model.'

Biddy stared an instant. 'Is he dead?'

Her brother laughed out at the candid cheerfulness, hopefulness almost, with which this inquiry broke from her. 'He's only dead to me. He has gone away.'

'Where has he gone?'

'I haven't the least idea.'

'Why, have you quarrelled?' Biddy asked.

'Quarrelled? For what do you take us? Does the nightingale quarrel with the moon?'

'I needn't ask which of you is the moon,' said Biddy.

'Of course I'm the nightingale. But, more literally,' Nick continued, 'Nash has melted back into the elements – he is part of the ambient air.' Then, as even with this literalness he saw that his sister was mystified, he added: 'I have a notion he has gone to India, and at the present moment is reclining on a bank of flowers in the vale of Cashmere.'

Biddy was silent a minute, after which she dropped: 'Julia will be glad – she dislikes him so.'

'If she dislikes him, why should she be glad he's in such a delightful situation?'

'I mean about his going away; she'll be glad of that.'

'My poor child, what has Julia to do with it?'

'She has more to do with things than you think,' Biddy replied, with some eagerness; but she had no sooner uttered the words than she perceptibly blushed. Hereupon, to attenuate the foolishness of her blush (only it had the opposite effect), she added: 'She thinks he has been a bad element in your life.'

Nick shook his head, smiling. 'She thinks, perhaps, but she doesn't think enough; otherwise, she would arrive at this thought – that she knows nothing whatever about my life.'

'Ah, Nick,' the girl pleaded, with solemn eyes, 'you don't imagine what an interest she takes in it. She has told me, many times – she has talked lots to me about it.' Biddy paused and then went on, with an anxious little smile shining through her gravity, as if she were trying cautiously how much her brother would take: 'She has a conviction that it was Mr Nash who made trouble between you.'

'My dear Biddy,' Nick rejoined, 'those are thoroughly second-rate ideas, the result of a perfectly superficial view. Excuse my possibly priggish tone, but they really attribute to Nash a part he's quite incapable of playing. He can neither make trouble nor take trouble; no trouble could ever either have come out of him or have gone into him. Moreover,' our young man continued, 'if Julia has talked to you so much about the matter, there's no harm in my talking to you a little. When she threw me over in an hour, it was on a perfectly definite occasion. That occasion was the presence in my studio of a dishevelled actress.'

'Oh Nick, she has not thrown you over!' Biddy protested. 'She has not – I have the proof.'

Nick felt, at this direct denial, a certain stir of indignation, and he looked at his sister with momentary sternness. 'Has she sent you here to tell me this? What do you mean by the proof?'

Biddy's eyes, at these questions, met her brother's with a strange expression, and for a few seconds, while she looked entreatingly into his own, she wavered there, with parted lips, vaguely stretching out her hands. The next minute she had burst into tears – she was sobbing on his breast. He said 'Hallo!' and soothed her; but it was very quickly over. Then she told him what she meant by her 'proof', and what she had had on her mind ever since she came into the room. It was a message from Julia, but not to say – not to say what he had asked her just before if she meant; though indeed Biddy, more familiar now, since her brother had had his arm round her, boldly expressed the hope that it might in the end come to the same thing. Julia simply wanted to know (she had instructed Biddy to sound him discreetly) if Nick would undertake her portrait; and the girl wound up this experiment in 'sounding' by the statement that their beautiful kinswoman was dying to sit.

'Dying to sit?' repeated Nick, whose turn it was, this time, to feel his colour rise.

'Any time you like after Easter, when she comes up to town. She wants a full-length, and your very best, your most splendid work.'

Nick stared, not caring that he had blushed. 'Is she serious?'

'Ah, Nick – serious!' Biddy reasoned tenderly. She came nearer to him, and he thought she was going to weep again. He took her by the shoulders, looking into her eyes.

'It's all right, if she knows *I* am. But why doesn't she come like any one else? I don't refuse people!'

'Nick, dearest Nick!' she went on, with her eyes conscious and pleading. He looked into them intently – as well as she, he could play at sounding – and for a moment, between these young persons, the air was lighted by the glimmer of mutual searchings and suppressed confessions. Nick read deep, and then, suddenly releasing his sister, he turned away. She didn't see his face in that movement, but an observer to whom it had been presented might have fancied that it denoted a foreboding which was not exactly a dread, yet was not exclusively a joy.

The first thing Nick made out in the room, when he could

distinguish, was Gabriel Nash's portrait, which immediately filled him with an unreasoning resentment. He seized it and turned it about; he jammed it back into its corner, with its face against the wall. This bustling transaction might have served to carry off the embarrassment with which he had finally averted himself from Biddy. The embarrassment however was all his own; none of it was reflected in the way Biddy resumed, after a silence in which she had followed his disposal of the picture:

'If she's so eager to come here (for it's here that she wants to sit, not in Great Stanhope Street – never!) how can she prove better that she doesn't care a bit if she meets Miss Rooth?'

'She won't meet Miss Rooth,' Nick replied, rather dryly.

'Oh, I'm sorry!' said Biddy. She was as frank as if she had achieved a sort of victory over her companion; and she seemed to regret the loss of a chance for Mrs Dallow to show magnanimity. Her tone made her brother laugh, but she went on, with confidence: 'She thought it was Mr Nash who made Miss Rooth come.'

'So he did, by the way,' said Nick.

'Well, then, wasn't that making trouble?'

'I thought you admitted there was no harm in her being here.'

'Yes, but he hoped there would be.'

'Poor Nash's hopes!' Nick laughed. 'My dear child, it would take a cleverer head than you or me, or even Julia, who must have invented that wise theory, to say what they were. However, let us agree, that even if they were perfectly devilish my good sense has been a match for them.'

'Oh, Nick, that's delightful!' chanted Biddy. Then she added, 'Do you mean she doesn't come any more?'

'The dishevelled actress? She hasn't been near me for months.'

'But she's in London – she's always acting? I've been away so much I've scarcely observed,' Biddy explained, with a slight change of note.

'The same part, poor creature, for nearly a year. It appears that that's success, in her profession. I saw her in the character several times last summer, but I haven't set foot in her theatre since.'

Biddy was silent a moment; then she suggested: 'Peter wouldn't have liked that.'

'Oh, Peter's likes!' sighed Nick, at his easel, beginning to work.

'I mean her acting the same part for a year.'

'I'm sure I don't know; he has never written me a word.'

'Nor me either,' Biddy returned.

There was another short silence, during which Nick brushed at a panel. It was terminated by his presently saying: 'There's one thing, certainly, Peter *would* like – that is simply to be here tonight. It's a great night – another great night – for the dishevelled one. She's to act Juliet for the first time.'

'Ah, how I should like to see her!' the girl cried.

Nick glanced at her; she sat watching him. 'She has sent me a stall; I wish she had sent me two. I should have been delighted to take you.'

'Don't you think you could get another?' asked Biddy.

'They must be in tremendous demand. But who knows, after all?' Nick added, at the same moment, looking round. 'Here's a chance – here's a quite extraordinary chance!'

His servant had opened the door and was ushering in a lady whose identity was indeed justly indicated in those words. 'Miss Rooth!' the man announced; but he was caught up by a gentleman who came next and who exclaimed, laughing and with a gesture gracefully corrective: 'No, no – no longer Miss Rooth!'

Miriam entered the place with her charming familiar grandeur, as she might have appeared, as she appeared every night, early in her first act, at the back of the stage, by the immemorial central door, presenting herself to the house, taking easy possession, repeating old movements, looking from one to the other of the actors before the footlights. The rich 'Good morning' that she threw into the air, holding out her hand to Biddy Dormer and then giving her left to Nick (as she might have given it to her own brother), had nothing to tell of intervals or alienations. She struck Biddy as still more terrible, in her splendid practice, than when she had seen her before – the practice and the splendour had now something almost royal. The girl had had occasion to make her courtesy to majesties and highnesses, but the flutter those effigies produced was nothing to the way in which, at the approach of this young lady, the agitated air seemed to recognize something supreme. So the deep, mild eyes that she bent upon Biddy were not soothing, though they were evidently intended to sooth. The girl wondered that Nick could have got so used to her (he joked at her as she came), and later in the day, still under the great impression of this incident, she even wondered that Peter could. It was true that Peter apparently hadn't.

'You never came – you never came,' said Miriam to Biddy, kindly, sadly; and Biddy recognizing the allusion, the invitation to visit the actress at home, had to explain how much she had been absent from London, and then even that her brother hadn't proposed to take her. 'Very true – he hasn't come himself. What is he doing now?' Miriam asked, standing near Biddy, but looking at Nick, who had immediately engaged in conversation with his other visitor, a gentleman whose face came back to the girl. She had seen this gentleman on the stage with Miss Rooth – that was it, the night Peter took her to the theatre with Florence Tressilian. Oh, that Nick would only do something of that sort now! This desire, quickened by the presence of the strange, expressive woman, by the way she scattered sweet syllables as if she were touching the piano-keys, combined with other things to make Biddy's head swim – other things too mingled to name, admiration and fear and dim divination and purposeless pride and curiosity and resistance, the impulse to go away and the determination not to go. The actress courted her with her voice (what was the matter with her and what did she want?) and Biddy tried in return to give an idea of what Nick was doing. Not succeeding very well she was going to appeal to her brother, but Miriam stopped her, saying it didn't matter; besides, Dashwood was telling Nick something – something they wanted him to know. 'We're in great excitement – he has taken a theatre,' Miriam added.

'Taken a theatre?' Biddy was vague.

'We're going to set up for ourselves. He's going to do for me altogether. It has all been arranged only within a day or two. It remains to be seen how it will answer,' Miriam smiled. Biddy murmured some friendly hope, and her interlocutress went on: 'Do you know why I've broken in here today, after a long absence – interrupting your poor brother, taking up his precious time! It's because I'm so nervous.'

'About your first night?' Biddy risked.

'Do you know about that – are you coming?' Miriam asked, quickly.

'No, I'm not coming – I haven't a place.'

'Will you come if I send you one?'

'Oh, but really, it's too beautiful of you,' stammered the girl.

'You shall have a box; your brother shall bring you. You can't squeeze in a pin, I'm told; but I've kept a box, I'll manage it. Only, if I do, you know, mind you come!' Miriam exclaimed, in suppliance, resting her hand on Biddy's.

'Don't be afraid! And may I bring a friend – the friend with whom I'm staying?'

Miriam looked at her. 'Do you mean Mrs Dallow?'

'No, no – Miss Tressilian. She puts me up, she has got a flat. Did you ever see a flat?' asked Biddy, expansively. 'My cousin's not in London.' Miriam replied that she might bring whom she liked, and Biddy broke out, to her brother: 'Fancy what kindness, Nick: we're to have a box tonight, and you're to take me!'

Nick turned to her, smiling, with an expression in his face which struck her even at the time as odd, but which she understood when the sense of it recurred to her later. Mr Dashwood interposed with the remark that it was all very well to talk about boxes, but that he didn't see where at that time of day any such luxury was to come from.

'You haven't kept one, as I told you?' Miriam demanded.

'As you told me, my dear? Tell the lamb to keep its tender mutton from the wolves!'

'You shall have one: we'll arrange it,' Miriam went on, to Biddy.

'Let me qualify that statement a little, Miss Dormer,' said Basil Dashwood. 'We'll arrange if it's humanly possible.'

'We'll arrange it even if it's inhumanly impossible – that's just the point,' Miriam declared, to the girl. 'Don't talk about trouble – what's he meant for but to take it? *Cela s'annonce bien*, you see,' she continued, to Nick: 'doesn't it look as if we should pull beautifully together?' And as he replied that he heartily congratulated her – he was immensely interested in what he had been told – she exclaimed, after resting her eyes on him a moment: 'What will you have? It seemed simpler! It was clear there had to be some one.' She explained further to Nick what had led her to come in at that moment, while Dashwood approached Biddy with civil assurances that they would see, they would leave no stone unturned, though he would not have taken it upon himself to promise.

Miriam reminded Nick of the blessing he had been to her nearly a year before, on her other first night, when she was fidgety and impatient: how he had let her come and sit there for hours – helped her to possess her soul till the evening and keep out of harm's way. The case was the same at present, with the aggravation indeed that he would understand – Dashwood's nerves as well as her own: they were a great deal worse than hers. Everything was ready for Juliet; they had been rehearsing for five months (it had kept her from going mad, with the eternity of the other piece), and *he* had occurred to her again in the last intolerable hours as the friend in need, the salutary stop-gap, no matter how much she bothered him. She shouldn't be turned out? Biddy broke away from Basil Dashwood: she must go, she must hurry off to Miss Tressilian with her news. Florence might make some stupid engagement for the evening: she must be warned in time. The girl took a flushed, excited leave, after having received a renewal of Miriam's pledge and even heard her say to Nick that he must now give back the stall that had been sent him – they would be sure to have another use for it.

51

THAT night, at the theatre, in the box (the miracle had been wrought, the treasure was found), Nick Dormer pointed out to his two companions the stall he had relinquished, which was close in front – noting how oddly, during the whole of the first act, it remained vacant. The house was magnificent, the actress was magnificent, everything was magnificent. To describe again so famous an occasion (it has been described repeatedly by other reporters) is not in the compass of the closing words of a history already too sustained. It is enough to say that this great night marked an era in contemporary art, and that for those who had a spectator's share in it the word 'triumph' acquired a new illustration. Miriam's Juliet was an exquisite image of young passion and young despair, expressed in the divinest, truest music that had ever poured from tragic lips. The great childish audience, gaping at her points, expanded there before her like a lap to catch flowers.

During the first interval our three friends in the box had plenty to talk about, and they were so occupied with it that for some time they failed to observe that a gentleman had at last come into the empty stall near the front. The discovery was presently formulated by Miss Tressilian, in the cheerful exclamation: 'Only fancy – there's Mr Sherringham!' This of course immediately became a high wonder – a wonder for Nick and Biddy, who had not heard of his return; and the marvel was increased by the fact that he gave no sign of looking for them, or even at them. Having taken possession of his place he sat very still in it, staring straight before him at the curtain. His abrupt reappearance contained mystifying elements both for Biddy and for Nick, so that it was mainly Miss Tressilian who had freedom of mind to throw off the theory that he had come back that very hour – had arrived from a long journey. Couldn't they see how strange he was and how brown, how burnt and how red, how tired and how worn? They all inspected him, though Biddy declined Miss Tressilian's glass; but he was evidently unconscious of observation, and finally Biddy, leaning back in her chair, dropped the fantastic words: 'He has come home to marry Juliet.'

Nick glanced at her; then he replied: 'What a disaster – to make such a journey as that and to be late for the fair!'

'Late for the fair?'

'Why, she's married – these three days. They did it very quietly; Miriam says because her mother hated it and hopes it won't be much known. All the same she's Basil Dashwood's wedded wife – he has come in just in time to take the receipts for Juliet. It's a good thing, no doubt, for there are at least two fortunes to be made out of her, and he'll give up the stage.' Nick explained to Miss Tressilian, who had inquired, that the gentleman in question was the actor who was playing Mercutio, and he asked Biddy if she had not known that this was what they were telling him in Rosedale Road in the morning. She replied that she had not understood, and she sank considerably behind the drapery of the box. From this cover she was able to launch creditably enough the exclamation –

'Poor Peter!'

Nick got up and stood looking at poor Peter. 'He ought to come round and speak to us, but if he doesn't see us I suppose he doesn't.' Nick quitted the box as if to go to the restored exile. I may add that as soon as he had done so Florence Tressilian bounded over to the dusky corner in which Biddy had nestled. What passed immediately between these young ladies need not concern us: it is sufficient to mention that two minutes later Miss Tressilian broke out –

'Look at him, dearest; he's turned his head this way!'

'Thank you, I don't care to look at him,' said Biddy; and she doubtless demeaned herself in the high spirit of these words. It nevertheless happened that directly afterwards she became aware that he had glanced at his watch, as if to judge how soon the curtain would rise again, and then had jumped up and passed quickly out of his place. The curtain had risen again without his coming back and without Nick's reappearing in the box. Indeed by the time Nick slipped in a good deal of the third act was over; and even then, even when the curtain descended, Peter Sherringham had not returned. Nick sat down in silence, to watch the stage, to which the breathless attention of his companions seemed to be attached, though Biddy after a moment threw back at him a single quick look. At the end of the act they were all occupied with the recalls, the applause and the responsive loveliness of Juliet as she was led out (Mercutio had to give her up to Romeo), and even for a few minutes after the uproar had subsided nothing was said among the three. At last Nick began:

'It's quite true, he has just arrived; he's in Great Stanhope Street. They've given him several weeks, to make up for the uncomfortable way they bundled him off (to arrive in time for some special business that had suddenly to be gone into) when he first went out: he tells me they promised that at the time. He got into Southampton only a few hours ago, rushed up by the first train he could catch and came off here without any dinner.'

'Fancy!' said Miss Tressilian; while Biddy asked if Peter might be in good health and had been happy. Nick replied that he said it was a beastly place, but he appeared all right. He was to be in England probably a month, he was awfully brown, he sent his love to Biddy. Miss Tressilian looked at his empty stall and was of the opinion that it would be more to the point for him to come in to see her.

'Oh, he'll turn up; we had a goodish talk in the lobby, where he met me. I think he went out somewhere.'

'How odd to come so many thousand miles for this, and then not to stay!' Biddy reflected.

'Did he come on purpose for this?' Miss Tressilian asked.

'Perhaps he's gone out to get his dinner!' joked Biddy.

Her friend suggested that he might be behind the scenes, but Nick expressed a doubt of this; and Biddy asked her brother if he himself were not going round. At this moment the curtain rose: Nick said he would go in the next interval. As soon as it came he quitted the box, remaining absent while it lasted.

All this time, in the house, there was no sign of Peter. Nick reappeared only as the fourth act was beginning, and uttered no word to his companions till it was over. Then, after a further delay produced by renewed evidences of the actress's victory, he described his visit to the stage and the wonderful spectacle of Miriam on the field of battle. Miss Tressilian inquired if he had found Mr Sherringham with her; to which he replied that, save across the footlights, she had not seen him. At this a soft exclamation broke from Biddy:

'Poor Peter! Where is he, then?'

Nick hesitated a moment. 'He's walking the streets.'

'Walking the streets?'

'I don't know – I give it up!' Nick replied; and his tone, for some minutes, reduced his companions to silence. But a little later Biddy said:

'Was it for him, this morning, she wanted that place, when she asked you to give yours back?'

'For him, exactly. It's very odd that she just managed to keep it, for all the use he makes of it! She told me just now that she heard from him, at his post, a short time ago, to the effect that he had seen in a newspaper a statement she was going to do

Juliet and that he firmly intended, though the ways and means were not clear to him (his leave of absence hadn't come out, and he couldn't be sure when it would come), to be present on her first night: therefore she must do him the service to keep a seat for him. She thought this a speech rather in the air, so that in the midst of all her cares she took no particular pains about the matter. She had an idea she had really done with him for a long time. But this afternoon what does he do but telegraph her from Southampton that he keeps his appointment and counts upon her for a stall? Unless she had got back mine she wouldn't have been able to accommodate him. When she was in Rosedale Road this morning she hadn't received his telegram; but his promise, his threat, whatever it was, came back to her; she had a sort of foreboding, and thought that, on the chance, she had better have something ready. When she got home she found his telegram, and she told me that he was the first person she saw in the house, through her fright, when she came on in the second act. It appears she was terrified this time, and it lasted half through the play.'

'She must be rather annoyed at his having gone away,' Miss Tressilian observed.

'Annoyed? I'm not so sure!' laughed Nick.

'Ah, here he comes back!' cried Biddy, behind her fan, as the absentee edged into his seat in time for the fifth act. He stood there a moment, first looking round the theatre; then he turned his eyes upon the box occupied by his relatives, smiling and waving his hand.

'After that he'll surely come and see you,' said Miss Tressilian.

'We shall see him as we go out,' Biddy replied: 'he must lose no more time.'

Nick looked at him with a glass; then he exclaimed: 'Well, I'm glad he has pulled himself together!'

'Why, what's the matter with him, since he wasn't disappointed in his seat?' Miss Tressilian demanded.

'The matter with him is that a couple of hours ago he had a great shock.'

'A great shock?'

'I may as well mention it at last,' Nick went on. 'I had to say something to him in the lobby there, when we met – something

I was pretty sure he couldn't like. I let him have it full in the face – it seemed to me better and wiser. I told him Juliet's married.'

'Didn't he know it?' asked Biddy, who, with her face raised, had listened in deep stillness to every word that fell from her brother.

'How should he have known it? It has only just happened, and they've been so clever, for reasons of their own (those people move among a lot of considerations that are absolutely foreign to us), about keeping it out of the papers. They put in a lot of lies, and they leave out the real things.'

'You don't mean to say Mr Sherringham wanted to *marry* her!' Miss Tressilian ejaculated.

'Don't ask me what he wanted – I dare say we shall never know. One thing is very certain: that he didn't like my news and that I shan't soon forget the look on his face as he turned away from me, slipping into the street. He was too much upset – he couldn't trust himself to come back; he had to walk about – he tried to walk it off.'

'Let us hope he *has* walked it off!'

'Ah, poor fellow – he couldn't hold out to the end; he has had to come back and look at her once more. He knows she'll be sublime in these last scenes.'

'Is he so much in love with her as that? What difference does it make, with an actress, if she *is* mar—?' But in this rash inquiry Miss Tressilian suddenly checked herself.

'We shall probably never know how much he has been in love with her nor what difference it makes. We shall never know exactly what he came back for, nor why he couldn't stand it out there any longer without relief, nor why he scrambled down here all but straight from the station, nor why, after all, for the last two hours, he has been roaming the streets. And it doesn't matter, for it's none of our business. But I'm sorry for him – she *is* going to be sublime,' Nick added. The curtain was rising on the tragic climax of the play.

Miriam Rooth was sublime; yet it may be confided to the reader that during these supreme scenes Bridget Dormer directed her eyes less to the inspired actress than to a figure in the stalls who sat with his own gaze fastened to the stage. It may further be intimated that Peter Sherringham, though he saw but a fragment of the performance, read clear at the last, in the intense light of genius that this fragment shed, that even so, after all, he had been rewarded for his formidable journey. The great trouble of his infatuation subsided, leaving behind it something tolerably deep and pure. This assuagement was far from being immediate, but it was helped on, unexpectedly to him, it began to dawn at least, the very next night he saw the play, when he sat through the whole of it. Then he felt somehow recalled to reality by the very perfection of the representation. He began to come back to it from a period of miserable madness. He had been baffled, he had got his answer; it must last him – that was plain. He didn't fully accept it the first week or the second; but he accepted it sooner than he would have supposed had he known what it was to be when he paced at night, under the southern stars, the deck of the ship that was bringing him to England.

It had been, as we know, Miss Tressilian's view, and even Biddy's, that evening, that Peter Sherringham would join them as they left the theatre. This view however was not confirmed by the event, for the gentleman in question vanished utterly (disappointingly crude behaviour on the part of a young diplomatist who had distinguished himself), before any one could put a hand on him. And he failed to make up for his crudity by coming to see any one the next day, or even the next. Indeed many days elapsed, and very little would have been known about him had it not been that, in the country, Mrs Dallow knew. What Mrs Dallow knew was eventually known to Biddy Dormer; and in this way it could be established in his favour that he had remained some extraordinarily small number of days in London, had almost directly gone over to Paris to see his old chief. He came back from Paris – Biddy knew this not from Mrs Dallow, but in a much more immediate way: she knew it by his pressing the little electric button at the door of Florence Tressilian's flat, one day when the good Florence was out and she herself was at home. He made on this occasion a very long visit. The good Florence knew it not much later, you may be sure (and how he had got their address from Nick), and she took an extravagant satisfaction in it. Mr Sherringham had never been to see her – the like of her – in his life: therefore it was clear what had made him begin. When he had once begun he kept it up, and Miss Tressilian's satisfaction increased.

Good as she was she could remember without the slightest relenting what Nick Dormer had repeated to them at the theatre about Peter's present post being a beastly place. However, she was not bound to make a stand at this if persons more nearly concerned, Lady Agnes and the girl herself, didn't mind it. How little *they* minded it, and Grace, and Julia Dallow, and even Nick, was proved in the course of a meeting that took place at Harsh during the Easter holidays. Mrs Dallow had a small and intimate party to celebrate her brother's betrothal. The two ladies came over from Broadwood; even Nick, for two days, went back to his old hunting-ground, and Miss Tressilian relinquished for as long a time the delights of her newly-arranged flat. Peter Sherringham obtained an extension of leave, so that he might go back to his legation with a wife. Fortunately, as it turned out, Biddy's ordeal, in the more or less torrid zone, was not cruelly prolonged, for the pair have already received a superior appointment. It is Lady Agnes's proud opinion that her daughter is even now shaping their destiny. I say 'even now', for these facts bring me very close to contemporary history. During those two days at Harsh Nick arranged with Julia Dallow the conditions, as they might be called, under which she should sit to him; and everyone will remember in how recent an exhibition general attention was attracted, as the newspapers said in describing the private view, to the noble portrait of a lady which was the final outcome of that arrangement. Gabriel Nash had been at many a private view, but he was not at that one.

These matters are highly recent however, as I say; so that in glancing about the little circle of the interests I have tried to evoke I am suddenly warned by a sharp sense of modernness. This renders it difficult for me, for example, in taking leave of our wonderful Miriam, to do much more than allude to the general impression that her remarkable career is even yet only in its early prime. Basil Dashwood has got his theatre, and his wife (people know now she *is* his wife) has added three or four new parts to her repertory; but every one is agreed that both in public and in private she has a great deal more to show. This is equally true of Nick Dormer, in regard to whom I may finally say that his friend Nash's predictions about his reunion with Mrs Dallow have not up to this time been justified. On the other hand, I must not omit to add, this lady has not, at the latest accounts, married Mr Macgeorge. It is very true there has been a rumour that Mr Macgeorge is worried about her – has even ceased to believe in her.

MORE ABOUT PENGUINS AND PELICANS

Penguinews, which appears every month, contains details of all the new books issued by Penguins as they are published. It is supplemented by our stocklist, which includes around 5,000 titles.

A specimen copy of *Penguinews* will be sent to you free on request. Please write to Dept EP, Penguin Books Ltd, Harmondsworth, Middlesex, for your copy.

In the U.S.A.: For a complete list of books available from Penguins in the United States write to Dept CS, Penguin Books, 625 Madison Avenue, New York, New York 10022.

In Canada: For a complete list of books available from Penguins in Canada write to Penguin Books Canada Ltd, 2801 John Street, Markham, Ontario L3R 1B4.

The Penguin English Library

JANE AUSTEN

MANSFIELD PARK

Edited by Tony Tanner

Mansfield Park (1814) is possibly the most profound and original of Jane Austen's novels. As its title suggests it is as much the story of a house as of the people who carry on their intrigues and affairs in and around it. Some modern readers have felt little sympathy with the patient, watchful heroine, Fanny: few, however, have failed to detect an underlying atmosphere which is all the more powerful for being suppressed. And certainly the greatest scenes show Jane Austen working at the very height of her artistic powers.

EMMA

Edited by Ronald Blythe

In planning *Emma*, which appeared in 1816, Jane Austen wrote: 'I am going to take a heroine whom no one but myself will much like.'

Yet, despite her manifest faults – her officiousness and her capacity for deluding herself – most readers will agree in liking Emma Woodhouse very much indeed. More complex and fully rounded than almost any of Jane Austen's other characters, her progress, through the mismanagement of other people's affairs to the crisis and resolution of her own, is a whole comedy of self-deceit and self-discovery.

Also published in Penguins:

PERSUASION

PRIDE AND PREJUDICE

SENSE AND SENSIBILITY

NORTHANGER ABBEY

LADY SUSAN / THE WATSONS / SANDITON

Penguin Modern Classics

JOSEPH CONRAD

'Conrad is among the very greatest novelists in the language – or any language' – F. R. Leavis in *The Great Tradition*

LORD JIM

The novel by which Conrad is most often remembered by perhaps a majority of readers, and the first considerable novel he wrote.

THE NIGGER OF THE NARCISSUS, TYPHOON *and Other Stories*

Conrad's first sea novel, together with 'Typhoon', 'Falk', 'Amy Foster', and 'Tomorrow'.

NOSTROMO: *A Tale of the Seaboard*

His story of revolution in South America, which Arnold Bennett regarded 'as one of the greatest novels of any age'.

HEART OF DARKNESS

T. S. Eliot's use of a quotation from *Heart of Darkness* as an epigraph to the original manuscript of *The Waste Land* was no doubt inspired by his belief that Mr Kurtz, the ambiguous hero of the story, stands at the dark heart of the twentieth century.

Also published in Penguins:
YOUTH *and* THE END OF THE TETHER
AN OUTCAST OF THE ISLANDS
TALES OF UNREST

ARNOLD BENNETT

THE CLAYHANGER TRILOGY

CLAYHANGER

Arnold Bennett's careful evocation of a boy growing to manhood during the last quarter of the nineteenth century, with its superb portrait of an autocratic father, stands on a literary level with *The Old Wives' Tale*.

HILDA LESSWAYS

Relating the early life of Hilda Lessways, before her marriage to Edwin Clayhanger. Her involvement with the enigmatic, self-made man, George Cannor, and his enterprises take her from the offices of an embryo newspaper in the Five Towns to a venture into the guest house business in Brighton.

THESE TWAIN

In what is in many ways the most accomplished novel of the trilogy, Bennett achieves a remarkably subtle and biting portrait of a marriage. Hilda and Edwin Clayhanger are living in Bursley, with Hilda's son by her disastrous 'marriage' to George Cannor. As they cope with immediate tensions and old wounds, they are forced continually to reassess their relationship.

Also published in Penguins:

THE CARD

THE GRAND BABYLON HOTEL

Penguin Modern Classics

HENRY JAMES

'He is as solitary in the history of the novel as Shakespeare in the history of poetry' – Graham Greene in *The Lost Childhood*

THE PORTRAIT OF A LADY

The poignant story of an American princess in Europe and her involvement in the toils of European guile.

THE BOSTONIANS

Intending to write 'a very American tale', Henry James drew attention, in *The Bostonians*, to 'the situation of women, the decline of the sentiment of sex, the agitation in their behalf'.

THE EUROPEANS

'This small book, written so early in James's career, is a masterpiece of major quality' – F. R. Leavis in *The Great Tradition*

Also published in Penguins:

WASHINGTON SQUARE

THE WINGS OF THE DOVE

WHAT MAISIE KNEW

THE AWKWARD AGE

THE GOLDEN BOWL

THE AMBASSADORS

SELECTED SHORT STORIES

THE TURN OF THE SCREW AND OTHER STORIES

THE PRINCESS CASAMASSIMA